Nigel Kalton's Lectures in Nonlinear Functional Analysis

UNIVERSITY LECTURE SERIES VOLUME 79

Nigel Kalton's Lectures in Nonlinear Functional Analysis

Adam Bowers

AMERICAN MATHEMATICAL SOCIETY
Providence, Rhode Island

EDITORIAL COMMITTEE

Christopher Bishop Robert Guralnick (Chair)
Panagiota Daskalopoulou Emily Riehl

2020 *Mathematics Subject Classification.* Primary 46-01, 46T99, 46B20, 46B80, 46B07, 46B04, 46B08.

For additional information and updates on this book, visit
www.ams.org/bookpages/ulect-79

Library of Congress Cataloging-in-Publication Data

Names: Bowers, Adam (Adam Roman), author.
Title: Nigel Kalton's lectures in nonlinear functional analysis / Adam Bowers.
Description: Providence, Rhode Island : American Mathematical Society, [2024] | Series: University lecture series, 1047-3998 ; volume 79 | Includes bibliographical references.
Identifiers: LCCN 2024032344 | ISBN 9781470473471 (paperback) | ISBN 9781470478797 (ebook)
Subjects: LCSH: Nonlinear functional analysis. | Kalton, Nigel J. (Nigel John), 1946–2010. | AMS: Functional analysis – Instructional exposition (textbooks, tutorial papers, etc.). | Functional analysis – Nonlinear functional analysis – None of the above, but in this section. | Functional analysis – Normed linear spaces and Banach spaces; Banach lattices – Geometry and structure of normed linear spaces. | Functional analysis – Normed linear spaces and Banach spaces; Banach lattices – Nonlinear classification of Banach spaces; nonlinear quotients. | Functional analysis – Normed linear spaces and Banach spaces; Banach lattices – Local theory of Banach spaces. | Functional analysis – Normed linear spaces and Banach spaces; Banach lattices – Isometric theory of Banach spaces. | Functional analysis – Normed linear spaces and Banach spaces; Banach lattices – Ultraproduct techniques in Banach space theory.
Classification: LCC QA321.5 .B68 2024 | DDC 515/.7248–dc23/eng20241004
LC record available at https://lccn.loc.gov/2024032344

University Lecture Series ISSN: 1047-3998 (print)
DOI: https://doi.org/10.1090/ulect/79

Copying and reprinting. Individual readers of this publication, and nonprofit libraries acting for them, are permitted to make fair use of the material, such as to copy select pages for use in teaching or research. Permission is granted to quote brief passages from this publication in reviews, provided the customary acknowledgment of the source is given.

Republication, systematic copying, or multiple reproduction of any material in this publication is permitted only under license from the American Mathematical Society. Requests for permission to reuse portions of AMS publication content are handled by the Copyright Clearance Center. For more information, please visit **www.ams.org/publications/pubpermissions**.

Send requests for translation rights and licensed reprints to reprint-permission@ams.org.

© 2024 by the American Mathematical Society. All rights reserved.
The American Mathematical Society retains all rights
except those granted to the United States Government.
Printed in the United States of America.

∞ The paper used in this book is acid-free and falls within the guidelines
established to ensure permanence and durability.
Visit the AMS home page at https://www.ams.org/

10 9 8 7 6 5 4 3 2 1   29 28 27 26 25 24

To the memory of Nigel J. Kalton (1946–2010)

Contents

Preface		ix
Notations and conventions		xi
Chapter 1.	Absolute Lipschitz retracts	1
Chapter 2.	Lipschitz extensions and Hilbert spaces	13
Chapter 3.	Convex subsets and selections	19
3.1.	The set of convex subsets	19
3.2.	The Michael selection theorem	24
Chapter 4.	Lipschitz classification of Banach spaces	27
Chapter 5.	Arens–Eells space	37
Chapter 6.	Differentiation and the isomorphism problem	49
	Additional comments	55
Chapter 7.	Differentiation and Haar-null sets	57
Chapter 8.	Property $\Pi(\lambda)$ and embeddings into c_0	75
Chapter 9.	Local complementation and the Heinrich–Mankiewicz theorem	89
Chapter 10.	The Lipschitz structure of c_0	103
Chapter 11.	Ultraproducts of Banach spaces	119
Chapter 12.	The uniform structure of Banach spaces	131
Chapter 13.	The unique uniform structure of sequence spaces	159
Chapter 14.	Uniform embeddings into a Hilbert space	185
14.1.	Theorem A	186
14.2.	Theorem B	200
14.3.	Conclusion	213
Chapter 15.	Uniform embeddings into reflexive spaces	215
	Some comments on unit balls	222
Chapter 16.	Exercises	225
16.1.	Examples I	225
16.2.	Examples II	227

Afterword (where to from here)	229
Appendix A. Vector integration	231
Appendix B. The Radon–Nikodym property	235
Appendix C. Gaussian measures	239
Appendix D. Notes on closest points	243
Bibliography	247
Index	253

Preface

This manuscript is based on the lectures given by Nigel Kalton during a Nonlinear Functional Analysis course during Fall Semester of 2008 at the University of Missouri at Columbia.

While typing these notes, I have endeavored to be as faithful as possible to the original source material, which is the collection of notes I took when attending the actual lectures. My objective when writing this manuscript was to share the lectures as they were given, so that others could benefit from the insight of Nigel Kalton and see the material presented in the way that he envisioned it. Of course, without his direct involvement, I cannot offer more than my best guess. I must confess to making some (in some cases considerable) changes to the original lectures, but in these cases, I follow the presentation in the published work of Nigel and his coauthors. (These changes are usually indicated in the text.)

When it was felt some additional explanation would be useful, it has been added in a "Comments" box, rather than placed in the narrative directly. It is hoped that in this way the editorial voice can be separated from the voice of the true lecturer—the master himself, Nigel Kalton.

I would like to offer my appreciation to the American Mathematical Society for publishing this work, and Ina Mette in particular for helping to make it happen. I would also like to show my appreciation to the reviewers for their excellent suggestions and corrections, which I was very glad to receive, and which led to a much better text. I would also like to thank Jennifer Kalton for her support.

I give my deepest gratitude to Gilles Godefroy. Without his expertise and support, and his patience and kindness, I would never have been able to complete this document.

Adam Bowers

Notations and conventions

Unless otherwise specified, all spaces (vector spaces, Banach spaces, etc.) are real spaces. We will usually begin a theorem with a line something like "Let X be a Banach space." In such cases, it should be assumed that X is a real Banach space. If X is a normed space, then we will denote the norm of x in X by $\|x\|_X$. Similarly, if X has a metric, we will use d_X to denote the metric. If there is little risk of ambiguity, however, we will usually write $\|\cdot\|$ and d (without the subscript).

Proofs will conclude with a square symbol ($\square$) to indicate the proof's conclusion. In several places, however, there are claims made within a proof, and the proof of such a claim will begin with the phrase "*Proof of claim*". Since these proofs are within other proofs, they will be concluded with a diamond ($\diamond$) to maintain a distinction between the main proof and the proof of the claim.

If X is a metric space, we will use the notation $B(x,r)$ to indicate the *closed* ball with center x and radius r. The closed *unit ball* $B(0,1)$ in X will usually be denoted by B_X. The unit *sphere* in X, which is the boundary of B_X will be denoted ∂B_X. There may be some variations of these notations used in the text, for specific purposes, and those will be commented upon within the text when needed.

In the literature, there are numerous conventions for denoting isomorphism between various types of spaces. In this document, we will use the symbol $\approx$ to indicate isomorphism. In certain cases, we will modify this symbol. For example, to indicate a Lipschitz isomorphism between spaces, we will use $\stackrel{\text{Lip}}{\approx}$.

If X is a metric space with a point 0, then the standard notation for the set of all real-valued Lipschitz functions f defined on X with the property that $f(0) = 0$ is $\text{Lip}_0(X)$. In this document, however, we will use $\text{Lip}(X)$. This should not be confused with $\text{Lip}(f)$, which denotes the smallest constant K for which

$$|f(x) - f(y)| \leq K d_X(x,y)$$

for all x and y in X, where d_X is the metric on X.

We will follow the standard convention of using ℓ_p to denote the classical sequence space of p-summable real sequences, where $1 \leq p < \infty$, and we will use ℓ_∞ to denote the set of bounded real sequences. For the space of p-summable sequences ($1 \leq p < \infty$) having entries in a normed space X (other than $\mathbb{R}$), we adopt the convention of writing $\ell_p(X)$. For these spaces, we use $\|\cdot\|_p$ to denote the norm, so that

$$\|(x_k)_{k=1}^\infty\|_p = \left(\sum_{k=1}^\infty \|x_k\|_X^p\right)^{1/p},$$

where $x_k \in X$ for each $k \in \mathbb{N}$ and $\|\cdot\|_X$ is the norm on X.

We will also follow the standard convention of using $L_p(a,b)$ to denote the set of p-integrable measurable functions on the interval $[a,b]$ when $1 \leq p < \infty$, and we will use $L_\infty(a,b)$ to denote the set of essentially bounded measurable functions on $[a,b]$. For these classical function spaces we use Lebesgue measure. Again, we will use $\|\cdot\|_p$ to denote the norms on these spaces, as long as there is no ambiguity with the norm on the sequence spaces. When we are interested in more general measurable spaces X with measures μ, we will write $L_p(X,\mu)$.

The set of real sequences converging to a limit will be denoted c, and the collection of real sequences converging to zero will be denoted c_0. We will use $c(X)$ and $c_0(X)$ when the elements of the sequence come from a space X other than $\mathbb{R}$. The notation $\mathcal{C}(K)$ will be used to denote the set of (real-valued) continuous functions on the (usually compact Hausdorff) topological set K. In all of these cases, we will use $\|\cdot\|_\infty$ to denote the supremum. We will write $\mathcal{M}(K)$ for the collection of measures on K having finite total variation, and we will use $\|\cdot\|_M$ to denote the value of that total variation. If $K = [a,b]$ is a closed interval, then we will write $\mathcal{C}[a,b]$ and $\mathcal{M}[a,b]$, without the enclosing parentheses, for notational simplicity.

When X is a Banach space, we denote the dual space by X^* and the second dual by X^{**}, and so forth. We will follow the convention of identifying X with its canonical isometric copy in X^{**}. Similarly, when a Banach space is reflexive, we will usually identify X with X^{**}, even though it would be more precise to say that the two spaces are isometrically isomorphic.

CHAPTER 1

Absolute Lipschitz retracts

Let (X, d_X) and (Y, d_Y) be metric spaces. A function $f : X \to Y$ is called *Lipschitz continuous* if there exists a constant $K \geq 0$ such that
$$d_Y(f(x), f(y)) \leq K\, d_X(x, y),$$
for x and y in X. Any such K is called a *Lipschitz constant* for f. The smallest Lipschitz constant for f is denoted $\mathrm{Lip}(f)$. A function that is Lipschitz continuous is said to be a *Lipschitz function*.

COMMENT 1.1. It is assumed that the reader is familiar with the definition of a metric space. Whenever M is a metric space, the metric on M will be denoted by d_M, unless stated otherwise. It is also assumed that the reader is familiar with the notions of Hilbert space and (more generally) Banach space.

The two metric spaces X and Y are called *Lipschitz isomorphic* if there exists a bijection $f : X \to Y$ such that both f and f^{-1} are Lipschitz functions. In such a case, there exist positive constants c and C such that
$$c\, d_X(x, y) \leq d_Y(f(x), f(y)) \leq C\, d_X(x, y),$$
for x and y in X.

A basic goal of nonlinear functional analysis is to answer the following question.

QUESTION 1.2. *If two Banach spaces are Lipschitz isomorphic, when are they also linearly isomorphic?*

Let X and Y be metric spaces and suppose $f : X \to Y$ is a function. The *modulus of continuity* of f is the function $\omega_f : [0, \infty] \to [0, \infty]$ defined by the formula
$$\omega_f(t) = \sup\left\{ d_Y(f(x), f(y)) \,:\, d_X(x, y) \leq t \right\},$$
where $t \geq 0$. Observe that f is Lipschitz continuous with Lipschitz constant K provided that $\omega_f(t) \leq Kt$ for all $t \geq 0$.

A function $f : X \to Y$ between two metric spaces is *uniformly continuous* if for every $\epsilon > 0$ there exists a $\delta > 0$ such that $d_Y(f(x), f(y)) < \epsilon$ whenever $d_X(x, y) < \delta$. In terms of the modulus of continuity, the function f is uniformly continuous if
$$\lim_{t \to 0^+} \omega_f(t) = 0.$$

DEFINITION 1.3. A metric space (X, d) is called *metrically convex* if given any x and y in X and any $t \in (0, 1)$, there exists a $z \in X$ such that
$$d(x, z) = t\, d(x, y) \quad \text{and} \quad d(z, y) = (1 - t)\, d(x, y).$$

EXAMPLE 1.4. A Banach space is metrically convex.

LEMMA 1.5. *Let $f : X \to Y$ be a function between two metric spaces. If X is metrically convex, then ω_f is subadditive:*
$$\omega_f(s+t) \leq \omega_f(s) + \omega_f(t)$$
for any s and t in $[0, \infty]$.

PROOF. Suppose $d_X(x,y) \leq s+t$. By assumption, X is metrically convex, and so there exists some $z \in X$ such that
$$d_X(x,z) = \frac{s}{s+t} d_X(x,y) \leq s \quad \text{and} \quad d_X(z,y) = \frac{t}{s+t} d_X(x,y) \leq t.$$
Therefore,
$$d_Y\big(f(x), f(z)\big) \leq \omega_f(s) \quad \text{and} \quad d_Y\big(f(z), f(y)\big) \leq \omega_f(t).$$
Consequently,
$$d_Y\big(f(x), f(y)\big) \leq \omega_f(s) + \omega_f(t),$$
from which the result follows. $\square$

REMARK 1.6. If ω_f is subadditive, then $\omega_f(t) < \infty$ for one value of t implies that $\omega_f(t) < \infty$ for all values of t.

Suppose X is a metric space with subset A. If Y is a metric space and $f_0 : A \to Y$ is Lipschitz continuous with Lipschitz constant K, is it possible to extend f_0 to a function $f : X \to Y$ such that f is also Lipschitz continuous? If a Lipschitz extension can be found, is it possible for the extension to also have Lipschitz constant K?

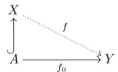

DEFINITION 1.7. A metric space Y is called a λ-*absolute Lipschitz retract* (or a λ-*ALR*) if whenever X is a metric space with subset A, any Lipschitz continuous function $f_0 : A \to Y$ can be extended to a Lipschitz continuous function $f : X \to Y$ with $\mathrm{Lip}(f) \leq \lambda \mathrm{Lip}(f_0)$, where $\lambda \geq 1$.

The next result can be thought of as a nonlinear Hahn-Banach theorem.

THEOREM 1.8. *The metric space $\mathbb{R}$, with the standard metric, is a 1-ALR.*

PROOF. Let (X, d) be a metric space with subset A and suppose $f_0 : A \to \mathbb{R}$ is a Lipschitz continuous function with $\mathrm{Lip}(f_0) = K$.

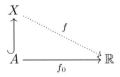

The goal is to find an extension f of f_0 that is Lipschitz continuous. Start by looking for a function $f : X \to \mathbb{R}$ that satisfies $|f(x) - f_0(a)| \leq K d(x,a)$ for all $x \in X$ and $a \in A$. Then, for each $x \in X$,
$$f_0(a) - K d(x,a) \leq f(x) \leq f_0(a) + K d(x,a),$$

for all $a \in A$. Since this is true for all $a \in A$, it follows that
$$\sup_{a \in A} \left\{ f_0(a) - Kd(x,a) \right\} \le f(x) \le \inf_{a \in A} \left\{ f_0(a) + Kd(x,a) \right\}.$$
It turns out that both the left and right sides of this inequality are Lipschitz extensions of f_0. Only one extension is required, so define $f : X \to \mathbb{R}$ by
$$f(x) = \sup_{a \in A} \left\{ f_0(a) - Kd(x,a) \right\}, \quad x \in X.$$
To show that f is an extension of f_0, suppose $x \in A$. Then, because f_0 is Lipschitz continuous with Lipschitz constant K, for all $a \in A$,
$$f_0(a) - Kd(x,a) \le f_0(x).$$
Computing the supremum over all $a \in A$, this results in the bound $f(x) \le f_0(x)$. However, there is a value for which this supremum is achieved (because $x \in A$), and thus $f(x) = f_0(x)$. It follows that $f|_A = f_0$, and so f is an extension of f_0.

It remains to show that f is Lipschitz continuous with Lipschitz constant K. Suppose $x \in X$ and $y \in X$, and let $a \in A$. Then, by the triangle inequality,
$$f_0(a) - Kd(x,a) \le \Big(f_0(a) - Kd(y,a) \Big) + Kd(x,y).$$
Computing the supremum over all $a \in A$, this becomes $f(x) \le f(y) + Kd(x,y)$, or $f(x) - f(y) \le Kd(x,y)$. Repeating the argument with the roles of x and y reversed, it follows that $|f(x) - f(y)| \le Kd(x,y)$, as required. $\square$

COMMENT 1.9. As was mentioned in the preceding proof (without justification), an alternate extension for f_0 is given by the formula
$$\tilde{f}(x) = \inf_{a \in A} \left\{ f_0(a) + Kd(x,a) \right\}, \quad x \in X.$$
The proof is similar to the one given to show $f|_A = f_0$. Since both f and $\tilde{f}$ are extensions of f_0, note that for all $x \in A$,
$$\sup_{a \in A} \left\{ f_0(a) - Kd(x,a) \right\} = f_0(x) = \inf_{a \in A} \left\{ f_0(a) + Kd(x,a) \right\}.$$

REMARK 1.10. The nonempty intervals $[a,b]$, $[c,\infty)$, and $(-\infty,d]$ are all 1-ALR. The proof in each case is the same as the proof of Theorem 1.8.

If Γ is an arbitrary set, the collection of all *bounded real-valued functions* on Γ is denoted $\ell_\infty(\Gamma)$. It can be shown that this is a Banach space with the norm $\|\xi\|_\infty = \sup_{\gamma \in \Gamma} |\xi(\gamma)|$.

THEOREM 1.11. *If Γ is an arbitrary set, then $\ell_\infty(\Gamma)$ is a 1-ALR.*

PROOF. Let X be a metric space with subset A and suppose $f_0 : A \to \ell_\infty(\Gamma)$ is a Lipschitz continuous function. For each $\gamma \in \Gamma$, the function $f_0^\gamma : A \to \mathbb{R}$ defined by $f_0^\gamma(a) = f_0(a)(\gamma)$ can be extended to a function $f^\gamma : X \to \mathbb{R}$, by Theorem 1.8. Then the function $f : X \to \ell_\infty(\Gamma)$ is defined so that $f(x)(\gamma) = f^\gamma(x)$ is a Lipschitz extension of f_0 and $\text{Lip}(f) = \text{Lip}(f_0)$. $\square$

PROPOSITION 1.12. *If X is a metric space, then X is isometric to a subset of $\ell_\infty(X)$.*

PROOF. Fix $a \in X$. Define a function $f : X \to \ell_\infty(X)$ so that for each $x \in X$, $f(x) \in \ell_\infty(X)$ is a function $f(x) : X \to \mathbb{R}$ given by the rule
$$f(x)(y) = d(x, y) - d(y, a), \quad y \in X.$$
Then, for each $y \in X$,
$$|f(x)(y)| = |d(x, y) - d(y, a)| \leq d(x, a),$$
and so $f(x) \in \ell_\infty(X)$ for each $x \in X$, and therefore f is well-defined. Now let x and x' be elements in X. For each $y \in X$,
$$|f(x)(y) - f(x')(y)| = |d(x, y) - d(x', y)| \leq d(x, x').$$
Consequently,
$$\|f(x) - f(x')\|_\infty = \sup_{y \in Y} \left\{ |f(x)(y) - f(x')(y)| \right\} \leq d(x, x').$$
Next, observe that
$$\|f(x) - f(x')\|_\infty \geq |f(x)(x') - f(x')(x')| = |d(x, x') - d(x', x')| = d(x, x').$$
It follows that $\|f(x) - f(x')\|_\infty = d(x, x')$, and so f is an isometry. $\square$

DEFINITION 1.13. Let Z be a metric space with subset Y. A function $r : Z \to Y$ is a *Lipschitz retraction* if it is Lipschitz continuous and $r(y) = y$ for all $y \in Y$.

COROLLARY 1.14. *A metric space Y is a λ-ALR if and only if for every metric space Z containing Y there exists a Lipschitz retraction $r : Z \to Y$ with $\mathrm{Lip}(r) \leq \lambda$.*

PROOF. Assume that Y is a λ-ALR and suppose that Z is a metric space such that $Y \subseteq Z$.

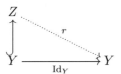

Since Y is a λ-ALR, there exists a Lipschitz continuous function $r : Z \to Y$ that extends the identity map Id_Y on Y (that is $r|_Y = \mathrm{Id}_Y$) such that $\mathrm{Lip}(r) \leq \lambda$.

Now assume that Y is a metric space such that for every metric space Z containing Y there exists a Lipschitz retraction $r : Z \to Y$ with $\mathrm{Lip}(r) \leq \lambda$. The goal is to show that Y is a λ-ALR. Suppose that A is a subset of the metric space X and suppose that $f_0 : A \to Y$ is a Lipschitz continuous function. The objective is to find a Lipschitz extension of f_0 from X to Y with a Lipschitz constant bounded by $\lambda \mathrm{Lip}(f_0)$.

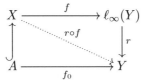

By Proposition 1.12, the metric space $\ell_\infty(Y)$ contains an isometric copy of Y. Without loss of generality, we may therefore assume that Y is a subset of $\ell_\infty(Y)$, and so (by assumption) there exists a Lipschitz retraction $r : \ell_\infty(Y) \to Y$ with $\mathrm{Lip}(r) \leq \lambda$.

Note that f_0 can be viewed as a function with codomain $\ell_\infty(Y)$. By Theorem 1.11, the metric space $\ell_\infty(Y)$ is a 1-ALR, and so $f_0 : A \to \ell_\infty(Y)$ has a Lipschitz extension $f : X \to \ell_\infty(Y)$ with $\operatorname{Lip}(f) = \operatorname{Lip}(f_0)$. The function $r \circ f : X \to Y$ is a Lipschitz extension of f_0 with $\operatorname{Lip}(r \circ f) \leq \lambda \operatorname{Lip}(f_0)$. Therefore, it follows that Y is a λ-ALR. $\square$

COMMENT 1.15. In some of the literature, a metric space Y is defined to be a λ-ALR precisely when, for every metric space Z containing Y, there exists a Lipschitz retraction $r : Z \to Y$ with $\operatorname{Lip}(r) \leq \lambda$. By Corollary 1.14, this definition and the one given in Definition 1.7 are equivalent.

DEFINITION 1.16. A metric space Y is said to have the *binary intersection property* (or Y has *BIP*) provided that whenever $\{B(x_i, r_i)\}_{i \in I}$ is a collection of closed balls in Y such that any two balls in the collection intersect nontrivially, then the intersection $\bigcap_{i \in I} B(x_i, r_i)$ is nonempty.

COMMENT 1.17. The set $B(x_i, r_i)$ appearing in Definition 1.16 is the closed ball centered at $x_i \in Y$ with radius $r_i > 0$.

EXAMPLE 1.18. The spaces $\mathbb{R}$ and ℓ_∞^2 have BIP where ℓ_∞^2 is the space $\mathbb{R}^2$ equipped with the supremum norm. If ℓ_2^2 is $\mathbb{R}^2$ equipped with the Euclidean norm, then ℓ_2^2 does not have BIP.[1]

THEOREM 1.19. *If Γ is an arbitrary set, then $\ell_\infty(\Gamma)$ has BIP.*

PROOF. Assume that $\{B(x_i, r_i)\}_{i \in I}$ is a collection of closed balls in $\ell_\infty(\Gamma)$ such that each pair of closed balls in the collection intersects nontrivially. Observe that

$$B(x_i, r_i) \cap B(x_j, r_j) \neq \emptyset \implies \|x_i - x_j\|_\infty \leq r_i + r_j.$$

Thus, by assumption, $\|x_i - x_j\|_\infty \leq r_i + r_j$ for each i and j in I.

Working backwards, the goal is to find a $z \in \ell_\infty(\Gamma)$ such that $\|z - x_i\|_\infty \leq r_i$ for all $i \in I$. That is,

$$-r_i \leq z(\gamma) - x_i(\gamma) \leq r_i, \quad \gamma \in \Gamma,\ i \in I.$$

This can be rewritten as

$$x_i(\gamma) - r_i \leq z(\gamma) \leq x_i(\gamma) + r_i, \quad \gamma \in \Gamma,\ i \in I.$$

Equivalently, we may say

$$\sup_{i \in I} \{x_i(\gamma) - r_i\} \leq \inf_{i \in I} \{x_i(\gamma) + r_i\}, \quad \gamma \in \Gamma.$$

An alternate way to write this is

$$x_i(\gamma) - r_i \leq x_j(\gamma) + r_j, \quad \{i,j\} \subseteq I,\ \gamma \in \Gamma,$$

or, in other words,

$$x_i(\gamma) - x_j(\gamma) \leq r_i + r_j, \quad \{i,j\} \subseteq I,\ \gamma \in \Gamma.$$

---

[1] In fact, if ℓ_p^2 denotes the space $\mathbb{R}^2$ equipped with the ℓ_p norm, then ℓ_p^2 fails to have BIP whenever $1 < p < \infty$. In general, a finite-dimensional normed space has BIP if and only if it is isomorphic to ℓ_∞^n, the space $\mathbb{R}^n$ equipped with the supremum norm. (See Theorem 3 in [**87**].)

Repeating the argument with i and j reversed reveals that
$$-(r_i + r_j) \leq x_i(\gamma) - x_j(\gamma), \quad \{i,j\} \subseteq I, \ \gamma \in \Gamma.$$
Therefore, the existence of $z \in B(x_i, r_i)$ for all $i \in I$ is equivalent to the condition $\|x_i - x_j\|_\infty \leq r_i + r_j$. This is known to be true (by assumption), and so the result follows. $\square$

THEOREM 1.20. *A metric space Y is a 1-ALR if and only if Y is (1) metrically convex and (2) has BIP.*

PROOF. Let Y be a metric space with metric d_Y. Start by assuming that Y is a 1-ALR and show that this assumption implies both (1) and (2).

Proof of (1) Let x and y be in Y such that $x \neq y$ and let $t \in (0,1)$. Define a function $f_0 : \{0,1\} \to Y$ by $f_0(0) = x$ and $f_0(1) = y$. Then f_0 is a Lipschitz continuous function and $\mathrm{Lip}(f_0) = d_Y(x,y)$. Since Y is a 1-ALR there exists a Lipschitz continuous function $f : [0,1] \to Y$ with $f(0) = x$ and $f(1) = y$ such that $\mathrm{Lip}(f) = \mathrm{Lip}(f_0) = d_Y(x,y)$. Let $z = f(t)$. Then
$$d_Y(x,z) = d_Y\big(f(0), f(t)\big) \leq \mathrm{Lip}(f)|0 - t| = t\, d_Y(x,y)$$
and
$$d_Y(y,z) = d_Y\big(f(1), f(t)\big) \leq \mathrm{Lip}(f)|1 - t| = (1-t)\, d_Y(x,y).$$
Since $d_Y(x,y) \leq d_Y(x,z) + d_Y(y,z)$, by the triangle inequality, the above inequalities must be equalities. This completes the proof of (1).

Proof of (2) Assume that $\{B(y_i, r_i)\}_{i \in I}$ is a collection of closed balls in Y such that any two intersect nontrivially. By Proposition 1.12, the metric space Y can be viewed as a subset of $\ell_\infty(Y)$. For each $i \in I$, let
$$B_{\ell_\infty(Y)}(y_i, r_i) = \big\{z \in \ell_\infty(Y) : d_{\ell_\infty(Y)}(y_i, z) \leq r_i\big\},$$
where $d_{\ell_\infty(Y)}$ is the metric on $\ell_\infty(Y)$ and by y_i we actually mean the element in $\ell_\infty(Y)$ identified to y_i by the isometry in Proposition 1.12. Then
$$B(y_i, r_i) = Y \cap B_{\ell_\infty(Y)}(y_i, r_i).$$
By Theorem 1.19, the space $\ell_\infty(Y)$ has BIP, and so there exists a $z \in \ell_\infty(Y)$ such that $z \in \bigcap_{i \in I} B_{\ell_\infty(Y)}(y_i, r_i)$.

By Corollary 1.14, because Y is assumed to be 1-ALR, there exists a Lipschitz retraction $\rho : \ell_\infty(Y) \to Y$ such that $\mathrm{Lip}(\rho) = 1$. Let $z' = \rho(z)$. Then
$$d_Y(z', y_i) = d_Y\big(\rho(z), \rho(y_i)\big) \leq d_{\ell_\infty(Y)}(z, y_i) \leq r_i.$$
Consequently, $z' \in B(y_i, r_i)$ for every $i \in I$, and so Y has BIP.

It remains to show that (1) and (2) together imply that the metric space Y is 1-ALR. Assume that Y is metrically convex and has BIP. Let X be a metric space with subset A and assume $f_0 : A \to Y$ is a Lipschitz continuous function such that $\mathrm{Lip}(f_0) = \lambda$.

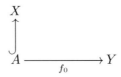

It suffices to consider the case $X\setminus A = \{x\}$. (The general case then follows from Zorn's lemma.) In the case where $X\setminus A$ contains the single point x, it is enough to find $z \in Y$ such that
$$d_Y(z, f_0(a)) \leq \lambda\, d_X(x,a), \quad a \in A.$$
If such a $z \in Y$ can be found, then $f_0 : A \to Y$ can be extended to $f : X \to Y$ by setting $f(x) = z$, and then it will be the case that $\operatorname{Lip}(f) \leq \lambda$.

The objective is to find
$$z \in \bigcap_{a \in A} B(f_0(a), \lambda\, d(x,a)).$$
Because Y has BIP (by assumption), it suffices to show that for any a and a' in A,
$$B(f_0(a), \lambda\, d(x,a)) \cap B(f_0(a'), \lambda\, d(x,a')) \neq \emptyset.$$
Observe that
$$d_Y(f_0(a), f_0(a')) \leq \lambda\, d_X(a,a') \leq \lambda[d_X(a,x) + d_X(a',x)].$$
Now let
$$t = \frac{d_X(x,a)}{d_X(x,a) + d_X(x,a')}.$$
Since Y is metrically convex (by assumption), there exists a $y \in Y$ such that
$$d_Y(y, f_0(a)) = t\, d_Y(f_0(a), f_0(a')), \quad d_Y(y, f_0(a')) = (1-t)\, d_Y(f_0(a), f_0(a')).$$
Thus,
$$d_Y(y, f_0(a)) \leq \lambda\, d_X(x,a) \quad \text{and} \quad d_Y(y, f_0(a')) \leq \lambda\, d_X(x,a').$$
Therefore, y is in the intersection of the two balls, and the proof is complete. $\square$

REMARK 1.21. The λ-ALR property in the study of Lipschitz continuous functions between metric spaces is analogous to the λ-injective property in the study of linear functions between Banach spaces.

A real Banach space Y is called λ-*injective* if whenever X is a Banach space with subspace E, any bounded linear function $T_0 : E \to Y$ can be extended to a bounded linear function $T : X \to Y$ with $\|T\| \leq \lambda\, \|T_0\|$, where $\lambda \geq 1$.

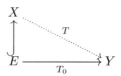

EXAMPLE 1.22. The space $\mathbb{R}$ is 1-injective by the Hahn-Banach theorem. If Γ is an arbitrary set, then $\ell_\infty(\Gamma)$ is 1-injective by applying the Hahn-Banach theorem to each coordinate separately.

THEOREM 1.23. *If Y is a real Banach space, then the following are equivalent:*
  (i) *Y is 1-injective.*
  (ii) *Y has BIP.*
  (iii) *Y is isometrically isomorphic to $C(\Omega)$, where Ω is a Stonean space.*

COMMENT 1.24. Recall that a *Stonean space* is an extremally disconnected compact Hausdorff space, where a topological space is *extremally disconnected* if the closure of every open set is open.

The proof of Theorem 1.23 is not given here, but a proof can be found in [**113**], as well as a brief discussion of its history. (See Theorem 2.1 in [**113**].)

For the next theorem, we recall that c_0 is the space of real-valued sequences that converge to zero. When equipped with the supremum norm, the sequence space c_0 is a Banach space.

THEOREM 1.25 (Lindenstrauss [**75**]). *The Banach space c_0 is a 2-ALR.*

PROOF. It suffices to show that there is a Lipschitz retraction $r : \ell_\infty \to c_0$ with $\mathrm{Lip}(r) = 2$. For each sequence $x = (x_n)_{n=1}^\infty$ in ℓ_∞, let
$$d(x) = \limsup_{n \to \infty} |x_n|.$$
Define the nth component of $r(x)$ in c_0 according to the rule
$$\bigl(r(x)\bigr)_n = \begin{cases} 0 & \text{if } |x_n| \leq d(x), \\ x_n - d(x)\,\mathrm{sign}(x_n) & \text{otherwise.} \end{cases}$$
If $x \in c_0$, then $d(x) = 0$, so $r(x) = x$.

Let x and y be sequences in ℓ_∞. Assume first that $|x_n| > d(x)$ and $|y_n| > d(y)$. Then
$$\bigl|(r(x))_n - (r(y))_n\bigr| = \left|\bigl(|x_n| - d(x)\bigr)\mathrm{sign}(x_n) - \bigl(|y_n| - d(y)\bigr)\mathrm{sign}(y_n)\right|.$$
If $\mathrm{sign}(x_n) = \mathrm{sign}(y_n)$, then
$$\bigl|(r(x))_n - (r(y))_n\bigr| \leq |x_n - y_n| + |d(x) - d(y)| \leq 2\|x - y\|_\infty.$$
If $\mathrm{sign}(x_n) = -\mathrm{sign}(y_n)$, then
$$\bigl|(r(x))_n - (r(y))_n\bigr| \leq \bigl(|x_n| - d(x)\bigr) + \bigl(|y_n| - d(y)\bigr) \leq |x_n| + |y_n| = |x_n - y_n|,$$
and so in this case
$$\bigl|(r(x))_n - (r(y))_n\bigr| \leq \|x - y\|_\infty.$$
Thus, either way, we have that
$$\bigl|(r(x))_n - (r(y))_n\bigr| \leq 2\|x - y\|_\infty.$$
Now assume that $|x_n| \leq d(x)$ and $|y_n| > d(y)$. (The argument is similar if the inequalities are switched.) In this case,
$$\bigl|(r(x))_n - (r(y))_n\bigr| = \left|\bigl(|y_n| - d(y)\bigr)\mathrm{sign}(y_n)\right| = |y_n| - d(y).$$
Observe that
$$|y_n| - d(y) = \bigl[|y_n| - d(x)\bigr] + \bigl[d(x) - d(y)\bigr].$$
It was assumed that $|x_n| \leq d(x)$, and so the above quantity is bounded by
$$\bigl[|y_n| - |x_n|\bigr] + \bigl[d(x) - d(y)\bigr] \leq |y_n - x_n| + |d(x) - d(y)| \leq 2\|x - y\|_\infty.$$
In every case,
$$\bigl|(r(x))_n - (r(y))_n\bigr| \leq 2\|x - y\|_\infty,$$

for all $n \in \mathbb{N}$, and so $\|r(x) - r(y)\|_\infty \leq 2\|x - y\|_\infty$. Therefore, the function r is a Lipschitz retraction with $\mathrm{Lip}(r) \leq 2$.

To show that $\mathrm{Lip}(r) = 2$, let $e = (1, 1, 1, \ldots)$ be the sequence where every term is 1, and let $e_n = (0, 0, 0, \ldots, 0, 1, 0, \ldots)$ be the sequence where the nth term is 1, but every other term is 0. Let $\rho : \ell_\infty \to c_0$ be a Lipschitz retraction with $\mathrm{Lip}(\rho) \leq \lambda$. Then
$$\|\rho(2e_n) - \rho(e)\|_\infty \leq \lambda \|2e_n - e\|_\infty = \lambda.$$
But $\rho|_{c_0} = \mathrm{Id}_{c_0}$. Thus,
$$\lambda \geq \|\rho(2e_n) - \rho(e)\|_\infty = \|2e_n - \rho(e)\|_\infty.$$
Therefore,
$$\left| (\rho(e))_n - 2 \right| \leq \lambda, \quad n \in \mathbb{N}.$$
Since $\rho(e) \in c_0$ (by assumption) and consequently $(\rho(e))_n \to 0$ as $n \to \infty$, it follows that $\lambda \geq 2$. □

COMMENT 1.26. In the proof of Theorem 1.25, it sufficed to show that there is a Lipschitz retraction $r : \ell_\infty \to c_0$ with $\mathrm{Lip}(r) = 2$ for the following reason: Let $i : c_0 \to \ell_\infty$ be the inclusion map. Then $i \circ f_0 : A \to \ell_\infty$ is a Lipschitz continuous function.

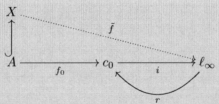

Since ℓ_∞ is a 1-ALR (Theorem 1.11), there exists a Lipschitz extension $\tilde{f} : X \to \ell_\infty$ with $\mathrm{Lip}(\tilde{f}) \leq \mathrm{Lip}(i \circ f_0) = \mathrm{Lip}(f_0)$. Then $f = r \circ \tilde{f} : X \to c_0$ is a Lipschitz extension of f_0 and $\mathrm{Lip}(f) \leq \mathrm{Lip}(r) \cdot \mathrm{Lip}(f_0)$.

The proof given above for Theorem 1.25 relied on the existence of a Lipschitz retraction from ℓ_∞ to c_0. The next result is a classical result that emphasizes the difference between the linear and nonlinear theory of Banach spaces.

THEOREM 1.27 (Phillips' Lemma). *There is no bounded linear projection of ℓ_∞ onto c_0.*

PROOF. If a map $P : \ell_\infty \to c_0$ is a bounded linear projection, then $P - \mathrm{Id}_{\ell_\infty}$ is a bounded linear operator from ℓ_∞ to itself such that $\ker(P - \mathrm{Id}_{\ell_\infty}) = c_0$. Thus, it suffices to show that there exists no bounded linear operator on ℓ_∞ with kernel c_0.

Therefore, begin by assuming that $S : \ell_\infty \to \ell_\infty$ is a bounded linear operator such that $\ker(S) = c_0$.

CLAIM 1. There is an uncountable family $\{\mathbb{A}_i\}_{i \in I}$ of infinite subsets of $\mathbb{N}$ such that $|\mathbb{A}_i \cap \mathbb{A}_j| < \infty$ whenever $i \neq j$.

Proof of claim. To prove this, use the fact that $\mathbb{N}$ can be put into one-to-one correspondence with $\mathbb{Q}$. Let $\phi : \mathbb{Q} \to \mathbb{N}$ be a bijection and let $I = \mathbb{R} \setminus \mathbb{Q}$ be the set of irrational numbers. For each $i \in I$, pick a sequence $(q_k^i)_{k=1}^\infty$ such that $q_k^i \in \mathbb{Q}$ for all $k \in \mathbb{N}$ and such that $q_k^i \to i$ as $k \to \infty$. Then the sets $\mathbb{A}_i = \{\phi(q_k^i) : k \in \mathbb{N}\}$ for $i \in I$ satisfy the requirements of the claim. ⋄

CLAIM 2. There exists some $i \in I$ such that $S|_{\ell_\infty(\mathbb{A}_i)} = 0$.

Proof of claim. Suppose the claim is not true. Then for each $i \in I$, there exists $\xi_i \in \ell_\infty(\mathbb{A}_i)$ such that $\|\xi_i\|_\infty = 1$ and $S\xi_i \neq 0$. Let $n_i \in \mathbb{N}$ be chosen so that $(S\xi_i)_{n_i} \neq 0$. There is some $m \in \mathbb{N}$ such that $n_i = m$ for uncountably many $i \in I$. Let $J = \{i \in I : n_i = m\}$. Then J is uncountable.

There exists a $\delta > 0$ such that the set $J' = \{i \in J : |(S\xi_i)_m| > \delta\}$ is still uncountable. By switching signs (if necessary), we assume that $(S\xi_i)_m$ is positive for all $i \in J'$, so that $(S\xi_i)_m > \delta$ for all $i \in J'$.

Pick a finite set $\{i_1, i_2, \ldots, i_N\} \subseteq J'$. Then
$$\|S\xi_{i_1} + S\xi_{i_2} + \cdots + S\xi_{i_N}\|_\infty \geq N\delta.$$
Because $|\mathbb{A}_{i_j} \cap \mathbb{A}_{i_k}| < \infty$ whenever $j \neq k$, the intersection of the support of any two sequences in the set $\{\xi_{i_1}, \xi_{i_2}, \ldots, \xi_{i_N}\}$ is finite. Therefore, it is possible to write
$$\xi_{i_1} + \xi_{i_2} + \cdots + \xi_{i_N} = \eta + \xi,$$
where $\|\eta\| \leq 1$ and ξ has finite support. In particular, $\xi \in c_0$, and so $S\xi = 0$. Thus,
$$N\delta \leq \|S(\xi_{i_1} + \xi_{i_2} + \cdots + \xi_{i_N})\|_\infty = \|S\eta\|_\infty \leq \|S\|.$$
This contradicts the assumption that S is a bounded linear operator. ⋄

From Claim 2, it follows that $\ell_\infty(\mathbb{A}_i) \subseteq \ker(S) = c_0$. This is a contradiction, because $\mathbb{A}_i$ is an infinite set. Therefore, there does not exist a bounded linear projection from ℓ_∞ onto c_0. □

Theorem 1.27 shows that there is no bounded linear projection of ℓ_∞ onto c_0, and so c_0 is not a complemented subspace of ℓ_∞.

COMMENT 1.28. We remind the reader that a closed subspace V of a Banach space X is said to be *complemented* in X provided there exists a bounded linear projection $P : X \to X$ such that $P(X) = V$. The name "complemented" comes from the fact that, in such a case, the Banach space X is linearly isomorphic to $V \oplus W$, where W is a closed linear subspace of X that is linearly isomorphic to X/V. (For more on this topic, see, for example, Theorem 4.42 in [21].)

REMARK 1.29. Theorem 1.27 is also known as the Phillips-Sobczyk lemma. It follows directly from the work of Phillips in 1940 [96], but was first formally observed by Sobczyk a year later, in 1941 [104]. The proof of Theorem 1.27 given above is based on the proof provided by Whitley in 1966 [111].

The next theorem, given without proof, is a fundamental result in the nonlinear theory of Banach spaces.

THEOREM 1.30 (Lindenstrauss). *If K is a compact metric space, then $\mathcal{C}(K)$ is a λ-ALR for some λ.*

REMARK 1.31. Theorem 1.30 was proved by Lindenstrauss in his 1964 paper [75]. His technique gave a value of $\lambda \leq 20$. However, in 2007, Kalton showed that $\mathcal{C}(K)$ is a 2-ALR [61].

COMMENT 1.32. Theorem 1.25 states that c_0 is a 2-ALR. On the other hand, Theorem 1.27 shows that c_0 is not λ-injective for any λ. In fact, it can be shown that there are no separable injective spaces. This was demonstrated by Rosenthal in 1970 when he showed that any infinite dimensional injective Banach space must contain a subspace linearly isomorphic to ℓ_∞ [102].

Although c_0 is not injective, it is what is known as *separably injective*. A Banach space Y is said to be λ-*separably injective* if whenever X is a *separable* Banach space with subspace E, any bounded linear function $T_0 : E \to Y$ can be extended to a bounded linear function $T : X \to Y$ with $\|T\| \leq \lambda \|T_0\|$. In 1941, Sobczyk showed that c_0 is 2-separably injective [104]. A very short proof of this theorem was provided by Veech in 1971 [109].

Not only is c_0 separably injective, it is the only infinite dimensional separable space that is separably injective. This was shown by Zippin in 1977 [112]. (A more detailed account of the development of these ideas can be found in [113].)

CHAPTER 2

Lipschitz extensions and Hilbert spaces

Let H be a Hilbert space with norm $\|\cdot\|$. If K is a nonempty closed convex subset of H, then for every $v \in H$ there is a unique point $\phi(v) \in K$ such that
$$\|v - \phi(v)\| = \inf_{w \in K} \|v - w\|.$$
That is, for every $v \in H$, there is a unique point $\phi(v)$ in K that is closest to x.[1] This defines a map $\phi : H \to K$ that is called the *closest-point map* for K.

PROPOSITION 2.1. *Let H be a Hilbert space and suppose K is a nonempty closed convex subset. The closest-point map $\phi : H \to K$ is a Lipschitz retraction with $\mathrm{Lip}(\phi) = 1$.*

PROOF. The closest-point map ϕ has the following property:

If $v \in H$, then $\langle v - \phi(v), \phi(v) - w \rangle \geq 0$ for all $w \in K$.

For x and y in H:
$$\|\phi(x) - \phi(y)\|^2 = \langle \phi(x) - \phi(y), \phi(x) - \phi(y) \rangle.$$
Expanding, this equals
$$\langle \phi(x) - \phi(y), x - y \rangle + \langle \phi(x) - \phi(y), \phi(x) - x \rangle - \langle \phi(x) - \phi(y), \phi(y) - y \rangle,$$
or
$$\langle \phi(x) - \phi(y), x - y \rangle - \langle x - \phi(x), \phi(x) - \phi(y) \rangle - \langle y - \phi(y), \phi(y) - \phi(y) \rangle.$$
The second and third inner products are nonnegative, by the above mentioned property of the closest-point map. (For the second inner product, let $v = x$ and $w = \phi(y)$. For the third, let $v = y$ and $w = \phi(y)$.) Therefore, by the Cauchy-Schwarz inequality,
$$\|\phi(x) - \phi(y)\|^2 \leq \langle \phi(x) - \phi(y), x - y \rangle \leq \|\phi(x) - \phi(y)\| \, \|x - y\|,$$
or
$$\|\phi(x) - \phi(y)\| \leq \|x - y\|.$$
This completes the proof. □

COMMENT 2.2. The property of the closest-point map ϕ used in the proof of Proposition 2.1 is not difficult to prove, but is by no means obvious. Suppose K is a closed convex subset of the Hilbert space H. Let $v \in H$ and $w \in K$. For each

---

[1]This fact is sometimes called the Hilbert projection theorem. For a proof, see (for example) Lemma 7.11 in [**21**], where it is called the "Closest Point Lemma".

$t \in (0,1)$, let $u_t = tw + (1-t)\phi(v)$. Then $u_t \in K$, because $\phi(v) \in K$ and K is a convex set. Since $\phi(v)$ is the point in K closest to v, it follows that
$$\|v - u_t\|^2 - \|v - \phi(v)\|^2 \geq 0.$$
Expanding this as an inner product, canceling terms, and ultimately dividing by t, this becomes
$$2\langle v - \phi(v), \phi(v) - w\rangle + t\|\phi(v) - w\|^2 \geq 0.$$
This is true for all $t \in (0,1)$, and so computing the limit as $t \to 0^+$ provides the result.

REMARK 2.3. If K is a closed subspace, then the closest-point map ϕ is the orthogonal projection.

THEOREM 2.4 (Kirszbraun's Theorem[2]). *Let H_1 and H_2 be Hilbert spaces. Suppose $A \subseteq H_1$ and let $f_0 : A \to H_2$ be a Lipschitz function. Then there is a Lipschitz extension $f : H_1 \to H_2$ with $\mathrm{Lip}(f) = \mathrm{Lip}(f_0)$.*

The proof of Theorem 2.4 will make use of the following proposition. The proposition is used to extend the Lipschitz function by one point and then Zorn's lemma is applied to extend to the entire space.[3]

PROPOSITION 2.5. *Suppose $\{B(x_i, r_i)\}_{i \in I}$ is a collection of closed balls in H_1 and $\{B(y_i, r_i)\}_{i \in I}$ is a collection of closed balls in H_2. If*
$$\|y_i - y_j\| \leq \|x_i - x_j\|, \quad \{i,j\} \subseteq I,$$
then
$$\bigcap_{i \in I} B(x_i, r_i) \neq \emptyset \implies \bigcap_{i \in I} B(y_i, r_i) \neq \emptyset.$$

PROOF. Suppose Proposition 2.5 is true for index sets I that are finite. Then if I is an infinite index set, the collection of balls $\{B(y_i, r_i)\}_{i \in I}$ has the finite intersection property. Thus, since all of the balls are weakly compact, there exists a point of intersection, and so the proof is complete. Therefore, it suffices to prove the proposition for I a finite set.

Assume $|I| < \infty$. Without loss of generality, assume that H_1 and H_2 are finite-dimensional Hilbert spaces. Let $x \in \bigcap_{i \in I} B(x_i, r_i)$, which is assumed to exist. If $x = x_{i_0}$ for some $i_0 \in I$, then $y_{i_0} \in B(y_i, r_i)$ for all $i \in I$. Thus, assume $x \neq x_i$ for any $i \in I$.

Define a function $r : H_2 \to \mathbb{R}$ by
$$r(z) = \max_{i \in I} \frac{\|y_i - z\|}{\|x_i - x\|}, \quad z \in H_2.$$
This function is continuous on H_2, and so r attains a minimum value at some point $y \in H_2$, since $\lim_{\|z\| \to \infty} r(z) = \infty$. Let $\lambda = r(y)$ and define
$$J = \left\{i \in I : \frac{\|y_i - y\|}{\|x_i - x\|} = \lambda\right\}.$$

---

[2]In a paper of 1934 [70], Kirszbraun proved the theorem for finite-dimensional Euclidean spaces. The theorem for Hilbert spaces was given by Valentine in 1945 [108]. Valentine seems to have been unaware of the work of Kirszbraun. (See also [107].)

[3]Kalton's exact words were "wave hands and say Zorn's lemma."

Note that $J \neq \emptyset$.

CLAIM. $y \in \operatorname{co}(\{y_i : i \in J\})$.

Proof of claim. Let $\phi(y)$ be the point in $\operatorname{co}(\{y_i : i \in J\})$ that is closest to y and assume $\phi(y) \neq y$. Observe that for each $i \in I$,
$$\langle y - \phi(y), y - y_i \rangle \geq \|y - \phi(y)\|^2,$$
because
$$\langle y - \phi(y), y - y_i \rangle - \langle y - \phi(y), y - \phi(y) \rangle = \langle y - \phi(y), \phi(y) - y_i \rangle \geq 0.$$
Let $\tau \in (0, 1)$ and let $y_\tau = (1 - \tau)y + \tau\phi(y)$ be a point on the line segment with endpoints at y and $\phi(y)$. Then for each $i \in J$:
$$\|y_\tau - y_i\|^2 = \|y_\tau - y + y - y_i\|^2 = \|y_\tau - y\|^2 + 2\langle y_\tau - y, y - y_i \rangle + \|y - y_i\|^2$$
$$= \| - \tau y + \tau\phi(y)\|^2 + 2\langle - \tau y + \tau\phi(y), y - y_i \rangle + \|y - y_i\|^2$$
$$= \tau^2 \|y - \phi(y)\|^2 - 2\tau\langle y - \phi(y), y - y_i \rangle + \|y - y_i\|^2.$$
Thus, making use of the earlier observation,
$$\|y_\tau - y_i\|^2 \leq \tau^2 \|y - \phi(y)\|^2 - 2\tau\|y - \phi(y)\|^2 + \|y - y_i\|^2 < \|y - y_i\|^2.$$
Then $r(y_\tau) < r(y)$, which contradicts the assumption that $r(y)$ is the minimum value of r. It follows that $\phi(y) = y$, and so $y \in \operatorname{co}(\{y_i : i \in J\})$. $\diamond$

Therefore, by the above claim, there is a finite sequence $(\alpha_i)_{i \in J}$ of positive real numbers such that $y = \sum_{i \in J} \alpha_i y_i$ and $\sum_{i \in J} \alpha_i = 1$. Consequently,
$$\sum_{i \in J} \alpha_i (y_i - y) = 0, \quad \text{and so} \quad \left\| \sum_{i \in J} \alpha_i (y_i - y) \right\|^2 = 0.$$
Observe that $\langle v, w \rangle = \frac{1}{2}(\|v\|^2 + \|w\|^2 - \|v - w\|^2)$. Thus,
$$0 = \left\| \sum_{i \in J} \alpha_i (y_i - y) \right\|^2 = \sum_{i \in J} \sum_{j \in J} \alpha_i \alpha_j \langle y_i - y, y_j - y \rangle$$
$$= \frac{1}{2} \sum_{i \in J} \sum_{j \in J} \alpha_i \alpha_j \Big(\|y_i - y\|^2 + \|y_j - y\|^2 - \|y_i - y_j\|^2 \Big)$$
$$\geq \frac{1}{2} \sum_{i \in J} \sum_{j \in J} \alpha_i \alpha_j \Big(\lambda^2 \|x_i - x\|^2 + \lambda^2 \|x_j - x\|^2 - \|x_i - x_j\|^2 \Big)$$
This last term can be rewritten as:
$$\frac{\lambda^2}{2} \sum_{i,j} \alpha_i \alpha_j \Big(\|x_i - x\|^2 + \|x_j - x\|^2 - \|x_i - x_j\|^2 \Big) + \frac{\lambda^2 - 1}{2} \sum_{i,j} \alpha_i \alpha_j \|x_i - x_j\|^2,$$
which equals
$$\lambda^2 \sum_{i \in J} \sum_{j \in J} \alpha_i \alpha_j \langle x_i - x, x_j - x \rangle + \frac{\lambda^2 - 1}{2} \sum_{i \in J} \sum_{j \in J} \alpha_i \alpha_j \|x_i - x_j\|^2.$$
Consequently,
$$0 \geq \lambda^2 \left\| \sum_{i \in J} \alpha_i (x_i - x) \right\|^2 + \frac{\lambda^2 - 1}{2} \sum_{i \in J} \sum_{j \in J} \alpha_i \alpha_j \|x_i - x_j\|^2.$$

This can only happen if $\lambda^2 \leq 1$ (because $\alpha_i > 0$ for all $i \in J$). As λ is nonnegative, it follows that $\lambda \leq 1$.

It has been established that
$$r(y) = \max_{i \in I} \frac{\|y_i - y\|}{\|x_i - x\|} \leq 1,$$
and hence
$$\|y_i - y\| \leq \|x_i - x\| \leq r_i, \quad i \in I.$$
Therefore, $y \in B(y_i, r_i)$ for each $i \in I$, as required. □

We now use Proposition 2.5 to prove Theorem 2.4 (Kirszbraun's theorem). As mentioned above, Proposition 2.5 will allow the extension of the Lipschitz function $f_0 : A \to H_2$ by one point, and then Zorn's lemma can be used to extend f_0 to the entire space H_1.

PROOF OF THEOREM 2.4. Without loss of generality, assume $\text{Lip}(f_0) = 1$. Let $x \in H_1$ be chosen so that $x \notin A$. It suffices to extend f_0 to a function $f : A \cup \{x\} \to H_2$ with $\text{Lip}(f) = 1$. (Then apply Zorn's lemma.)

By assumption,
$$\|f_0(a) - f_0(a')\| \leq \|a - a'\|, \quad \{a, a'\} \subseteq A.$$
Therefore, by Proposition 2.5,
$$x \in \bigcap_{a \in A} B(a, \|x - a\|) \implies \bigcap_{a \in A} B(f_0(a), \|x - a\|) \neq \emptyset.$$
Let $y \in \bigcap_{a \in A} B(f_0(a), \|x - a\|)$ and put $f(x) = y$. This completes the proof. □

REMARK 2.6. An alternate way of stating Theorem 2.4 is if $\omega_{f_0}(t) \leq Kt$, then f_0 can be extended to a a function $f : H_1 \to H_2$ such that $\omega_f(t) \leq Kt$. In fact, more can be said. If θ is a concave increasing function with $\lim_{t \to 0^+} \theta(t) = 0$, and if $\omega_{f_0}(t) \leq \theta(t)$, then there exists an extension $f : H_1 \to H_2$ such that $\omega_f(t) \leq \theta(t)$.[4]

THEOREM 2.7 (Minty[5]). Let (M, d) be a metric space with subset A and suppose H is a Hilbert space. If $f_0 : A \to H$ is a function for which there is a constant $K > 0$ such that
$$\|f_0(a) - f_0(b)\|_H \leq K \sqrt{d(a,b)}, \quad \{a,b\} \subseteq A,$$
then there exists an extension $f : M \to H$ such that
$$\|f(a) - f(b)\|_H \leq K \sqrt{d(a,b)}, \quad \{a,b\} \subseteq M.$$

COMMENT 2.8. If $g : X \to Y$ is a function between metric spaces such that $d_Y(g(x), g(y)) \leq K[d_X(x,y)]^\alpha$ for all x and y in X, then g is called *Hölder continuous of class α* (or simply *Hölder class α*) with constant K. In Theorem 2.7, the function f_0 is Hölder class $1/2$, and the theorem tells us that it can be extended to a function f that is also Hölder class $1/2$, and each has the same constant K.

---

[4] The extension of Theorem 2.4 to uniformly continuous functions is due to Grünbaum and Zarantonello [43]. For a proof, see Theorem 1.12 in [13].

[5] This theorem is a result of Minty from 1970 [85], although the original proof is different than the one presented here.

2. LIPSCHITZ EXTENSIONS AND HILBERT SPACES

PROOF. Without loss of generality, assume $M = A \cup \{x\}$, where $x \notin A$. (Apply Zorn's Lemma to extend in the general case.) Also without loss of generality, assume $K = 1$. The goal is to find an element $y \in H$ such that
$$\|y - f_0(a)\|_H \leq \sqrt{d(x,a)}, \quad a \in A,$$
and then let $y = f(x)$. Therefore, it suffices to show $\bigcap_{a \in A} B\big(f_0(a), \sqrt{d(x,a)}\big) \neq \emptyset$.

Consider the Hilbert space
$$\ell_2(A) = \Big\{h : A \to \mathbb{R} \;:\; \sum_{a \in A} |h(a)|^2 < \infty\Big\}.$$

Let
$$e_a(b) = \begin{cases} 1 & \text{if } b = a, \\ 0 & \text{if } b \neq a. \end{cases}$$

Note that
$$0 \in \bigcap_{a \in A} B_{\ell_2(A)}\big(\sqrt{d(x,a)}\, e_a, \sqrt{d(x,a)}\big).$$

In particular, this collection of closed balls in the Hilbert space $\ell_2(A)$ has nonempty intersection. Furthermore, observe that for any $\{a, a'\} \subseteq A$,
$$\|f_0(a) - f_0(a')\|_H \leq d(a,a')^{1/2} \leq \big(d(a,x) + d(a',x)\big)^{1/2}.$$

Therefore,
$$\|f_0(a) - f_0(a')\|_H \leq \big\| \sqrt{d(x,a)}\, e_a - \sqrt{d(x,a')}\, e_{a'} \big\|_{\ell_2(A)}.$$

Consequently, by Proposition 2.5, the intersection $\bigcap_{a \in A} B\big(f_0(a), \sqrt{d(x,a)}\big)$ is nonempty. Choose y in this intersection and let $f(x) = y$. This completes the proof. $\square$

COMMENT 2.9. In the above proof, $B_{\ell_2(A)}(x, r)$ is the closed ball centered at x with radius r in the metric space $\ell_2(A)$.

CHAPTER 3

Convex subsets and selections

3.1. The set of convex subsets

Let X be a Banach space and denote the metric induced by the norm by d_X. If K is a closed convex subset of X, then $\mathcal{H}(K)$ is the collection of all nonempty closed convex bounded subsets of K. If A and B are sets in $\mathcal{H}(K)$, then let

$$d(A,B) = \max\left\{\sup_{a\in A} d_X(a,B),\ \sup_{b\in B} d_X(b,A)\right\}.$$

Then d is a complete metric (called the *Hausdorff metric*) on $\mathcal{H}(K)$. Observe that the single-point set $\{x\}$ is in $\mathcal{H}(K)$ for each $x \in K$. In this way, K can be viewed as a subset of $\mathcal{H}(K)$. (Note that K is a member of $\mathcal{H}(K)$ only if K is bounded.)

COMMENT 3.1. Recall that if A is a set in the metric space (X, d_X), then $d_X(x, A) = \inf_{a \in A} d_X(x, a)$ is the distance between the point x and the set A.

THEOREM 3.2 (Lindenstrauss, 1964[1]). *If K is a closed convex subset of a real Banach space, then $\mathcal{H}(K)$ is an 8-ALR.*

QUESTION 3.3. *Can the constant 8 in Theorem 3.2 be improved?*

The proof of Theorem 3.2 will be given later (as Corollary 3.17) after some preliminary results.

PROPOSITION 3.4. *If K is a closed and convex subset of a Banach space X, then there exists a Lipschitz function $\phi : X \to \mathcal{H}(K)$ such that $\phi(x) = \{x\}$ for all $x \in K$ and with $\mathrm{Lip}(\phi) \leq 8$.*

PROOF. Denote the norm on X by $\|\cdot\|$ and the corresponding metric by d_X. Define $\phi : X \to \mathcal{H}(K)$ by

$$\phi(x) = B\bigl(x, 3\,d_X(x,K)\bigr) \cap K, \quad x \in X.$$

Let x and y be distinct elements in X. If x and y are both in K, then nothing needs to be shown, so assume (without loss of generality) that $x \notin K$. It suffices to estimate $d_X\bigl(u, \phi(x)\bigr)$, where $u \in \phi(y)$. Let $\theta = d_X(x,y)$. Then $\theta > 0$, by assumption.

---

[1]The origin of this theorem is in a 1961 paper by Isbell, who showed that $\mathcal{H}(X)$ is an absolute uniform retract. In particular, Isbell showed that if Y is a metric space containing $\mathcal{H}(X)$ as a subspace, then there is a uniformly continuous projection from Y onto $\mathcal{H}(X)$. (See Lemma 3.2 of [48].) In 1964, Lindenstrauss observed that Isbell's projection was actually Lipschitz continuous with a Lipschitz constant less than $\lambda < 12$ [75]. The bound 8 appearing in this theorem was discovered by Przesławski and Yost in 1995. (See Theorem 7 in [98].)

Let $\delta = d_X(x, K)$, which is assumed to be positive. Because $u \in \phi(y)$, and in particular $u \in B(y, 3\,d_X(y, K))$, it follows (using the triangle inequality) that
$$\|u - y\| \le 3\,d_X(y, K) \le 3\bigl(d_X(y, x) + d_X(x, K)\bigr) \le 3(\theta + \delta).$$
Consequently,
$$\|u - x\| \le \|u - y\| + \|y - x\| \le 3(\theta + \delta) + \theta = 4\theta + 3\delta.$$

Now let η be a real number such that $0 < \eta < \delta$, and pick an element $v \in K$ such that $\|v - x\| < \delta + \eta$. For each $t \in (0, 1)$, define $w_t = (1 - t)v + tu$. Note that $w_t \in K$ for each $t \in (0, 1)$, by the convexity of K. Observe that
$$\|w_t - x\| \le (1 - t)\|v - x\| + t\|u - x\| \le (1 - t)(\delta + \eta) + t(4\theta + 3\delta).$$
Pick
$$t = \frac{2\delta - \eta}{2\delta + 4\theta - \eta}.$$
A straightforward calculation shows that the rightmost term in the inequality evaluates to 3δ, which means that $\|w_t - x\| \le 3\delta$ for this choice of t, and so $w_t \in \phi(x)$ for t as given above.

Since the chosen value of t is in the interval $(0, 1)$, the element $w_t \in K$, and so the distance $d_X(u, K)$ can be estimated by $\|w_t - u\|$. To that end, observe that
$$\|w_t - u\| = \|(1 - t)v + tu - u\| = (1 - t)\|v - u\| = \frac{4\theta}{2\delta + 4\theta - \eta}\|v - u\|.$$
Note that
$$\|v - u\| \le \|v - x\| + \|x - u\| \le (\delta + \eta) + (4\theta + 3\delta) = 4\delta + 4\theta + \eta.$$
Thus, because $w_t \in \phi(x)$, it follows that
$$d_X\bigl(u, \phi(x)\bigr) \le \|u - w_t\| \le \frac{4\theta}{2\delta + 4\theta - \eta} \cdot (4\delta + 4\theta + \eta).$$
This bound is independent of the choice of η, and so (letting $\eta \to 0^+$) the bound becomes
$$d_X\bigl(u, \phi(x)\bigr) \le \frac{4\theta}{2\delta + 4\theta} \cdot (4\delta + 4\theta) = \frac{8\theta(\delta + \theta)}{\delta + 2\theta} \le 8\theta.$$
Therefore,
$$d\bigl(\phi(x), \phi(y)\bigr) \le 8\|x - y\|,$$
where d is the Hausdorff metric on $\mathcal{H}(K)$, and so $\mathrm{Lip}(\phi) \le 8$, as required. $\square$

DEFINITION 3.5. Let X be a real Banach space with closed convex subset K. If A is a bounded subset of K, then the *support functional* of A is the function $P_A : X^* \to \mathbb{R}$ defined by
$$P_A(x^*) = \sup_{x \in A} x^*(x), \quad x^* \in X^*.$$

REMARK 3.6. The real-valued function P_A is a sublinear functional on X^* with $|P_A(x^*)| \le c\|x^*\|$, where $c = \sup_{x \in A} \|x\|$, and c is finite because A is a bounded set.

Observe that if K is a closed convex subset of a Banach space X, then the map $A \mapsto P_A|_{\partial B_{X^*}}$ defines a function from $\mathcal{H}(K)$ into $\ell_\infty(\partial B_{X^*})$.

COMMENT 3.7. For notational simplicity, it is assumed that the support functional P_A is restricted to the set ∂B_{X^*} for the remainder of this chapter.

LEMMA 3.8. *If K is a closed and convex subset of a real Banach space X, then $\|P_A - P_B\|_{\ell_\infty(\partial B_{X^*})} = d(A, B)$ for all A and B in $\mathcal{H}(K)$.*

PROOF. Suppose $\|x^*\| = 1$. Let $\eta > 0$ be arbitrary and choose an $x \in A$ such that $x^*(x) > P_A(x^*) - \eta$.

There exists a $y \in B$ such that
$$\|x - y\| < d_X(x, B) + \eta \leq d(A, B) + \eta.$$
Then
$$x^*(x) - x^*(y) < d(A, B) + \eta.$$
Consequently,
$$d(A, B) > x^*(x) - x^*(y) - \eta > P_A(x^*) - x^*(y) - 2\eta.$$
Since $x^*(y) \leq P_B(x^*)$, it follows that
$$d(A, B) > P_A(x^*) - P_B(x^*) - 2\eta.$$
The choice of η was arbitrary, and so
$$d(A, B) \geq P_A(x^*) - P_B(x^*).$$
The argument can be repeated with the roles of A and B reversed, and so it is also true that $d(A, B) \geq P_B(x^*) - P_A(x^*)$. Therefore,
$$d(A, B) \geq |P_A(x^*) - P_B(x^*)|.$$
This is true for all $x^* \in \partial B_{X^*}$, and so $d(A, B) \geq \|P_A - P_B\|_{\ell_\infty(\partial B_{X^*})}$.

It remains to show the reverse inequality. Without loss of generality, assume $A \neq B$, so that $d(A, B) \neq 0$. Let η be a real number such that $0 < \eta < \frac{1}{2}d(A, B)$. Recall that
$$d(A, B) = \max\left\{\sup_{x \in B} d_X(x, A), \sup_{x \in A} d_X(x, B)\right\}.$$
Suppose that $d(A, B) = \sup_{x \in B} d_X(x, A)$. Then there exists an element $x \in B$ such that $d(A, B) < d_X(x, A) + \eta$. Let δ be any positive number for which we have the inequality $\delta < d_X(x, A) - \eta$. Then $x \notin \overline{A + \delta B_X}$. By the Hahn–Banach separation theorem,[2] there exists some element $x^* \in X^*$ with $\|x^*\| = 1$ such that $x^*(x) > x^*(a) + \delta x^*(u)$ for all $a \in A$ and $u \in B_X$. Then we have
$$x^*(x) > \sup_{a \in A} x^*(a) + \delta \sup_{u \in B_X} x^*(u) = P_A(x^*) + \delta\|x^*\| = P_A(x^*) + \delta.$$
Since $P_B(x^*) \geq x^*(x)$, it follows that
$$\delta < P_B(x^*) - P_A(x^*) \leq \|P_A - P_B\|_{\ell_\infty(\partial B_{X^*})}.$$
This is true for any $\delta < d_X(x, A) - \eta$, and so
$$\|P_A - P_B\|_{\ell_\infty(\partial B_{X^*})} \geq d_X(x, A) - \eta > d(A, B) - 2\eta.$$

---

[2]There are several formulations of the Hahn–Banach theorems. The version used here is the following: *Let E be a real locally convex topological vector space and let C be a closed nonempty convex subset of E. If $x \notin C$, then there exists a continuous linear functional x^* on E such that $x^*(x) > \sup_{c \in C} x^*(c)$.* (See Theorem 5.20 in [**21**].)

If $d(A,B) = \sup_{x \in A} d_X(x, B)$, then the above argument can be repeated by reversing the roles of A and B. In either case, the same inequality is obtained. The choice of η was arbitrary, and so $\|P_A - P_B\|_{\ell_\infty(\partial B_{X^*})} \geq d(A,B)$. The result follows. $\square$

LEMMA 3.9. *If K is a closed and convex subset of a real Banach space X and $t \in (0,1)$, then $tP_A + (1-t)P_B = P_{\overline{tA+(1-t)B}}$ for all A and B in $\mathcal{H}(K)$.*

PROOF. Let $C = tA + (1-t)B$. Let $x \in C$, so that $x = ta + (1-t)b$ for some $a \in A$ and $b \in B$. Then for $x^* \in \partial B_{X^*}$,
$$x^*(x) \leq \sup_{c \in C} x^*(c) \leq \sup_{c \in \overline{C}} x^*(c) = P_{\overline{C}}(x^*).$$
Computing directly,
$$x^*(x) = x^*\big(ta + (1-t)b\big) = tx^*(a) + (1-t)x^*(b).$$
Therefore,
$$tx^*(a) + (1-t)x^*(b) \leq P_{\overline{C}}(x^*).$$
This is true for every $a \in A$ and $b \in B$, and so
$$t \sup_{a \in A} x^*(a) + (1-t) \sup_{b \in B} x^*(b) \leq P_{\overline{C}}(x^*),$$
or
$$tP_A(x^*) + (1-t)P_B(x^*) \leq P_{\overline{C}}(x^*).$$
To show the reverse inequality let $q(x^*) = tP_A(x^*) + (1-t)P_B(x^*)$. If $c \in C$, then $c = ta + (1-t)b$ for some $a \in A$ and $b \in B$, and so
$$x^*(c) = tx^*(a) + (1-t)x^*(b) \leq tP_A(x^*) + (1-t)P_B(x^*) = q(x^*).$$
Now, let $x \in \overline{C}$ and choose $c_n \in C$ such that $\|x - c_n\| \to 0$ as $n \to \infty$. Then
$$x^*(x) \leq \sup_{n \in \mathbb{N}} x^*(c_n) \leq \sup_{n \in \mathbb{N}} q(x^*) = q(x^*).$$
This is true for all $x \in \overline{C}$, and consequently
$$P_{\overline{C}}(x^*) = \sup_{x \in \overline{C}} x^*(x) \leq q(x^*).$$
This completes the proof. $\square$

REMARK 3.10. Observe that the function $q = tP_A + (1-t)P_B$ from Lemma 3.9 is weak* lower semicontinuous.

REMARK 3.11. If X^* is a *reflexive* Banach space, then $tA + (1-t)B$ is weakly compact, and hence closed (in the norm topology) when A and B are in $\mathcal{H}(K)$. Thus, in this circumstance, $tP_A + (1-t)P_B = P_{tA+(1-t)B}$.

PROPOSITION 3.12. *If K is a closed and convex subset of a real Banach space X, then $\{P_A : A \in \mathcal{H}(K)\}$ is a closed and convex subset of $\ell_\infty(\partial B_{X^*})$.*

PROOF. Closure follows from Lemma 3.8 and convexity from Lemma 3.9. $\square$

If K is a closed convex subset of a real Banach space X, then Proposition 3.12 shows that, with the appropriate interpretation,
$$K \subseteq \mathcal{H}(K) \subseteq \ell_\infty(\partial B_{X^*}).$$

COMMENT 3.13. By Proposition 3.12, the set $\mathcal{H}(K)$ is a closed and convex subset of the Banach space $\ell_\infty(\partial B_{X^*})$. Consequently, $\mathcal{H}(\mathcal{H}(K))$ is a well defined object.

COROLLARY 3.14. *There exists a Lipschitz map* $\phi : \ell_\infty(\partial B_{X^*}) \to \mathcal{H}(\mathcal{H}(K))$ *such that* $\phi(P_A) = \{P_A\}$ *for* $A \in \mathcal{H}(K)$ *and such that* $\mathrm{Lip}(\phi) \leq 8$.

PROOF. By Proposition 3.12, the set $\{P_A : A \in \mathcal{H}(K)\}$ is a closed and convex subset of the Banach space $\ell_\infty(\partial B_{X^*})$. Therefore, by Proposition 3.4, there exists a Lipschitz function
$$\phi : \ell_\infty(\partial B_{X^*}) \longrightarrow \mathcal{H}\Big(\{P_A : A \in \mathcal{H}(K)\}\Big)$$
for which $\phi(P_A) = \{P_A\}$ for all $A \in \mathcal{H}(K)$ and $\mathrm{Lip}(\phi) \leq 8$. Since $\{P_A : A \in \mathcal{H}(K)\}$ is identifiable with the set $\mathcal{H}(K)$, the result follows. $\square$

LEMMA 3.15. *There exists a 1-Lipschitz retraction* $\psi : \mathcal{H}(\mathcal{H}(K)) \to \mathcal{H}(K)$ *such that the following diagram commutes.*

$$\begin{array}{ccc} \ell_\infty(\partial B_{X^*}) & \xrightarrow{\phi} & \mathcal{H}(\mathcal{H}(K)) \\ \uparrow & \swarrow_{\psi} & \\ \mathcal{H}(K) & & \end{array}$$

PROOF. For $\mathcal{C} \in \mathcal{H}(\mathcal{H}(K))$, define $\psi(\mathcal{C}) = \sup_{P_A \in \mathcal{C}} P_A$. This is a Lipschitz function with $\mathrm{Lip}(\psi) \leq 1$ and $\psi(\{P_A\}) = P_A$ for all $A \in \mathcal{H}(K)$. $\square$

COMMENT 3.16. To expand on the above proof, if
$$\mathcal{C} \in \mathcal{H}(\mathcal{H}(K)) = \mathcal{H}\Big(\{P_A : A \in \mathcal{H}(K)\}\Big),$$
then $\mathcal{C}$ is a closed bounded convex subset of $\{P_A : A \in \mathcal{H}(K)\}$. It follows that $\mathcal{C} = \{P_A : A \in \mathcal{C}_K\}$, where $\mathcal{C}_K$ is a family of closed bounded convex subsets of K. Then,
$$\psi(\mathcal{C}) = \sup_{P_A \in \mathcal{C}} P_A = \sup_{C \in \mathcal{C}_K} P_C \in \ell_\infty(\partial B_{X^*}).$$
Therefore, for each $x^* \in \partial B_{X^*}$,
$$\psi(\mathcal{C})(x^*) = \sup_{C \in \mathcal{C}_K} P_C(x^*) = \sup_{C \in \mathcal{C}_K} \Big(\sup_{x \in C} x^*(x)\Big) = \sup_{x \in D} x^*(x) = P_D(x^*),$$
where $D = \overline{\bigcup_{C \in \mathcal{C}_K} C}$, and so $\psi(\mathcal{C}) \in \mathcal{H}(K)$.

Note that the set D is a closed bounded convex set in K because $\mathcal{C}$ is a closed bounded convex set in $\mathcal{H}(K)$.

COROLLARY 3.17 (Theorem 3.2 restated). *If K is a closed convex subset of a real Banach space, then $\mathcal{H}(K)$ is an 8-ALR.*

PROOF. This follows from Corollary 3.14 and Lemma 3.15. $\square$

COMMENT 3.18. The argument used to prove Corollary 3.17 is similar to the one described in Comment 1.26. To prove that $\mathcal{H}(K)$ is a 8-ALR, it is necessary to show that whenever M is a metric space with subset M_0, any Lipschitz function $f_0 : M_0 \to \mathcal{H}(K)$ can be extended to a Lipschitz function $f : M \to \mathcal{H}(K)$ with $\mathrm{Lip}(f) \leq 8\,\mathrm{Lip}(f_0)$. Consider the following diagram:

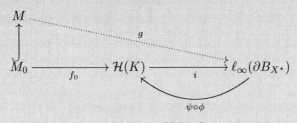

In this diagram, the map $g : M \to \ell_\infty(\partial B_{X^*})$ is a Lipschitz extension of the Lipschitz function $g_0 : M_0 \to \ell_\infty(\partial B_{X^*})$ defined by the composition $g_0 = i \circ f_0$, where $i : \mathcal{H}(K) \to \ell_\infty(\partial B_{X^*})$ is the inclusion map. It is known that $\ell_\infty(\partial B_{X^*})$ is a 1-ALR (by Theorem 1.11), and so it is known that the extension g of g_0 exists with $\mathrm{Lip}(g) \leq \mathrm{Lip}(g_0)$. Now let $f = \psi \circ \phi \circ g$ and the proof is complete.

The result of Corollary 3.17 can be generalized by the following theorem.

THEOREM 3.19. *Let S be a set and let A be a closed and convex subset of $\ell_\infty(S)$. If $\sup_{i \in I} a_i \in A$ whenever $\{a_i\}_{i \in I}$ is a collection of elements in A that is bounded above, then A is an 8-ALR.*

PROOF. Define $r : \ell_\infty(S) \to A$ by
$$r(x) = \sup \{a \in A : a(s) \leq x(s) + 3\,d_X(x, A) \text{ for all } s \in S\}, \quad x \in \ell_\infty(S).$$
Then r is a Lipschitz retraction with $\mathrm{Lip}(r) \leq 8$. (The arguments verifying this are similar to those seen leading up to the proof of Corollary 3.17.) □

REMARK 3.20. To see how Corollary 3.17 follows from Theorem 3.19, observe that the collection $\{C_i\}_{i \in I}$ of sets in $\mathcal{H}(K)$ corresponds to the subset $\{P_{C_i} : i \in I\}$ of $\ell_\infty(\partial B_{X^*})$. If this set is bounded above, then
$$\sup_{i \in I} P_{C_i} = P_C, \quad \text{where} \quad C = \overline{\mathrm{co}}\Big(\bigcup_{i \in I} C_i\Big).$$
Therefore, the set $\mathcal{H}(K)$ is an 8-ALR, by Theorem 3.19.

COMMENT 3.21. In the preceding remark, $S = \partial B_{X^*}$, $A = \{P_X : C \in \mathcal{H}(K)\}$, and the subset $\{P_{C_i} : i \in I\}$ of $\ell_\infty(\partial B_{X^*})$ is playing the role of the sequence $\{a_i\}_{i \in I}$. It is worth mentioning that the set C in the remark is the closed convex hull of the union of the sets in $\{C_i : i \in I\}$, rather than just the closure of the union (as in the proof of Lemma 3.15), because $\{P_{C_i} : i \in I\}$ is not assumed to come from $\mathcal{H}(A)$.

3.2. The Michael selection theorem

Suppose X is a Banach space and M is a metric space. If $\phi : M \to 2^X$, so that $\phi(a) \subset X$ for each $a \in M$, then a common question is "Does there exist a continuous selection function $f : M \to X$ such that $f(a) \in \phi(a)$ for each $a \in M$?"

A typical problem occurs when $M = Y$, where Y is a Banach space with closed subspace E, and $\phi : Y \to Y/E$ is the quotient map. When is it possible for each $y \in Y$ to continuously select a point $f(y)$ in $y + E$?

THEOREM 3.22 (Michael selection theorem). *Assume M is a metric space and X is a Banach space. If $\phi : M \to 2^X$ is a lower semi-continuous function such that $\phi(a)$ is a nonempty closed convex set for each $a \in M$, then there exists a continuous function $f : M \to X$ such that $f(a) \in \phi(a)$ for each $a \in M$.*

REMARK 3.23. The function $\phi : M \to 2^X$ is lower semi-continuous if and only if for each open set U in X, the set $\{a : \phi(a) \cap U \neq \emptyset\}$ is open in M.

REMARK 3.24. In the hypotheses of Theorem 3.22, it is assumed that M is a metric space; however, it is sufficient to assume only that M is paracompact, as this is the only metric space property of M that will be used in the proof of Theorem 3.22.

COMMENT 3.25. We recall that a topological space M (that is not necessarily a metric space) is called *paracompact* if every open cover has an open refinement that is locally finite. It may be worthwhile to recall what these terms mean.

A *refinement* of an open cover $\{U_i\}_{i \in I}$ of M is another open cover $\{V_j\}_{j \in J}$ of M for which each V_j is contained in some U_i. That is, for each $j \in J$, there exists at least one $i \in I$ so that $V_j \subseteq U_i$.

When we say that the refinement $\{V_j\}_{j \in J}$ is a *locally finite* cover, we mean that each point in M has an open neighborhood that intersects only finitely many sets in the cover. That is, for each x in M, there is an open set W containing x for which $W \cap V_j \neq \emptyset$ for only finitely may $j \in J$.

Any compact set is paracompact, since any open cover of a compact set admits a finite subcover. A theorem of Stone from 1948 shows that every metric space is paracompact [**105**]. (An accessible proof of this nontrivial fact can be found in [**103**].)

For the proof of Theorem 3.22, we will use an important property of paracompact Hausdorff spaces related to partitions of unity. Suppose that M is a paracompact Hausdorff space and let $\{U_i\}_{i \in I}$ be an open cover of M. Then there exists a partition of unity $\{\psi_\alpha\}_{\alpha \in A}$ where ψ_α is a real-valued continuous function on M with $0 \leq \psi_\alpha \leq 1$ for each $\alpha \in A$, such that:

(i) For each $\alpha \in A$, there is some $i \in I$ such that $\mathrm{supp}(\psi_\alpha) \subset U_i$.

(ii) $\displaystyle\sum_{\alpha \in A} \psi_\alpha(a) = 1$ for each $a \in M$.

(iii) For each $a \in M$, there exists an open neighborhood V of a such that the set $\{\alpha \in A : \mathrm{supp}(\psi_\alpha) \cap V \neq \emptyset\}$ is finite.

When a partition of unity $\{\psi_\alpha\}_{\alpha \in A}$ satisfies the properties given above, it is said to be *subordinate* to the open covering $\{U_i\}_{i \in I}$. Consequently, paracompact Hausdorff spaces (which include metric spaces) admit partitions of unity subordinate to any open cover. (For a proof of this, see Theorem 41.7 in [**86**].)

PROOF OF THEOREM 3.22. Since M is paracompact, given any open cover $\{U_i\}_{i \in I}$ of M, there exists a partition of unity that is subordinate to $\{U_i\}_{i \in I}$.

That is, there exists a collection $\{\psi_\alpha\}_{\alpha \in A}$ of functions, where ψ_α is a real-valued continuous function on M with $0 \le \psi_\alpha \le 1$ for each $\alpha \in A$, such that:

(i) For each $\alpha \in A$, there is some $i \in I$ such that $\mathrm{supp}(\psi_\alpha) \subset U_i$.
(ii) $\displaystyle\sum_{\alpha \in A} \psi_\alpha(a) = 1$ for each $a \in M$.
(iii) For each $a \in M$, there exists an open neighborhood V of a such that the set $\{\alpha \in A : \mathrm{supp}(\psi_\alpha) \cap V \ne \emptyset\}$ is finite.

(See the remarks at the end of Comment 3.25.)

CLAIM. *Let $\theta : M \to 2^X$ be a lower semi-continuous function such that $\theta(a)$ is a nonempty closed convex set for each $a \in M$. Given $\epsilon > 0$, there exists a continuous function $g : M \to X$ such that $d_X(g(a), \theta(a)) < \epsilon$ for all $a \in M$.*

Proof of claim. Let $\epsilon > 0$ be given. For each $v \in X$, define
$$G_v = \Big\{ a \in M : \mathrm{int}\,[B_{d_X}(v, \epsilon)] \cap \theta(a) \ne \emptyset \Big\}.$$
The collection $\{G_v\}_{v \in X}$ is an open covering of M. Therefore, there exists a partition of unity subordinate to this covering, say $\{\psi_\alpha\}_{\alpha \in A}$. To each $\alpha \in A$, assume that v_α is chosen so that $\mathrm{supp}(\psi_\alpha) \subset G_{v_\alpha}$. Let
$$g(a) = \sum_{\alpha \in A} \psi_\alpha(a)\, v_\alpha.$$
The function g is continuous. Furthermore,
$$a \in G_{v_\alpha} \implies d_X(v_\alpha, \theta(a)) < \epsilon \implies d_X(g(a), \theta(a)) < \epsilon,$$
which is as required. ◇

By the preceding claim, using ϕ in place of θ, there exists a continuous function $f_0 : M \to X$ such that $d_X(f_0(a), \phi(a)) < 1$ for all $a \in M$. Let
$$\phi_1(a) = \phi(a) \cap \mathrm{int}\,\Big[B_{d_X}(f_0(a), 1) \Big].$$
Similarly, applying the claim to ϕ_1 in place of θ, there is a continuous function $f_1 : M \to X$ such that $d_X(f_1(a), \phi_1(a)) < \tfrac{1}{2}$ for all $a \in M$. Let
$$\phi_2(a) = \phi(a) \cap \mathrm{int}\,\Big[B_{d_X}(f_1(a), \tfrac{1}{2}) \Big].$$
Continuing inductively, for each $n \in \mathbb{N}$, there is a continuous function $f_n : M \to X$ such that $d_X(f_n(a), \phi_n(a)) < \tfrac{1}{2^n}$ for all $a \in M$, where
$$\phi_n(a) = \phi(a) \cap \mathrm{int}\,\Big[B_{d_X}\Big(f_{n-1}(a), \tfrac{1}{2^{n-1}}\Big) \Big].$$

Since $\|f_n(a) - f_{n-1}(a)\|_X < \tfrac{3}{2^n}$ for each $n \in \mathbb{N}$, the limit $f(a) = \lim_{n \to \infty} f_n(a)$ exists uniformly in a, and hence the function f is continuous. It follows that $d(f(a), \phi(a)) = 0$, and therefore $f(a) \in \phi(a)$, because the set $\phi(a)$ is assumed to be a closed set. □

CHAPTER 4

Lipschitz classification of Banach spaces

A general problem in nonlinear functional analysis, known as the Lipschitz isomorphism problem, is to answer the following question.

QUESTION 4.1 (Lipschitz isomorphism problem). *If two Banach spaces are Lipschitz isomorphic, when are they also linearly isomorphic?*

COMMENT 4.2. Recall that X and Y are Lipschitz isomorphic if there exists a bijection $f : X \to Y$ such that both f and f^{-1} are Lipschitz functions. The two spaces are linearly isomorphic if such an f exists so that both f and f^{-1} are linear.

THEOREM 4.3 (Mazur–Ulam Theorem [**82**]). *Suppose that X and Y are real Banach spaces. If $f : X \to Y$ is a surjective isometry and $f(0) = 0$, then f is linear.*

The proof of Theorem 4.3 is given below, after some preliminary results.

REMARK 4.4. Suppose $f : X \to Y$ is a surjective isometry between the real Banach spaces X and Y, but $f(0) \neq 0$. If we define a new function $\widetilde{f} : X \to Y$ by the rule $\widetilde{f}(x) = f(x) - f(0)$ for all $x \in X$, then this new function $\widetilde{f}$ is a surjective isometry and $\widetilde{f}(0) = 0$. Consequently, we can apply Theorem 4.3 to this new function, and conclude that $\widetilde{f}$ is linear. It follows that $f = \widetilde{f} + f(0)$, and so f is an affine function, as it is the composition of a linear function with a translation.

REMARK 4.5. A Banach space Y is called *strictly convex* if whenever u and v are points in Y such that $\|u\| = 1$ and $\|v\| = 1$, the equality $\|u + v\| = 2$ implies that $u = v$. If Y is strictly convex, then any isometry $f : X \to Y$ between Banach spaces is necessarily affine, even if f is not a surjection. To see why, we recall that a function $f : X \to Y$ is affine if

(4.1) $$f\big((1-t)x + ty\big) = (1-t)f(x) + tf(y)$$

for all x and y in X and $t \in [0, 1]$. Now suppose that u and v are points in Y. If z is a point in Y such that $\|z - u\| = \|z - v\| = \frac{1}{2}\|u - v\|$, then z is called a *metric midpoint* of u and v. If a Banach space is strictly convex, then the point $z = \frac{u+v}{2}$ is the unique metric midpoint of u and v.

If $f : X \to Y$ is an isometry between Banach spaces, then f maps metric midpoints of x and y in X to metric midpoints of $f(x)$ and $f(y)$ in Y. In particular, $f\left(\frac{x+y}{2}\right)$ is a metric midpoint of $f(x)$ and $f(y)$ in Y. Consequently, if Y is assumed to be strictly convex, then $f\left(\frac{x+y}{2}\right) = \frac{f(x)+f(y)}{2}$, by uniqueness. This is sufficient to show that f is affine, because repeated application of the equality shows that (4.1) holds for any dyadic rational number t, and then continuity allows extension to all real numbers in the interval $[0, 1]$.

COMMENT 4.6. The Banach spaces ℓ_p and $L_p([0,1])$ are strictly convex if $1 < p < \infty$, and so metric midpoints are unique in these spaces. In fact, the spaces ℓ_p and $L_p([0,1])$ are uniformly convex if $1 < p < \infty$, and any uniformly convex space is necessarily strictly convex. A Banach space X is called *uniformly convex* if given $\epsilon > 0$, there exists a $\delta > 0$ such that if $\|u\| = \|v\| = 1$ and $\|u - v\| \geq \epsilon$, then

$$\left\| \frac{u+v}{2} \right\| \leq 1 - \delta.$$

(We will see this notion again in Definition 11.23.) The fact that a uniformly convex space is strictly convex follows immediately from the definitions. The nontrivial fact that ℓ_p and $L_p([0,1])$ are uniformly convex if $1 < p < \infty$ follows from Clarkson's inequalities. For more on this, see Example 11.24 (and the comment following it).

The spaces $L_1([0,1])$ and $L_\infty([0,1])$ are not strictly convex, and metric midpoints are plentiful in these spaces. For $L_\infty([0,1])$, this is perhaps apparent. For $L_1([0,1])$, it may not be so clear, and so we give the following proposition.

PROPOSITION 4.7. *Suppose that f and g are two distinct functions in $L_1(0,1)$ (so that $f \neq g$ almost everywhere). There is an infinite sequence $(h_k)_{k=1}^\infty$ of metric midpoints of f and g such that $\|h_j - h_k\| = \frac{1}{2}\|f - g\|$ whenever $j \neq k$.*

PROOF. First, we consider the case where $g = -f$. We will construct an infinite sequence $(f_k)_{k=1}^\infty$ of metric midpoints of f and $-f$ such that $\|f_j - f_k\| = \|f\|$ whenever $j \neq k$. For each $k \in \mathbb{N}$, we may (using induction) divide the interval $[0,1]$ into 2^k disjoint subsets $\{A_{k,1}, A_{k,2} \ldots A_{k,2^k}\}$ so that $A_{k,i} = A_{k+1,2i-1} \cup A_{k+1,2i}$ and such that $\int_{A_{k,i}} |f(x)| \, dx = 2^{-k}$ for each $i \in \{1, \ldots, 2^k\}$. Next, for each $k \in \mathbb{N}$, define a function $f_k : [0,1] \to \mathbb{R}$ so that $f_k(x) = (-1)^i f(x)$ for $x \in A_{k,i}$ and $i \in \{1, \ldots, 2^k\}$. Then $(f_k)_{k=1}^\infty$ is the sequence we wanted to construct.

Now we assume that f and g are two functions in $L_1(0,1)$ which are not equal almost everywhere. Then $v = \frac{f-g}{2}$ is a nonzero function in $L_1(0,1)$ and so (by what we showed in the first paragraph of this proof) there exists a sequence $(v_k)_{k=1}^\infty$ of metric midpoints of v and $-v$ for which $\|v_j - v_k\| = \|v\|$ whenever $j \neq k$. For each $k \in \mathbb{N}$, let

$$h_k = \frac{f+g}{2} + v_k.$$

Then h_k is a metric midpoint of f and g. Furthermore, whenever $j \neq k$, we have that $\|h_j - h_k\| = \|v\| = \frac{1}{2}\|f - g\|$, and so $(h_k)_{k=1}^\infty$ is a sequence of metric midpoints of f and g satisfying the desired property. $\square$

The sequence spaces ℓ_1 and ℓ_∞ are also not strictly convex, which is easy to demonstrate directly. For example, if $e_1 = (1, 0, 0, \ldots)$ and $e_2 = (0, 1, 0, \ldots)$, then $\|e_1\|_1 = \|e_2\|_1 = 1$ and $\|e_1 + e_2\|_1 = 2$, but $e_1 \neq e_2$. That shows that ℓ_1 is not strictly convex. For ℓ_∞, consider e_1 and $e_1 + e_2$. In this case, we have $\|e_1\|_\infty = \|e_1 + e_2\|_\infty = 1$ and $\|e_1 + (e_1 + e_2)\|_\infty = 2$, but $e_1 \neq e_1 + e_2$.

REMARK 4.8. Theorem 4.3 is not true if the Banach spaces are allowed to be *complex* Banach spaces. To see this, let X be a complex Banach space and define a new Banach space $\overline{X}$ with the same elements and norm as X, but with scalar multiplication $\otimes$ given by $\lambda \otimes x = \overline{\lambda} x$ for $\lambda \in \mathbb{C}$ and $x \in X$, where juxtaposition denotes the multiplication in X. The conclusion then follows from the following theorem, given here without proof.

THEOREM 4.9 (Bourgain, 1986 [18]). *There exists a complex Banach space X such that X and $\overline{X}$ are not linearly isomorphic.*

Before providing the proof of Theorem 4.3, some preliminary definitions and results will be given.

DEFINITION 4.10. Let S be a symmetric (so that $S = -S$) and bounded subset of a Banach space X. Then

$$\operatorname{diam}(S) = \sup\{\|s - s'\| : s \in S,\ s' \in S\},$$

$$\operatorname{rad}(S) = \inf\{r > 0 : \exists s_0 \in S \text{ with } \|s - s_0\| \leq r,\ \forall s \in S\}, \text{ and}$$

$$\operatorname{Cent}(S) = \{s_0 \in S : \|s - s_0\| \leq \operatorname{rad}(S),\ \forall s \in S\}$$

are the *diameter*, the *radius*, and the set of *centers* of S, respectively.

LEMMA 4.11. *If S is a symmetric and bounded subset of the Banach space X, and if $0 \in S$, then*
  (i) $\operatorname{rad}(S) = \frac{1}{2}\operatorname{diam}(S)$,
  (ii) $0 \in \operatorname{Cent}(S)$,
  (iii) $\operatorname{diam}(\operatorname{Cent}(S)) \leq \operatorname{rad}(S)$,
  (iv) $\operatorname{Cent}(S) = -\operatorname{Cent}(S)$.

PROOF. (i) Suppose that $M = \sup\{\|s\| : s \in S\}$. By the triangle inequality, $\operatorname{diam}(S) \leq 2M$. Since S is symmetric,

$$\operatorname{diam}(S) \geq \sup\{\|s - (-s)\| : s \in S\} = 2M.$$

Consequently, $\operatorname{diam}(S) = 2M$. It remains to show that $M = \operatorname{rad}(S)$. By choosing $s_0 = 0$ in the definition of the radius, it follows that $\operatorname{rad}(S) \leq M$. To show the reverse inequality, let $\delta > 0$ be given and choose $s_0 \in S$ such that

$$\sup_{s \in S} \|s - s_0\| < \operatorname{rad}(S) + \delta.$$

Observe that

$$2\|s\| = \|2s\| = \|s - (-s)\| \leq \|s - s_0\| + \|(-s) - s_0\| < 2\operatorname{rad}(S) + 2\delta.$$

Since the choice of δ is arbitrary, it follows that $\|s\| \leq \operatorname{rad}(S)$ for all $s \in S$, and so $M \leq \operatorname{rad}(S)$. This completes the proof of (i).
  (ii) This follows from the fact that $\|s\| \leq M$ for each $s \in S$ and $M = \operatorname{rad}(S)$.
  (iii) If s_0 and s_1 are in $\operatorname{Cent}(S)$, then $\|s_1 - s_0\| \leq \operatorname{rad}(S)$. Since this is true for every s_0 and s_1 in $\operatorname{Cent}(S)$, the result follows.
  (iv) Suppose that $s_0 \in \operatorname{Cent}(S)$. Then for any $s \in S$,

$$\|s - (-s_0)\| = \|s + s_0\| = \|(-s) - s_0\| \leq \operatorname{rad}(S),$$

because $-s \in S$, since S is symmetric. Thus, the point $-s_0$ is in $\operatorname{Cent}(S)$ whenever s_0 is in $\operatorname{Cent}(S)$, and so $\operatorname{Cent}(S)$ is symmetric. □

COMMENT 4.12. In the literature, it is common to define rad(S) and Cent(S) by
$$\operatorname{rad}(S) = \inf\left\{r > 0 : \exists x_0 \in X \text{ with } \|s - x_0\| \leq r,\ \forall s \in S\right\} \text{ and}$$
$$\operatorname{Cent}(S) = \left\{x_0 \in X : \|s - x_0\| \leq \operatorname{rad}(S),\ \forall s \in S\right\}.$$
(The definition of diam(S) is unchanged.) In this context, diam(S), rad(S), and Cent(S) are sometimes known as the *Chebyshev diameter*, the *Chebyshev radius*, and the set of *Chebyshev centers* of S, respectively. In [**13**], they are called (respectively) the *relative radius* and the *relative Chebyshev cnter*. (See Definition 1.23.) With these definitions, the conclusions in (i), (ii), and (iv) of Lemma 4.11 remain true (because S is assumed to be bounded), but it is not necessary to assume $0 \in S$. The conclusion in (iii) is weakened to diam(Cent(S)) $\leq$ diam(S).

LEMMA 4.13. *Let X and Y be Banach spaces and assume $f : X \to Y$ is a surjective isometry. If there exist points $x \in X$ and $y \in Y$ such that $f(x) = y$ and $f(-x) = -y$, then $f(0) = 0$.*

PROOF. Let $S_0 = B(x, \|x\|) \cap B(-x, \|x\|)$. Then S_0 is a bounded symmetric subset of X and $0 \in S_0$. Let
$$S_1 = \operatorname{Cent}(S_0),\ S_2 = \operatorname{Cent}(S_1),\ \ldots,\ S_n = \operatorname{Cent}(S_{n-1}),\ \ldots$$
Because f is a surjective isometry,
$$f(S_0) = B(y, \|y\|) \cap B(-y, \|y\|).$$
Furthermore,
$$f(\operatorname{Cent}(S_n)) = \operatorname{Cent}(f(S_n))\quad \text{for all}\quad n \in \{0, 1, 2, \ldots\}.$$
Note that diam(S_0) $\leq 2\|x\|$. Therefore,
$$\operatorname{diam}(S_n) \leq \frac{2\|x\|}{2^n} = \frac{\|x\|}{2^{n-1}}.$$
Since $S_0 \supseteq S_1 \supseteq S_2 \supseteq \cdots$, it follows that
$$\bigcap_{n=0}^{\infty} S_n = \{0\}\quad \text{and}\quad \bigcap_{n=0}^{\infty} f(S_n) = \{0\},$$
and so $f(0) = 0$. □

PROOF OF THEOREM 4.3. It suffices to show that $f\left(\frac{x_1+x_2}{2}\right) = \frac{f(x_1)+f(x_2)}{2}$ for each x_1 and x_2 in X. To that end, let x_1 and x_2 be elements in X and define a function $g : X \to Y$ (that depends on x_1 and x_2) by the rule
$$g(x) = f\left(\frac{x_1 + x_2}{2} + x\right) - \frac{1}{2}\Big[f(x_1) + f(x_2)\Big].$$
Observe that g is a surjective isometry because f is a surjective isometry. We will show that there exists a point $\tilde{x} \in X$ such that $g(-\tilde{x}) = -g(\tilde{x})$ and then use Lemma 4.13 to conclude that $g(0) = 0$.

Let $\tilde{x} = \frac{x_1 - x_2}{2}$. Then
$$g(\tilde{x}) = g\left(\frac{x_1 - x_2}{2}\right) = \frac{1}{2}f(x_1) - \frac{1}{2}f(x_2)$$

and
$$g(-\tilde{x}) = g\left(\frac{x_2 - x_1}{2}\right) = \frac{1}{2}f(x_2) - \frac{1}{2}f(x_1).$$
We observe that $g(-\tilde{x}) = -g(\tilde{x})$, and so it follows that $g(0) = 0$, by Lemma 4.13. Since $g(0) = 0$, we now have that
$$g(0) = f\left(\frac{x_1 + x_2}{2}\right) - \frac{1}{2}\Big[f(x_1) + f(x_2)\Big] = 0,$$
and so
$$f\left(\frac{x_1 + x_2}{2}\right) = \frac{1}{2}\Big[f(x_1) + f(x_2)\Big] = \frac{f(x_1) + f(x_2)}{2}.$$
This suffices to show that f is linear, because we can extend the above equality to dyadic rationals and then use the continuity of f to extend the equality to all real numbers. $\square$

In Theorem 4.3, the assumption that the isometry $f : X \to Y$ is a surjection cannot be relaxed if Y is not strictly convex.

EXAMPLE 4.14. Let $f : \mathbb{R} \to \ell_\infty^2$ be defined by the formula $f(x) = (x, \sin x)$. This map, which is not linear, is an isometry *into* the codomain, but not *onto* it.

We will not be able to answer Question 4.1, so instead we will address the following question that is related to Question 4.1.

QUESTION 4.15. *If a Banach space X isometrically embeds in a Banach space Y, does X linearly isometrically embed in Y?*

COMMENT 4.16. Question 4.15 is not asking if a given isometric embedding $f : X \to Y$ is necessarily linear. The question is asking the following: If $f : X \to Y$ is an isometric embedding of Banach spaces, does there exist a linear isometric embedding $g : X \to Y$? The function g will in general be a different function from the original function f.

THEOREM 4.17 (Figiel [36]). *Let $f : X \to Y$ be an isometry and let $[f(X)]$ denote the closed linear span of $f(X)$ in Y. If $[f(X)] = Y$ and $f(0) = 0$, then there is a linear quotient map $Q : Y \to X$ with $\|Q\| \le 1$ such that $(Q \circ f)(x) = x$ for all $x \in X$.*

EXAMPLE 4.18. The linear quotient map for f in Example 4.14 is given by the formula $Q(x, y) = x$.

The key to the proof of Theorem 4.17 is to show that for each $x^* \in X^*$, there exists a $y^* \in Y^*$ with
$$y^* \circ f = x^* \quad \text{and} \quad \|y^*\| \le \|x^*\|.$$
Such a y^* must be unique (by the density of $f(X)$ in Y) and it must be that $\|y^*\| = \|x^*\|$ (because f is an isometry). This result, which will be used to construct the linear quotient map Q, will be proved in multiple steps. The first step, which is Lemma 4.19 is to prove the result in a very specific case, which will then be extended.

LEMMA 4.19. *Let $f : X \to Y$ be an isometry of Banach spaces such that $f(0) = 0$, and assume $E \subset X$ is finite dimensional. If e_0^* is an element of E^* for which there exists some $e_0 \in E$ such that $\|e_0\| = 1$ and*
$$\|e_0 + tv\| = \|e_0\| + te_0^*(v) + o(|t|),$$
for all $t \in \mathbb{R}$ and $v \in E$, then there exists a $y^* \in Y^*$ such that $\|y^*\| \leq \|e_0^*\|$ and such that $(y^* \circ f)(v) = e_0^*(v)$ for all $v \in E$.

PROOF. Let $t > 0$. Because f is an isometry and $\|e_0\| = 1$,
$$\|f(te_0) - f(-te_0)\| = 2t.$$
For each $t > 0$, let $y_t^* \in Y^*$ be chosen as a norming element for $f(te_0) - f(-te_0)$, so that $\|y_t^*\| = 1$ and
$$y_t^*\Big(f(te_0) - f(-te_0)\Big) = 2t.$$

Suppose that a and b are real numbers such that $-t < a < b < t$. Then, again using the assumption that f is an isometry, together with the assumptions that $\|e_0\| = 1$ and $\|y_t^*\| = 1$,
$$y_t^*\Big(f(te_0) - f(be_0)\Big) \leq t - b \quad \text{and} \quad y_t^*\Big(f(ae_0) - f(-te_0)\Big) \leq a + t.$$
Consequently,
$$2t = y_t^*\Big(f(te_0) - f(-te_0)\Big)$$
$$= \underbrace{y_t^*\Big(f(te_0) - f(be_0)\Big)}_{\leq t-b} + y_t^*\Big(f(be_0) - f(ae_0)\Big) + \underbrace{y_t^*\Big(f(ae_0) - f(-te_0)\Big)}_{\leq a+t},$$
and so
$$y_t^*\Big(f(be_0) - f(ae_0)\Big) \geq 2t - (t - b) - (a + t) = b - a.$$
The reverse inequality is necessarily true, and so it follows that
$$y_t^*\Big(f(be_0) - f(ae_0)\Big) = b - a,$$
for each $t \in (0, \infty)$. Let y^* be a weak*-cluster point of the set $\{y_t^* : t \in (0, \infty)\}$ as $t \to \infty$. Then $\|y^*\| = 1$ and
$$y^*\Big(f(be_0) - f(ae_0)\Big) = b - a,$$
for all real numbers a and b.

The objective is to show that $(y^* \circ f)(v) = e_0^*(v)$ for all $v \in E$ for this choice of y^*. Let $v \in E$. By the assumptions on e_0^*, for all $t \in \mathbb{R}$,
$$\|e_0 + tv\| = 1 + te_0^*(v) + o(t).$$
Consequently,
$$\|te_0 - v\| = t\left\|e_0 - \frac{v}{t}\right\| = t\left(1 - \frac{e_0^*(v)}{t} + \frac{\epsilon(t)}{t}\right) = t - e_0^*(v) + \epsilon(t),$$
where $\epsilon(t) \to 0$ as $t \to \infty$. Thus,
$$y^*\Big(f(te_0) - f(v)\Big) \leq t - e_0^*(v) + \epsilon(t).$$
However,
$$y^*\Big(f(te_0) - f(v)\Big) = y^*\Big(f(te_0) - f(0e_0)\Big) - y^*f(v) = (t - 0) - y^*f(v),$$

and so
$$t - y^*f(v) \leq t - e_0^*(v) + \epsilon(t).$$
Simplifying this inequality and computing the limit as $t \to \infty$ results in the new inequality $y^*f(v) \geq e_0^*(v)$.

By a similar argument,
$$\|-te_0 - v\| = t + e_0^*(v) + \epsilon(t),$$
where $\epsilon(t) \to 0$ as $t \to \infty$, and so
$$y^*f(v) + t = y^*\Big(f(v) - f(-te_0)\Big) \leq t + e_0^*(v) + \epsilon(t).$$
It follows that $y^*f(v) \leq e_0^*(v)$. Therefore, $y^*f(v) = e_0^*(v)$, as required. $\square$

COMMENT 4.20. In the hypothesis of Lemma 4.19, it was assumed that e_0^* is an element of E^* for which there exists some $e_0 \in E$ such that $\|e_0\| = 1$ and
$$\|e_0 + tv\| = \|e_0\| + te_0^*(v) + o(t),$$
for all $t \in \mathbb{R}$ and all $v \in E$. When this equation is satisfied, it is said that the norm on E is *Gâteaux differentiable* at e_0 and e_0^* is called the *Gâteaux derivative* of the norm at e_0. (Gâteaux differentiability will be treated in more detail in Chapter 6.)

Lemma 4.19 states that there exists a $y^* \in Y^*$ such that $y^* \circ f = e_0^*$ on a finite dimensional subspace E provided that e_0^* is a Gâteaux derivative of the norm at some point e_0 in E. This may seem like too strong an assumption to make on e_0^*; however, the next theorem (given without proof) shows that there are a sufficient number of Gâteaux derivatives (at points having norm one) to extend the conclusion of Lemma 4.19 to a larger collection of linear functionals.

We remark here that Theorem 4.21 applies to separable Banach spaces. Theorem 4.17 (Figiel's Theorem) does not make the assumption that X is separable, and it is for this reason that we restrict our attention to finite dimensional subspaces of X in Lemmas 4.19 and 4.22. In the proof of Theorem 4.17, we will use the finite intersection property to extend to the entire Banach space X.

THEOREM 4.21 (Mazur [81]). *If X is a separable Banach space, then the set of $x_0 \in \partial B_X$ such that the norm is Gâteaux differentiable at x_0 is a dense G_δ-set.*

LEMMA 4.22. *Let $f : X \to Y$ be an isometry of Banach spaces and assume $E \subset X$ is finite dimensional. If $e^* \in E^*$, then there exists a $y^* \in Y^*$ such that $\|y^*\| \leq \|e^*\|$ and $(y^* \circ f)(v) = e^*(v)$ for all $v \in E$.*

PROOF. Without loss of generality, assume $\|e^*\| \leq 1$. Let V be the set of all $e_0^* \in B_{E^*}$ such that e_0^* is a Gâteaux derivative of the norm (on E) at a point $e_0 \in E$ for which $\|e_0\| = 1$.

CLAIM. $e^* \in \overline{\mathrm{co}}(V)$, where $\overline{\mathrm{co}}(V)$ is the closed convex hull of V.

Proof of claim. Suppose to the contrary that $e^* \notin \overline{\mathrm{co}}(V)$. Then there exists a point $e \in E$ with $\|e\| = 1$ such that
$$e^*(e) > \sup_{e_0^* \in V} e_0^*(e).$$

By Theorem 4.21, given $\epsilon > 0$, there exists a point $e_0 \in E$ with $\|e_0\| = 1$ such that $\|e - e_0\| < \epsilon$ and such that the norm is Gâteaux differentiable at e_0 with derivative $e_0^* \in V$. Then
$$\epsilon > \|e - e_0\| = \|e_0 + (-1)e\| = \|e_0\| + (-1)e_0^*(e) + o(|t|) \geq 1 - e_0^*(e).$$
Therefore,
$$e_0^*(e) > 1 - \epsilon.$$
Since the choice of $\epsilon > 0$ was arbitrary, it follows that $\sup_{e_0^* \in V} e_0^*(e) \geq 1$, and so
$$e^*(e) > \sup_{e_0^* \in V} e_0^*(e) \geq 1.$$
This contradicts the assumption that e and e^* have norm 1, and so $e^* \in \overline{\mathrm{co}}(V)$, as claimed. $\diamond$

We now return to the proof of Lemma 4.22. Suppose that $e^* \in \mathrm{co}(V)$. Then $e^* = \sum_{j=1}^m \lambda_j e_j^*$, where $e_j^* \in V$ and $\lambda_j \in [0,1]$ for each $j \in \{1, \ldots, m\}$ and $\sum_{j=1}^m \lambda_j = 1$. By Lemma 4.19, there exists for each $j \in \{1, \ldots, m\}$ a bounded linear functional $y_j^* \in Y^*$ such that $y_j^* \circ f = e_j^*$ on E. Let $y^* = \sum_{j=1}^m \lambda_j y_j^*$. Then $\|y^*\| \leq 1$ and, on E,
$$y^* \circ f = \left(\sum_{j=1}^m \lambda_j y_j^*\right) \circ f = \sum_{j=1}^m \lambda_j e_j^* = e^*.$$
Passing to limits, it follows that if $e^* \in E^*$, then there exists a bounded linear functional $y^* \in Y^*$ such that $\|y^*\| \leq 1$ and $y^* \circ f = e^*$ on E. $\square$

PROOF OF THEOREM 4.17 (FIGIEL'S THEOREM). Let $\mathcal{F}$ be the collection of all finite dimensional subspaces of X. Let $x^* \in X^*$ and without loss of generality assume that $\|x^*\| = 1$. Then $x^*|_E \in E^*$ for any $E \in \mathcal{F}$, and so (by Lemma 4.22) there exists a $y_E^* \in Y^*$ such that $\|y_E^*\| \leq \|x^*\|$ and such that $y_E^* \circ f = x^*|_E$ on E.

Let
$$A_E = \{y^* : \|y^*\| \leq 1 \text{ and } y^* \circ f = x^*|_E \text{ on } E\}.$$
It has been shown that A_E is not empty. Furthermore, the set A_E is weak*-closed, and hence weak*-compact. Suppose that $\{E_1, \ldots, E_m\}$ is a collection of sets from $\mathcal{F}$. Then
$$A_{E_1} \cap \cdots \cap A_{E_m} \supseteq A_{E_1 + \cdots + E_m}.$$
Thus, by the finite intersection property, there exists a $y^* \in \bigcap_{E \in \mathcal{F}} A_E$ such that $\|y^*\| = 1$ and $y^* \circ f = x^*$ on X.

For $\{\alpha_1, \ldots, \alpha_m\} \subset \mathbb{R}$ and $\{x_1, \ldots, x_m\} \subset X$, define $Q : f(X) \to X$ by
$$Q\Big(\sum_{j=1}^m \alpha_j f(x_j)\Big) = \sum_{j=1}^m \alpha_j x_j.$$
We claim that
$$\Big\|\sum_{j=1}^m \alpha_j x_j\Big\| \leq \Big\|\sum_{j=1}^m \alpha_j f(x_j)\Big\|.$$
In order to show this, let $x^* \in X^*$ be chosen so that $\|x^*\| = 1$ and
$$x^*\Big(\sum_{j=1}^m \alpha_j x_j\Big) = \Big\|\sum_{j=1}^m \alpha_j x_j\Big\|.$$

Next, choose $y^* \in Y^*$ such that $\|y^*\| = 1$ and such that $y^* \circ f = x^*$. Then

$$x^*\Big(\sum_{j=1}^m \alpha_j x_j\Big) = \sum_{j=1}^m \alpha_j\, x^*(x_j) = \sum_{j=1}^m \alpha_j\, y^*(f(x_j)) \leq \Big\|\sum_{j=1}^m \alpha_j f(x_j)\Big\|.$$

Consequently,

$$\Big\|Q\Big(\sum_{j=1}^m \alpha_j f(x_j)\Big)\Big\| = \Big\|\sum_{j=1}^m \alpha_j x_j\Big\| = x^*\Big(\sum_{j=1}^m \alpha_j x_j\Big) \leq \Big\|\sum_{j=1}^m \alpha_j f(x_j)\Big\|.$$

Therefore, the map Q extends to a map $Q : Y \to X$ (because Y is the closed linear span of $f(X)$) with $\|Q\| \leq 1$ such that $Q(f(x)) = x$. $\square$

COMMENT 4.23. Theorem 4.17 does not directly provide an answer to Question 4.15. In Chapter 6, however, it will be seen that Theorem 4.17 is a key tool in providing an answer to Question 4.15.

REMARK 4.24. Question 4.1 asks if it is possible to determine the linear structure of a Banach space given its nonlinear structure. That is:

When are two Lipschitz isomorphic Banach spaces also linearly isomorphic?

This question is a natural followup to an earlier question posed by Fréchet in 1928 [**37**]:

When are separable Banach spaces homeomorphic as metric spaces?

It was not until 1965 that Kadets [**55**] answered this question by showing that *any two separable infinite-dimensional Banach spaces are homeomorphic*. This was later extended by Anderson [**6**], who showed that any two separable locally convex Fréchet spaces are homeomorphic, and later still extended even further by Toruńczyk [**106**], who showed that two locally convex Fréchet spaces are homeomorphic so long as they have the same density character.

While the homeomorphic theory of Banach spaces (and indeed locally convex spaces) is understood, there is more to say when the locally convex assumption is relaxed. Indeed, in 1994, Cauty showed the existence of a separable F-space (complete metric linear space) that is not homeomorphic to a separable Banach space. In particular, Cauty showed that there is a separable F-space that is not an absolute retract [**22**].

CHAPTER 5

Arens–Eells space

QUESTION 5.1. *If a Banach space X Lipschitz embeds into a Banach space Y, is it true that X linearly embeds into Y?*

The above question, which was one of the early motivating questions in nonlinear functional analysis will be answered in this chapter.[1] The method used here (following closely the development by Godefroy and Kalton in [**39**]) involves a new space, called an Arens–Eells space. In order to define this space, several concepts must be introduced.

A metric space (M, d) is called *pointed* provided that there is a point 0 in M. For a pointed metric space M, let $\mathrm{Lip}(M)$ denote the collection of Lipschitz functions $f : M \to \mathbb{R}$ for which $f(0) = 0$. The set $\mathrm{Lip}(M)$ is a Banach space when given the Lipschitz norm:

$$\|f\|_{\mathrm{Lip}(M)} = \sup\left\{ \frac{|f(x) - f(y)|}{d(x, y)} \; : \; x \in M, \; y \in M, \; x \neq y \right\}.$$

The Banach space $\mathrm{Lip}(M)$ is also known as the *Lipschitz dual* of M.

REMARK 5.2. It is common in the literature to let $\mathrm{Lip}(M)$ denote the collection of all real-valued Lipschitz functions on M, and to denote the Lipschitz dual of M by $\mathrm{Lip}_0(M)$, which emphasizes the fact that $f(0) = 0$ for each f in $\mathrm{Lip}_0(M)$. The Lipschitz dual of M is also sometimes denoted $M^\sharp$.

If $x \in M$, then the evaluation functional $\delta_x : \mathrm{Lip}(M) \to \mathbb{R}$ is given by the formula $\delta_x(f) = f(x)$ for each $f \in \mathrm{Lip}(M)$. This defines a bounded linear functional on $\mathrm{Lip}(M)$, and so for each $x \in M$, the function δ_x is in $\mathrm{Lip}(M)^*$. Because $\delta_0 = 0$ and $\|\delta_x - \delta_y\| = d(x, y)$, the map $x \mapsto \delta_x$ is an isometric embedding of M into $\mathrm{Lip}(M)^*$ that maps 0 to 0. This embedding is denoted δ, so that $\delta : M \to \mathrm{Lip}(M)^*$ is given by the formula $\delta(x) = \delta_x$ for each $x \in M$.

DEFINITION 5.3. Let M be a pointed metric space. The closed linear span of $\delta(M)$ in $\mathrm{Lip}(M)^*$ is denoted $\text{Æ}(M)$ and is called the *Arens–Eells space* of M.

REMARK 5.4. The Arens–Eells space is named after Arens and Eells, who introduced the space in the 1950s [**7**], although the basic idea was due to Kantorovitch [**68**]. The name "Arens–Eells space" and the use of the symbol Æ were introduced by N. Weaver [**110**]. The Arens–Eells space of M is also called the *Lipschitz-free space over M*, in which case it is denoted $\mathcal{F}(M)$. The "free space" terminology and notation were popularized by Godefroy and Kalton [**39**], but can also be found (in a slightly different form) in the work of Pestov ([**94**], for example).

---

[1]We already asked a similar question in Question 4.15, but in that case we considered isometric embeddings. We will address Question 5.1 in this chapter and Question 4.15 in Chapter 6.

COMMENT 5.5. In the remainder of this chapter, δ will always be used to denote the natural embedding $\delta : M \to \AE(M)$ of the pointed metric space M into it's associated Arens–Eells space. If there is any risk of ambiguity due to the presence of multiple metric spaces, the embedding will be denoted δ_M to emphasize that M is the metric space.

If it is not apparent that $\delta : M \to \AE(M)$ is an isometry, observe that for fixed x and y in M, we have

$$\|\delta(x) - \delta(y)\| = \sup\left\{|(\delta_x - \delta_y)(f)| \,:\, \|f\|_{\mathrm{Lip}(M)} \le 1\right\}$$
$$= \sup\left\{|f(x) - f(y)| \,:\, \|f\|_{\mathrm{Lip}(M)} \le 1\right\} \le d(x, y).$$

In order to show that this inequality is actually an equality, it will suffice to find an example of a Lipschitz continuous function f with Lipschitz constant 1 such that $f(0) = 0$ and such that $|(\delta_x - \delta_y)(f)| = d(x, y)$, where x and y are given. One such function is

$$f(w) = d(w, y) - d(y, 0), \quad w \in M.$$

We will call $\delta : M \to \AE(M)$ the *Dirac lifting* or the *natural isometric embedding* of M into $\AE(M)$.

PROPOSITION 5.6. *If M is a pointed metric space, then $\AE(M)^* = \mathrm{Lip}(M)$.*

PROOF. Define a map $T : \AE(M)^* \to \mathrm{Lip}(M)$ so that for each $z^* \in \AE(M)^*$, the image $T(z^*)$ is a real valued function on M given by $T(z^*)(x) = z^*(\delta_x)$. Then for each x and y in M, we have that

$$|T(z^*)(x) - T(z^*)(y)| = |z^*(\delta_x) - z^*(\delta_y)| \le \|z^*\| \|\delta(x) - \delta(y)\| = \|z^*\| d(x, y),$$

where we have used the fact that δ is an isometry. It follows that the function $T(z^*)$ is a member of $\mathrm{Lip}(M)$ and $\|T(z^*)\|_{\mathrm{Lip}(M)} \le \|z^*\|$.

To show the reverse inequality, recall that $\AE(M)$ is the closed linear span of $\delta(M)$ in $\mathrm{Lip}(M)^*$ and let $\langle \cdot, \cdot \rangle$ denote the dual action of $\mathrm{Lip}(M)^*$ on $\mathrm{Lip}(M)$. Now let $\mu = \sum_{i=1}^n \alpha_i \delta_{x_i}$ where $\alpha_i \in \mathbb{R}$ and $x_i \in M$ for each i. Then $\mu \in \mathrm{Lip}(M)^*$ and

$$\langle \mu, g \rangle = \sum_{i=1}^n \alpha_i g(x_i)$$

for each $g \in \mathrm{Lip}(M)$. Consequently, for each $z^* \in \AE(M)^*$, we have

$$\langle \mu, T(z^*) \rangle = \sum_{i=1}^n \alpha_i T(z^*)(x_i) = \sum_{i=1}^n \alpha_i z^*(\delta_{x_i}) = z^*\left(\sum_{i=1}^n \alpha_i \delta_{x_i}\right) = z^*(\mu).$$

From this, we conclude that

(5.1) $$\langle \mu, T(z^*) \rangle = z^*(\mu)$$

for all $\mu \in \AE(M)$. Therefore, taking the supremum over such μ having norm 1 in $\AE(M)$, we have that

$$\|z^*\| = \sup_{\|\mu\|_{\AE(M)} = 1} |\langle \mu, T(z^*) \rangle| \le \|T(z^*)\|_{\mathrm{Lip}(M)},$$

as required.

It remains to show that T is a bijection. First, if $T(z_1^*) = T(z_2^*)$, then (5.1) tells us that $z_1^*(\mu) = z_2^*(\mu)$ for all $\mu \in \text{Æ}(M)$, and so $z_1^* = z_2^*$. It follows that T is a one-to-one mapping. To see that T is also a surjection, suppose that $f \in \text{Lip}(M)$. Define $z^* \in \text{Æ}(M)^*$ by $z^*(\mu) = \langle \mu, f \rangle$ for each $\mu \in \text{Æ}(M)$, where as before $\langle \cdot, \cdot \rangle$ denotes the dual action of $\text{Lip}(M)^*$ on $\text{Lip}(M)$. Then

$$T(z^*)(x) = z^*(\delta_x) = \langle \delta_x, f \rangle = f(x).$$

This completes the proof. $\square$

LEMMA 5.7. *Let M and M' be two pointed metric spaces. If $\phi : M \to M'$ is a Lipschitz function such that $\phi(0) = 0$, then there exists a unique linear function $T_\phi : \text{Æ}(M) \to \text{Æ}(M')$ such that $T_\phi \circ \delta_M = \delta_{M'} \circ \phi$ (see the diagram below)*

$$\begin{array}{ccc} \text{Æ}(M) & \xrightarrow{T_\phi} & \text{Æ}(M') \\ \delta_M \uparrow & & \uparrow \delta_{M'} \\ M & \xrightarrow{\phi} & M' \end{array}$$

and such that $\|T_\phi\| = \text{Lip}(\phi)$, where $\text{Lip}(\phi)$ is the Lipschitz constant for ϕ.

PROOF. For the given Lipschitz function $\phi : M \to M'$, define a new function $\phi^\sharp : \text{Lip}(M') \to \text{Lip}(M)$ by $\phi^\sharp(f) = f \circ \phi$ for $f \in \text{Lip}(M')$. The linear map $\phi^\sharp$ is weak*-to-weak* continuous, and hence there exists a bounded linear map T_ϕ between the preduals for which $\phi^\sharp$ is the Banach space operator adjoint. That is, there exists an operator $T_\phi : \text{Æ}(M) \to \text{Æ}(M')$ such that $(T_\phi)^* = \phi^\sharp$. In order to show that $T_\phi \circ \delta_M = \delta_{M'} \circ \phi$, let $x \in M$ and let $f \in \text{Lip}(M')$. Then,

$$\big((T_\phi \circ \delta_M)(x)\big)(f) = \big(T_\phi(\delta_M)_x\big)(f) = (\delta_M)_x\big(T_\phi^*(f)\big) = (\delta_M)_x(f \circ \phi)$$
$$= (f \circ \phi)(x) = f\big(\phi(x)\big) = (\delta_{M'})_{\phi(x)}(f) = \big((\delta_{M'} \circ \phi)(x)\big)(f).$$

Uniqueness follows directly.

For the final part of the proof, we show that $\|T_\phi\| = \text{Lip}(\phi)$. Since $\phi^\sharp$ is the adjoint of T_ϕ, we have that $\|T_\phi\| = \|(T_\phi)^*\| = \|\phi^\sharp\|$. To show that $\|\phi^\sharp\| = \text{Lip}(\phi)$, we first observe that

$$\|\phi^\sharp\| = \sup\{\|\phi^\sharp(f)\|_{\text{Lip}(M)} : \text{Lip}(f) \leq 1\} = \sup\{\|f \circ \phi\|_{\text{Lip}(M)} : \text{Lip}(f) \leq 1\}.$$

For each $f \in \text{Lip}(M')$ and each x and y in M, we have that

$$|(f \circ \phi)(x) - (f \circ \phi)(y)| \leq \text{Lip}(f) \text{Lip}(\phi) \, d_M(x, y),$$

and so $\|\phi^\sharp\| \leq \text{Lip}(\phi)$.

It remains to show the reverse inequality. Since the Lipschitz constant of ϕ is obtained as a supremum, for each $n \in \mathbb{N}$, we may pick x_n and y_n in M such that

$$d_{M'}(\phi(x_n), \phi(y_n)) > \text{Lip}(\phi) - \frac{1}{n}.$$

Since x_n and y_n are in M, it follows that $\delta_{M'}(\phi(x_n)) - \delta_{M'}(\phi(y_n)) \in \text{Æ}(M')$, and so we may choose some $z_n^* \in \text{Æ}(M')^*$ having norm 1 so that

$$z_n^*(\delta_{M'}(\phi(x_n)) - \delta_{M'}(\phi(y_n))) = \|\delta_{M'}(\phi(x_n)) - \delta_{M'}(\phi(y_n))\|_{\text{Æ}(M')}.$$

The map $\delta_{M'}$ is an isometry, and so we have that

$$|(z_n^* \circ \delta_{M'} \circ \phi)(x_n) - (z_n^* \circ \delta_{M'} \circ \phi)(y_n)| = d_{M'}(\phi(x_n), \phi(y_n)) > \text{Lip}(\phi) - \frac{1}{n}.$$

Consequently, for each $n \in \mathbb{N}$,
$$\|z_n^* \circ \delta_{M'} \circ \phi\|_{\mathrm{Lip}(M)} \geq \mathrm{Lip}(\phi) - \frac{1}{n}.$$
If for each $n \in \mathbb{N}$ we let $f_n = z_n^* \circ \delta_{M'}$, then $f_n \in \mathrm{Lip}(M')$ and $\mathrm{Lip}(f_n) = 1$. Therefore,
$$\|\phi^\sharp\| = \sup\{\|f \circ \phi\|_{\mathrm{Lip}(M)} : \mathrm{Lip}(f) \leq 1\} \geq \mathrm{Lip}(\phi).$$
This gives the reverse inequality, and so the proof is complete. $\square$

COROLLARY 5.8. *Let M and M' be two pointed metric spaces. If $M \subseteq M'$, then $\AE(M) \subseteq \AE(M')$.*

Suppose that X is a Banach space. Define the *barycenter map* $\beta : \AE(X) \to X$ by setting
$$\beta\left(\sum_{i=1}^n \alpha_i \delta_{x_i}\right) = \sum_{i=1}^n \alpha_i x_i,$$
for finite sequences of points $(x_i)_{i=1}^n$ in X and scalars $(\alpha_i)_{i=1}^n$ in $\mathbb{R}$, and then extending to $\AE(X)$ using density. The fact that we may extend β in this way follows from the next lemma.

LEMMA 5.9. *If X is a Banach space, then the barycenter map $\beta : \AE(X) \to X$ is a linear quotient map and $\beta\delta = \mathrm{Id}_X$.*

PROOF. Suppose $(x_i)_{i=1}^n$ is a finite sequence of points in X and $(\alpha_i)_{i=1}^n$ is a finite sequence of scalars in $\mathbb{R}$. let x^* be an element in X^* such that $\|x^*\| = 1$ and such that x^* is a norming element for $\sum_{i=1}^n \alpha_i x_i$ in the norm on X. Then
$$\left\|\sum_{i=1}^n \alpha_i x_i\right\|_X = x^*\left(\sum_{i=1}^n \alpha_i x_i\right) = \sum_{i=1}^n \alpha_i x^*(x_i) = \sum_{i=1}^n \alpha_i \delta_{x_i}(x^*) \leq \left\|\sum_{i=1}^n \alpha_i \delta_{x_i}\right\|_{\AE(X)}.$$
Therefore, the function β extends to a bounded linear operator from $\AE(X)$ onto X. The rest follows directly. $\square$

COROLLARY 5.10. *Let M be a pointed metric space and let X be a Banach space. If $\phi : M \to X$ is a Lipschitz function such that $\phi(0) = 0$, then there exists a unique linear function $T : \AE(M) \to X$ such that $\|T\| = \mathrm{Lip}(\phi)$ and $T \circ \delta = \phi$.*

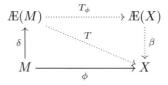

PROOF. This follows from Lemmas 5.7 and 5.9. (See the diagram above.) $\square$

Of particular interest will be the case when X and Y are Banach spaces such that $X \subseteq Y$. In that case, the diagram is as below, where $i : X \to Y$ is the inclusion map.

$$\begin{array}{ccc} Æ(X) & \xrightarrow{T_i} & Æ(Y) \\ {\scriptstyle \delta_X}\uparrow & & \uparrow{\scriptstyle \delta_Y} \\ X & \xrightarrow{i} & Y \end{array}$$

In this case, the linear map T_i is an isometric embedding, so that it can be seen that $Æ(X) \subseteq Æ(Y)$.

LEMMA 5.11. *Let M be a pointed metric space and suppose $\mu \in Æ(M)$. For any $\epsilon > 0$, there exists a sequence of real numbers $(a_n)_{n=1}^\infty$ and sequences of points $(x_n)_{n=1}^\infty$ and $(y_n)_{n=1}^\infty$ in M such that*

$$\mu = \sum_{n=1}^\infty a_n [\delta(x_n) - \delta(y_n)],$$

where $\sum_{n=1}^\infty |a_n| d(x_n, y_n) \leq \|\mu\|_{Æ(M)} + \epsilon$.

PROOF. Let $\langle \cdot, \cdot \rangle$ denote the dual action of $\mathrm{Lip}(M)$ on $Æ(M)$. Observe that

$$B_{Æ(M)} = \overline{\mathrm{co}}\left\{ \frac{1}{d(x,y)}[\delta(x) - \delta(y)] \ : \ x \in M, y \in M, x \neq y \right\}.$$

If this were not the case, then given $\mu \in Æ(M)$ with $\|\mu\| \leq 1$, there would exist a function $f \in \mathrm{Lip}(M)$ with $f(0) = 0$ such that

$$\|f\|_{\mathrm{Lip}(M)} \geq \langle \mu, f \rangle > \sup_{\substack{x \in M, y \in M \\ x \neq y}} \left\langle \frac{1}{d(x,y)}[\delta(x) - \delta(y)], f \right\rangle = \|f\|_{\mathrm{Lip}(M)},$$

which is a contradiction. The result follows from this fact. $\square$

PROPOSITION 5.12. *Let X be a Banach space. If $T : \ell_\infty \to X$ is a linear function such that $T|_{c_0}$ is an isomorphism, then X contains a copy of ℓ_∞ and, in particular, X is not separable.*

PROOF. Let $E = T(c_0)$. There exists a sequence $(e_n^*)_{n=1}^\infty$ of elements in E^* with $\sup_{n \in \mathbb{N}} \|e_n^*\| = \|T^{-1}\|$ such that, for each $e \in E$,

$$T^{-1}(e) = \left(e_n^*(e) \right)_{n=1}^\infty.$$

For each $n \in \mathbb{N}$, use the Hahn-Banach theorem to extend $e_n^* \in E^*$ to a linear functional $x^* \in X^*$ with $\|x_n^*\| = \|e_n^*\|$. Define a linear operator $S : X \to \ell_\infty$ by

$$S(x) = \left(x_n^*(x) \right)_{n=1}^\infty.$$

Then $ST : \ell_\infty \to \ell_\infty$ and $ST|_{c_0} = I$, where I is the identity operator on ℓ_∞. It follows that $ST - I : \ell_\infty \to \ell_\infty$ is a bounded linear operator on ℓ_∞ for which $c_0 \subseteq \ker(ST - I)$.

Recall the proof of Phillips' lemma (Theorem 1.27). In the proof, we showed that a bounded linear operator on ℓ_∞ cannot have c_0 as its kernel. However, we actually showed more than this. Indeed, we showed that if W is a bounded linear operator on ℓ_∞ for which $c_0 \subseteq \ker(W)$, then there exists a subspace Y of ℓ_∞ such that Y is linearly isomorphic to ℓ_∞ and such that $Y \subseteq \ker(W)$. That is, if $\ker(W)$ contains a copy of c_0, then it must contain a copy of ℓ_∞.

Now observe that $ST - I : \ell_\infty \to \ell_\infty$ is a bounded linear operator on ℓ_∞ for which $c_0 \subseteq \ker(ST - I)$. Consequently, there exists a subspace Y in ℓ_∞ such that Y is linearly isomorphic to ℓ_∞ and $Y \subseteq \ker(ST - I)$. It follows that $ST|_Y = I$, and so $T(Y)$ is a linear subspace of X that is linearly isomorphic to ℓ_∞. $\square$

COROLLARY 5.13. *If an injective Banach space contains a copy of c_0, then it must also contain a copy of ℓ_∞.*

PROOF. Let X be an injective Banach space and let Z be a subspace of X that is linearly isomorphic to c_0. There exists a map $T_0 : c_0 \to X$ that is an isomorphism onto the image $T_0(c_0) = Z$. Since X is injective, T_0 can be extended to a map $T : \ell_\infty \to X$ such that $T|_{c_0}$ is an isomorphism. Therefore, by Proposition 5.12, X contains a copy of ℓ_∞. $\square$

DEFINITION 5.14. A Banach space X is said to have the *separable complementation property* (or *SCP*, for short) provided that whenever Y is a separable subspace of X, there exists a separable subspace Z of X such that $Y \subseteq Z$ and Z is complemented in X.

COMMENT 5.15. We remind the reader that a subspace Z of a Banach space X is said to be *complemented* in X provided there exists a bounded linear projection $P : X \to X$ such that $P(X) = Z$.

We remark that, since there exists no bounded linear projection of ℓ_∞ onto c_0, by Phillips' lemma (Theorem 1.27), the subspace c_0 is not complemented in ℓ_∞.

PROPOSITION 5.16. *If X has SCP, then X does not contain a subspace isomorphic to ℓ_∞.*

PROOF. Suppose X has SCP and suppose X contains a subspace W isomorphic to ℓ_∞. The subspace W contains a subspace Y that is isomorphic to c_0, which is separable. Consequently, because X has SCP, there exists a separable subspace Z of X that is complemented in X and that contains Y (which we recall is a copy of c_0). It follows that the subspace $Z \cap W$ is complemented in W and contains a copy of c_0.

Since W is isomorphic to ℓ_∞, it is injective. A complemented subspace of an injective space is also injective, and so $Z \cap W$ is an injective space that contains a copy of c_0. By Corollary 5.13, the injective space $Z \cap W$ contains a copy of ℓ_∞. This contradicts the fact that Z is separable. $\square$

PROPOSITION 5.17. *Æ(ℓ_∞) has SCP.*

PROOF. Let Y be a separable subspace of Æ(ℓ_∞). Then there exists a separable subspace S of ℓ_∞ such that $Y \subseteq$ Æ(S) and such that S is isometrically isomorphic to some space $\mathcal{C}(K)$, where K is a compact metric space. It follows that S is a 2-ALR (by Remark 1.31), and so there exists a Lipschitz retraction $r : \ell_\infty \to S$

such that $\text{Lip}(r) \leq 2$ (by Corollary 1.14).

$$\begin{array}{ccc} \text{Æ}(\ell_\infty) & \xrightarrow{T_r} & \text{Æ}(S) \\ {\scriptstyle \delta_{\ell_\infty}}\big\uparrow & & \big\uparrow {\scriptstyle \delta_S} \\ \ell_\infty & \xrightarrow{r} & S \end{array}$$

Therefore, by Lemma 5.7, there exists a linear operator $T_r : \text{Æ}(\ell_\infty) \to \text{Æ}(S)$ such that $\|T_r\| \leq 2$ and such that $T_r \circ \delta_{\ell_\infty} = \delta_S \circ r$. Consequently, T_r is a projection onto $\text{Æ}(S)$, and so $\text{Æ}(S)$ is complemented in $\text{Æ}(\ell_\infty)$. $\square$

COMMENT 5.18. In the proof of Proposition 5.17, Y is a separable subspace of $\text{Æ}(\ell_\infty)$, which is the closed linear span of the set $\{\delta(x) : x \in \ell_\infty\}$. Thus, there exists a countable subset $\{x_n : n \in \mathbb{N}\}$ of ℓ_∞ such that $Y \subseteq [\delta(x_n)]_{n=1}^\infty$. Let $S = [x_n]_{n=1}^\infty$. Then S is a separable subspace of ℓ_∞ and $Y \subseteq \text{Æ}(S)$. If S is not isometrically isomorphic to a $\mathcal{C}(K)$ space, it may be necessary to enlarge it slightly, by adding the constant sequence $e = (1, 1, 1, \ldots)$ and then looking at the algebraic closure. More precisely, consider the smallest closed linear subspace of ℓ_∞ that contains S and e that is also closed under pointwise multiplication.

A characterization of real Banach spaces that are isometrically isomorphic to $\mathcal{C}(K)$ spaces can be found in [5] (Theorem 4.2.1).

COROLLARY 5.19. ℓ_∞ embeds isometrically into $\text{Æ}(\ell_\infty)$, but not linearly isomorphically.

PROOF. The map $\delta : \ell_\infty \to \text{Æ}(\ell_\infty)$ is an isometric embedding; however, the space $\text{Æ}(\ell_\infty)$ has SCP (by Proposition 5.17), and so cannot contain a copy of ℓ_∞ as a linear subspace (by Proposition 5.16). $\square$

COROLLARY 5.20. If $\beta : \text{Æ}(\ell_\infty) \to \ell_\infty$ is the barycenter map for ℓ_∞, then $\text{Æ}(\ell_\infty)$ is Lipschitz isomorphic to $\ker(\beta) \oplus \ell_\infty$. In particular, there are Lipschitz isomorphic Banach spaces that are not also linearly isomorphic.

PROOF. Let $\delta : \ell_\infty \to \text{Æ}(\ell_\infty)$ be the natural embedding of ℓ_∞ into $\text{Æ}(\ell_\infty)$. Define a map $f : \text{Æ}(\ell_\infty) \to \ker(\beta) \oplus \ell_\infty$ by

$$f(\mu) = \Big(\mu - (\delta \circ \beta)(\mu), \beta(\mu)\Big), \quad \mu \in \text{Æ}(\ell_\infty).$$

The map f is the desired Lipschitz isomorphism. It follows that $\text{Æ}(\ell_\infty)$ is Lipschitz isomorphic to $\ker(\beta) \oplus \ell_\infty$, but $\text{Æ}(\ell_\infty)$ is not linearly isomorphic to $\ker(\beta) \oplus \ell_\infty$, by Corollary 5.19. $\square$

COMMENT 5.21. It will be convenient to denote that two spaces A and B are Lipschitz isomorphic using the notation

$$A \stackrel{\text{Lip}}{\approx} B.$$

A linear isomorphism between A and B will be denoted by $A \approx B$.

Corollary 5.19 shows that there exists a Banach space that Lipschitz embeds into another Banach space (isometrically, in fact), but does not linearly embed into that Banach space. This answers Question 5.1 negatively. Furthermore, with Corollary 5.20, we have shown that there exist Banach spaces which are Lipschitz isomorphic, but not linearly isomorphic, providing a partial answer to Question 4.1.

Whether or not Lipschitz isomorphic Banach spaces are necessarily linearly isomorphic was an open question for a long time, until it was first answered by Aharoni and Lindenstrauss in 1978 [**3**]. The method used by Aharoni and Lindenstrauss was quite different than the one used here.

Recall that there exists an uncountable family $\{\mathbb{A}_i\}_{i \in I}$ of infinite subsets of $\mathbb{N}$ such that $|\mathbb{A}_i \cap \mathbb{A}_j| < \infty$ whenever $i \neq j$. (See the proof of Theorem 1.27, which is Phillips' lemma.) Let $q: \ell_\infty \to \ell_\infty/c_0$ be the quotient map. Then $\{q(\chi_{\mathbb{A}_i})\}_{i \in I}$ is a basis for $c_0(I)$, where $\chi_{\mathbb{A}_i}$ is the characteristic function for $\mathbb{A}_i$.

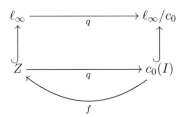

Aharoni and Lindenstrauss showed that if Z is the closed linear subspace of ℓ_∞ spanned by c_0 and $\{\chi_{\mathbb{A}_i}\}_{i \in I}$, then there exists a Lipschitz function $f: c_0(I) \to Z$ such that $q \circ f = \mathrm{Id}_{c_0(I)}$. It follows that

$$Z \stackrel{\mathrm{Lip}}{\approx} c_0 \oplus c_0(I) = c_0(I),$$

and so $c_0(I)$ is Lipschitz isomorphic to a subspace of ℓ_∞. However, $c_0(I)$ cannot be linearly embedded in ℓ_∞, which provides the desired result.

The Arens–Eells spaces provide many examples of spaces that are Lipschitz isomorphic, but not linearly isomorphic. The following theorem is an example of that.

THEOREM 5.22. *If R is a nonseparable reflexive Banach space, then R cannot be linearly embedded in $Æ(R)$.*

In order to prove Theorem 5.22, several preliminary results will be needed.

LEMMA 5.23. *Let X be a separable Banach space. If W is the subset of $\mathrm{Lip}(X) = Æ(X)^*$ consisting of weakly continuous functions on X (that is, continuous with respect to the weak topology on X), then W is a norming set. That is,*

$$\|\mu\|_{Æ(X)} = \sup\left\{\langle \mu, f \rangle \ : \ f \in W \ \text{and} \ \|f\|_{\mathrm{Lip}(X)} \leq 1\right\},$$

where $\langle \mu, f \rangle$ represents the dual action of $f \in \mathrm{Lip}(X)$ on $\mu \in Æ(X)$.

PROOF. View X as a closed subspace of $\mathcal{C}[0,1]$, which is a space that has a monotone basis. A Lipschitz function f in $\mathrm{Lip}(X)$ can be extended to a Lipschitz function $\tilde{f}$ in $\mathrm{Lip}(\mathcal{C}[0,1])$ because $\mathbb{R}$ is 1-ALR (Theorem 1.8). If P_n is the nth basis projection for each $n \in \mathbb{N}$, then $\tilde{f} \circ P_n$ is weakly continuous and $\tilde{f} \circ P_n \to \tilde{f}$ pointwise as $n \to \infty$. The result follows. □

COMMENT 5.24. It may be useful to make a few remarks about the proof of Lemma 5.23. We may view X as a closed subspace of $\mathcal{C}[0,1]$, by the Banach–Mazur theorem, but only because X is assumed to be separable. (For a proof of the Banach–Mazur theorem, see Theorem 1.4.4 in [5].)

The projections $(P_n)_{n=1}^\infty$ are the partial sum projections associated to the monotone basis of $\mathcal{C}[0,1]$. To be more general, assume that a separable Banach space E has a (Schauder) basis $(e_n)_{n=1}^\infty$ with associated biorthogonal functionals $(e_n^*)_{n=1}^\infty$ in E^*. Then for each $n \in \mathbb{N}$, the nth *partial sum projection* is defined to be

$$P_n(x) = \sum_{k=1}^{n} e_k^*(x)\, e_k, \quad x \in E.$$

The constant $K = \sup_{n\in\mathbb{N}} \|P_n\|$ is called the *basis constant*, and the basis is called *monotone* if $K = 1$.

In the case we are considering in Lemma 5.23, we have a bounded linear projection $P_n : \mathcal{C}[0,1] \to \mathcal{C}[0,1]$, so that $\tilde{f} \circ P_n : \mathcal{C}[0,1] \to \mathbb{R}$ is a Lipschitz function for each $n \in \mathbb{N}$. For every $g \in \mathcal{C}[0,1]$, we have

$$\lim_{n\to\infty} (\tilde{f} \circ P_n)(g) = \lim_{n\to\infty} \tilde{f}(P_n(g)) = \tilde{f}(g).$$

So when we say that $\tilde{f} \circ P_n \to \tilde{f}$ pointwise as $n \to \infty$, we mean as real-valued functions on $\mathcal{C}[0,1]$. It remains only to observe that $\tilde{f} \circ P_n$ is weakly continuous for each $n \in \mathbb{N}$.

REMARK 5.25. In light of Lemma 5.11, it is convenient to view $\mathcal{E}(X)$ as the completion of the set of Borel measures on X with finite support under the norm

$$\|\mu\|_{\mathcal{E}(X)} = \sup\left\{ \int f\, d\mu \,:\, \|f\|_{\mathrm{Lip}(X)} \le 1 \right\},$$

where a measure μ with finite support can be written as

$$\mu = \sum_{k=1}^{n} a_k\, [\delta(x_k) - \delta(y_k)],$$

with $(a_k)_{k=1}^n$ a finite sequence in $\mathbb{R}$ and $(x_k)_{k=1}^n$ and $(y_k)_{k=1}^n$ finite sequences in X.

LEMMA 5.26. *Let X be a Banach space and let W be a weakly compact subset of X. Suppose that $(\mu_n)_{n=1}^\infty$ is a sequence of Borel measures on X having finite support in W and suppose that there is some $K > 0$ such that $|\mu_n|(W) \le K$ for each $n \in \mathbb{N}$. If there is a number $\epsilon > 0$ such that*

$$\inf\left\{ \|\mu_n - \mu_m\|_{\mathcal{E}(X)} \,:\, n \in \mathbb{N} \text{ and } m \in \mathbb{N},\, n \ne m \right\} > \epsilon,$$

then there exists a function $f \in \mathrm{Lip}(X)$ with $\|f\|_{\mathrm{Lip}(X)} \le 1$ and an infinite subset $\mathbb{M}$ of $\mathbb{N}$ such that

(5.2) $$\left| \int f\, d\mu_n \right| \ge \frac{\epsilon}{8} \quad \text{for all } n \in \mathbb{M}.$$

In particular, $\mu_n \not\to 0$ weakly.

PROOF. It is sufficient to prove the lemma for X a separable Banach space. Let $A = \sup_{x\in W} \|x\|$. Then A is finite and $\sup_{n\in\mathbb{N}} \|\mu_n\|_{\mathcal{E}(X)} \le AK$, by assumption.

We will construct a subsequence of $(\mu_n)_{n=1}^\infty$ so that elements in the subsequence satisfy the inequality in (5.2), and $\mathbb{M}$ will be the collection of indices used in the subsequence. Start by constructing an initial subsequence $(\nu_k)_{k=1}^\infty$ of $(\mu_k)_{k=1}^\infty$ inductively, according to the procedure described below.

Let c be a number such that $\epsilon > c > \frac{\epsilon}{2}$. Let n be in $\mathbb{N}$ and suppose measures $\nu_1, \ldots, \nu_{n-1}$ have been chosen from the sequence $(\mu_k)_{k=1}^\infty$. Let

$$S_{n-1} = \overline{\mathrm{co}}\Big(\mathrm{supp}(\nu_1) \cup \cdots \cup \mathrm{supp}(\nu_{n-1})\Big),$$

where $\mathrm{supp}(\nu)$ is the support of the measure ν, and let

$$E_{n-1} = \overline{\mathrm{co}}\Big(\{\pm \delta(x) : x \in S_{n-1}\}\Big) \subseteq \mathrm{Æ}(X).$$

Since the convex hull of a compact set is compact, observe that E_{n-1} is compact in the norm topology on $\mathrm{Æ}(X)$. Let $b = \frac{16AK^2}{\epsilon}$ and choose ν_n from the sequence $(\mu_k)_{k=1}^\infty$ so that

$$d_{\mathrm{Æ}(X)}\Big(\nu_n, 2^n b E_{n-1}\Big) > c > \frac{\epsilon}{2}.$$

By Lemma 5.23, there exists a weakly continuous function $f_n \in \mathrm{Lip}(X)$ such that $\|f_n\|_{\mathrm{Lip}(X)} \leq 1$ and such that

(5.3) $$\int f_n \, d\nu_n > c + 2^n b \int f_n \, d\gamma,$$

for all $\gamma \in E_{n-1}$. Note that this implies that

$$\left|\int f_n \, d\gamma\right| \leq \frac{AK}{2^n b}, \quad \gamma \in E_{n-1}.$$

Consequently, for all $k < n$,

$$\int |f_n| \, d|\nu_k| = \int_{\{f_n > 0\}} f_n \, d|\nu_k| - \int_{\{f_n < 0\}} f_n \, d|\nu_k|$$

$$\leq 2 \cdot \frac{AK}{2^n b} \cdot K = \frac{2AK^2}{2^n b},$$

because restricting $|\nu_k|$ to either $\{f_n > 0\}$ or $\{f_n < 0\}$ gives an element of the set KE_{n-1} (hence the extra factor of K and the end of the above inequality). On the other hand (for the case where $k = n$),

$$\int f_n \, d\nu_n > c,$$

because of (5.3).

By passing to a further subsequence, if necessary, it may be assumed that $(|\nu_k|)_{k=1}^\infty$ converges in the weak* topology to some measure $\nu \in \mathcal{C}(W)^*$, where W inherits the weak topology from X.

Let $S = \bigcup_{n=1}^\infty S_n$. Then S is a closed convex subset of X and ν is supported in S, by Mazur's theorem (see, for example, Theorem 5.45 in [21]). Observe that

$$|f_n(x)| = \left|\int f_n \, d\delta(x)\right| \leq \frac{AK}{2^n b}, \quad x \in S_{n-1}.$$

It follows that $f_n(x) \to 0$ as $n \to \infty$ for all $x \in S$. Thus, since each of the functions is bounded on the support of ν by A,

$$\lim_{n \to \infty} \int |f_n| \, d\nu = 0,$$

by the bounded convergence theorem. Additionally, because f_n is weakly continuous for each $n \in \mathbb{N}$,

$$\lim_{k \to \infty} \int |f_n| \, d|\nu_k| = \int |f_n| \, d\nu.$$

Therefore, it is possible to pass to a further subsequence (if necessary), so that

$$\int |f_n| \, d|\nu_k| < 2 \cdot \frac{AK^2}{2^n b}, \quad k \neq n.$$

Let $\mathbb{M}$ be the collection of all indices in the subsequence chosen above. For each $n \in \mathbb{M}$, let $h_n = \max\{f_n, 0\}$ or $h_n = \max\{-f_n, 0\}$, so that h_n is nonnegative and

$$\left| \int h_n \, d\nu_n \right| \geq \frac{c}{2} > \frac{\epsilon}{4}.$$

Now let $f = \sup_{k \in \mathbb{N}} h_k$. Then $\|f\|_{\text{Lip}(X)} \leq 1$ and for $n \in \mathbb{N}$,

$$0 \leq f - h_n \leq \sum_{k \neq n} h_k,$$

from which it can be seen that

$$\left| \int f \, d\nu_n - \int h_n \, d\nu_n \right| \leq \sum_{k \neq n} \int |h_k| \, d|\nu_n| \leq \frac{\epsilon}{8}.$$

The desired result follows. $\square$

COMMENT 5.27. In the proof of Lemma 5.26, we commented that the functions f_n are bounded on the support of ν by A. This follows from the fact that each f_n is a Lipschitz continuous function with Lipschitz constant 1, together with the fact that the support of ν is in W and $A = \sup_{x \in W} \|x\|$. That is, for each $n \in \mathbb{N}$ and $x \in W$, we have

$$|f_n(x)| = |f_n(x) - f_n(0)| \leq \|f\|_{\text{Lip}(X)} \|x - 0\| = \|x\| \leq A.$$

PROOF OF THEOREM 5.22. Note that B_R is weakly compact because R is reflexive. Assume that R linearly embeds in $\AE(R)$. Let $\epsilon > 0$ be given. There exists an uncountable family of elements $\{\nu_i\}_{i \in I}$ of $\AE(R)$ that forms a relatively weakly compact subset in $\AE(R)$ such that $\|\nu_i - \nu_j\| > \epsilon$ for all distinct i and j in I. For each $i \in I$, pick a measure μ_i having finite support so that $\|\nu_i - \mu_i\| < \frac{\epsilon}{26}$. Observe that if $i \neq j$, then

$$\|\mu_i - \mu_j\| > \epsilon - \frac{\epsilon}{13} = \frac{12\epsilon}{13}.$$

For each $m \in \mathbb{N}$, let

$$C_m = \left\{ \mu \; : \; \mu \text{ is supported on } mB_R \text{ and } |\mu|(mB_R) \leq m \right\}.$$

There exists an integer m_0 such that $\mu_i \in C_{m_0}$ for countably many $i \in I$. Let $\{i_1, i_2, i_3, \ldots\}$ be the collection of indices i for which $\mu_i \in C_{m_0}$. By passing to a subsequence (if necessary), it can be assumed that the sequence $(\nu_{i_n})_{n\in\mathbb{N}}$ converges weakly to an element ν in $\cyr{E}(R)$. Let $\mu \in C_m$ be a measure such that $\|\nu - \mu\| \leq \frac{\epsilon}{26}$.

By Lemma 5.26, there exists a function $f \in \operatorname{Lip}(R)$ with $\|f\|_{\operatorname{Lip}(R)} \leq 1$ and an infinite subset $\mathbb{M}$ of indices from $(i_n)_{n=1}^\infty$ such that

$$\left| \int f \, d(\mu_i - \mu) \right| \geq \frac{1}{8}\left(\frac{12\epsilon}{13}\right) = \frac{3\epsilon}{26},$$

for all $i \in \mathbb{M}$.

However, since $\nu_{i_n} \to \nu$ weakly, it follows that

$$\lim_{n\to\infty} \int f \, d(\nu_{i_n} - \nu) = 0,$$

and thus

$$\limsup_{n\to\infty} \left| \int f \, d(\mu_{i_n} - \mu) \right| \leq \frac{\epsilon}{13}.$$

This is a contradiction, and so R cannot embed linearly into $\cyr{E}(R)$. □

COMMENT 5.28. In the proof of Theorem 5.22, Lemma 5.26 is applied to the sequence $(\mu_{i_n} - \mu)_{n=1}^\infty$. Note that $\|(\mu_i - \mu) - (\mu_j - \mu)\| = \|\mu_i - \mu_j\|$.

Theorem 5.22 shows that it is possible to have two Banach spaces that are Lipschitz isomorphic, but not linearly isomorphic. The proof, however, requires the Banach spaces in question to be nonseparable. It is still an open question whether or not two Lipschitz isomorphic separable spaces must necessarily be linearly isomorphic. This is the essence of the following question.

QUESTION 5.29 (Lipschitz isomorphism problem). *If X and Y are separable Banach spaces such that $X \stackrel{\operatorname{Lip}}{\approx} Y$, is it true that $X \approx Y$?*

CHAPTER 6

Differentiation and the isomorphism problem

The current objective is to answer the following question, which is Question 4.15 for separable Banach spaces.

QUESTION 6.1. *If a separable Banach space X isometrically embeds into a Banach space Y, does X linearly isometrically embed in Y?*

In Corollary 5.19, we saw that it is possible for a nonseparable Banach space to isometrically embed into a Banach space, but not linearly embed into that Banach space. Consequently, in Question 6.1, we have restated Question 4.15 for the case where X is a separable Banach space. We will answer Question 6.1 (positively) in Corollary 6.14. We will follow the approach of Kalton and Godefroy in [**39**]. We begin by giving some definitions.

DEFINITION 6.2. A function $f : X \to Y$ between Banach spaces is called *Gâteaux differentiable* at $x_0 \in X$ if there is a bounded linear map $T : X \to Y$ such that
$$\lim_{t \to 0} \frac{f(x_0 + tx) - f(x_0)}{t} = T(x) \quad \text{for all} \quad x \in X.$$
That is,
$$f(x_0 + tx) = f(x_0) + tT(x) + o(|t|),$$
for $t \in \mathbb{R}$ and $x \in X$. (The limit is in the topology on Y.) The linear map T is called the *Gâteaux derivative* of f at x_0 and is denoted $Df(x_0)(x)$.

Another type of differentiability, which is formally stronger than Gâteaux differentiability, is called *Fréchet* differentiability.

DEFINITION 6.3. A function $f : X \to Y$ between Banach spaces is called *Fréchet differentiable* at $x_0 \in X$ if there is a bounded linear map $T : X \to Y$ such that
$$\lim_{x \to 0} \frac{\|f(x_0 + x) - f(x_0) - T(x)\|_Y}{\|x\|_X} = 0.$$
That is,
$$f(x_0 + x) = f(x_0) + T(x) + o(\|x\|_X),$$
for $x \in X$. The linear map T is called the *Fréchet derivative* of f at x_0 and is denoted $Df(x_0)(x)$.

REMARK 6.4. Any function that is Fréchet differentiable is Gâteaux differentiable (and in such a case, the two derivatives coincide), but there exist functions that are Gâteaux differentiable and not Fréchet differentiable.

When a function $f : X \to Y$ is differentiable in either the Fréchet sense or the Gâteaux sense, we use the notation $Df(x_0)$ to denote the linear map T that is the derivative of f at x_0. In particular, $T(x)$ will be denoted $Df(x_0)(x)$.

EXAMPLE 6.5. A Lipschitz function $f : \mathbb{R}^m \to \mathbb{R}^n$ is Fréchet differentiable almost everywhere (and the Fréchet derivative is the derivative in the usual sense of the Jacobian).

Example 6.5 is known as *Rademacher's theorem*. We will revisit Rademacher's theorem in the next chapter. (See Lemma 7.22.)

EXAMPLE 6.6. Define a function $f : [0,1] \to L_1(0,1)$ by $f(x) = \chi_{[0,x)}$, where $\chi_{[0,x)}$ is the characteristic function of the set $[0,x)$, for $x \in [0,1]$. Then f is a Lipschitz function. However, for $x \in [0,1)$ and $t > 0$,
$$\frac{f(x+t) - f(x)}{t} = \frac{\chi_{[0,x+t)} - \chi_{[0,x)}}{t} = \frac{\chi_{[x,x+t)}}{t}.$$
The limit does not exist in $L_1(0,1)$ as $t \to 0^+$. A similar argument holds for $x \in (0,1]$ and $t < 0$, and consequently f does not have a Gâteaux derivative at x for any $x \in [0,1]$.

EXAMPLE 6.7. Define a Lipschitz function $f : \mathbb{R} \to c_0$ by
$$f(x) = \left(\frac{\sin(nx)}{n}\right)_{n=1}^\infty,$$
for $x \in \mathbb{R}$. If f had a Gâteaux derivative at x, it would be the sequence $\bigl(\cos(nx)\bigr)_{n=1}^\infty$, but that sequence is not in c_0. Therefore, f is a Lipschitz function that does not have a Gâteaux derivative at x for any $x \in \mathbb{R}$.

COMMENT 6.8. The definitions of Fréchet and Gâteaux differentiability coincide for functions $f : \mathbb{R} \to Y$. (That is, when $X = \mathbb{R}$.) In this case then, it is common (and unambiguous) to refer to a function as differentiable (or not differentiable, as the case may be). Examples 6.6 and 6.7 provide functions with domain $\mathbb{R}$ that are not differentiable. These examples were given by Clarkson in 1936 [24], although not exactly in the form presented here.

It is certainly true that a function $f : \mathbb{R} \to Y$ is Fréchet differentiable precisely when it is Gâteaux differentiable. What may not be obvious is that if a function $f : X \to Y$ is a *Lipschitz* function and X is *finite dimensional*, then f is Fréchet differentiable at a point if and only if it is Gâteaux differentiable there. (See Proposition 4.3 in [13].)

In light of the previous paragraph, it is difficult (that is, impossible) to find examples of Lipschitz continuous functions $f : \mathbb{R}^m \to \mathbb{R}^n$ that are Gâteaux differentiable but not Fréchet differentiable. It is, however, possible for a noncontinuous function $f : \mathbb{R}^m \to \mathbb{R}^n$ to have a Gâteaux derivative at a point and not be Fréchet differentiable there. A simple example given by Benyamini and Lindenstrauss (in [13]) is the function $f : \mathbb{R}^2 \to \mathbb{R}$ defined by
$$f(x,y) = \begin{cases} \frac{x^4 y}{x^6 + y^3} & \text{if } (x,y) \neq (0,0), \\ 0 & \text{if } (x,y) = (0,0). \end{cases}$$
This function is not continuous at $(0,0)$, but it does have a Gâteaux derivative there. Indeed, it is a simple matter to show that, for this f,
$$\lim_{t \to 0} \frac{f(tx,ty) - f(0,0)}{t} = 0,$$

for each $(x,y) \in \mathbb{R}^2$, which means that the Gâteaux derivative of f at $(0,0)$ is 0. However, a function which is Fréchet differentiable at x is necessarily continuous at x, and hence the discontinuity of f at $(0,0)$ gives us the desired conclusion.

The previous examples (Examples 6.6 and 6.7) show that not every Lipschitz function between Banach spaces is Gâteaux differentiable. However, under reasonable conditions, Gâteaux differentiable functions between Banach spaces are plentiful, as the following theorem (given here without proof) makes clear.

THEOREM 6.9. *Let X be a separable Banach space. Any Lipschitz function $f : X \to \mathbb{R}$ can be approximated uniformly by Lipschitz functions that are Gâteaux differentiable everywhere.*

In fact, Theorem 6.9 can be generalized to Lipschitz functions from a separable Banach space X into a Banach space Y, provided Y has something called the Radon–Nikodym property. (See Definition 7.1 for the precise definition of the Radon–Nikodym property.) A proof of the more general version of Theorem 6.9 can be found in the appendix. (See Theorem C.7.)

It is sometimes convenient to compute a Gâteaux derivative on a dense subset. Lemma 6.10 shows that this is sufficient.

LEMMA 6.10. *Suppose $f : X \to Y$ is a Lipschitz function between Banach spaces. If $T : X \to Y$ is a linear function such that*

$$\lim_{t \to 0} \frac{f(x_0 + tw) - f(x_0)}{t} = T(w),$$

for all w in a dense subset W of X, then f is Gâteaux differentiable at x_0 in X.

PROOF. Suppose $v \in X$. Without loss of generality, suppose that $\|v\| \leq 1$. Given $\epsilon > 0$, there exists a $w \in W$ such that $\|w\| \leq 1$ and $\|v - w\| < \epsilon$ and

$$\lim_{t \to 0} \frac{f(x_0 + tw) - f(x_0)}{t} = T(w).$$

Observe that

$$\left\| \frac{f(x_0 + tv) - f(x_0)}{t} - \frac{f(x_0 + tw) - f(x_0)}{t} \right\| = \frac{\|f(x_0 + tv) - f(x_0 + tw)\|}{|t|}$$

$$\leq \frac{\mathrm{Lip}(f) \|(x_0 + tv) - (x_0 + tw)\|}{|t|} = \mathrm{Lip}(f) \|v - w\|.$$

By the triangle inequality,

$$\left\| \frac{f(x_0 + tv) - f(x_0)}{t} - T(v) \right\|$$

is bounded above by

$$\left\| \frac{f(x_0 + tv) - f(x_0)}{t} - \frac{f(x_0 + tw) - f(x_0)}{t} \right\|$$
$$+ \left\| \frac{f(x_0 + tw) - f(x_0)}{t} - T(w) \right\| + \left\| T(w) - T(v) \right\|.$$

The middle term tends to zero as $t \to 0$, because $w \in W$, and therefore,
$$\lim_{t \to 0} \left\| \frac{f(x_0 + tv) - f(x_0)}{t} - T(v) \right\| \leq \Big(\text{Lip}(f) + \|T\| \Big) \epsilon.$$
The choice of $\epsilon > 0$ was arbitrary, and hence this limit is zero. It follows that f is Gâteaux differentiable at x_0. $\square$

We will use Gâteaux differentiability techniques to show that separable Banach spaces have something called the *lifting property*.

DEFINITION 6.11. A Banach space X is said to have the *lifting property* if there exists a continuous linear map $T : X \to \text{Æ}(X)$ such that $\beta \circ T = \text{Id}_X$, where $\text{Æ}(X)$ is the Arens–Eells space for X and $\beta : \text{Æ}(X) \to X$ is the barycenter map. Furthermore, if the map T has norm one, then we say X has the *isometric lifting property*.

THEOREM 6.12. *Separable Banach spaces have the isometric lifting property.*

PROOF. Suppose X is a separable Banach space and let $(v_n)_{n=1}^\infty$ be a linearly independent sequence of vectors in X such that $\sum_{n=1}^\infty \|v_n\| \leq 1$ and for which X is the closed linear span of $(v_n)_{n=1}^\infty$. Let V be the span of the elements in $(v_n)_{n=1}^\infty$, so that $X = \overline{V}$.

Let $(\eta_n)_{n=1}^\infty$ be a sequence of independent scalar valued random variables, each uniformly distributed on $[0,1]$, and let
$$\xi = \sum_{n=1}^\infty \eta_n v_n.$$
Let $\Omega = [0,1]^\mathbb{N}$ be the Hilbert cube with $\mathbb{P}$ the product measure, where each factor $[0,1]$ is given Lebesgue measure λ. Then ξ is a random variable on Ω.

For each $n \in \mathbb{N}$, let Ω_n be a copy of Ω with the nth factor removed. That is, let $\Omega_n = [0,1]^{\mathbb{N} \setminus \{n\}}$. Furthermore, for each $n \in \mathbb{N}$, let $\pi_n : \Omega \to \Omega_n$ be the natural projection of Ω onto Ω_n, and let $\mathbb{P}_n = \pi_n(\mathbb{P})$. Define a random variable ξ_n on $(\Omega_n, \mathbb{P}_n)$ by
$$\xi_n = \sum_{k \neq n}^\infty \eta_k v_k.$$
For each $n \in \mathbb{N}$, define $\phi_n \in \text{Æ}(X)$ by the Bochner integral[1]
$$\phi_n = \int_{\Omega_n} \Big[\delta(v_n + \xi_n(\omega)) - \delta(\xi_n(\omega)) \Big] \mathbb{P}_n(d\omega),$$
where $\delta : X \to \text{Æ}(X)$ is the isometric embedding of X in $\text{Æ}(X)$ given by $\delta(x) = \delta_x$ for each $x \in X$, where δ_x is the evaluation functional given by $\delta_x(f) = f(x)$ for each $f \in \text{Lip}(X)$. (See Comment 5.5 and the comments before Definition 5.3.)

CLAIM 1. For each $n \in \mathbb{N}$, we have $\beta(\phi_n) = v_n$.

---

[1] See Appendix A for the definition and properties of the Bochner integral.

Proof of claim. In order to prove Claim 1, we compute directly, using the fact that β is a continuous linear map:

$$\beta(\phi_n) = \int_{\Omega_n} \beta\Big(\delta(v_n + \xi_n(\omega)) - \delta(\xi_n(\omega))\Big)\,\mathbb{P}_n(d\omega)$$

$$= \int_{\Omega_n} \Big[(\beta\delta)(v_n + \xi_n(\omega)) - (\beta\delta)(\xi_n(\omega))\Big]\,\mathbb{P}_n(d\omega).$$

Since $\beta\delta = \mathrm{Id}_X$ (by Lemma 5.9),

$$\beta(\phi_n) = \int_{\Omega_n} \Big(v_n + \xi_n(\omega) - \xi_n(\omega)\Big)\,\mathbb{P}_n(d\omega) = \int_{\Omega_n} v_n\,\mathbb{P}_n(d\omega) = v_n\,\mathbb{P}_n(\Omega_n).$$

The claim follows from the fact that $(\Omega_n, \mathbb{P}_n)$ is a probability space. $\diamond$

Define a linear map $T_V : V \to \cancel{E}(X)$ so that $T_V(v_n) = \phi_n$. It follows from Claim 1 that $\beta T_V = \mathrm{Id}_V$.

CLAIM 2. If $f \in \mathrm{Lip}(X)$ is a Gâteaux differentiable function on X, then for each $n \in \mathbb{N}$ and $\omega \in \Omega_n$, we have

$$f(v_n + \xi_n(\omega)) - f(\xi_n(\omega)) = \int_0^1 Df(tv_n + \xi_n(\omega))(v_n)\,dt,$$

where $Df(tv_n + \xi_n(\omega))(v_n)$ is the Gâteaux derivative of f at $tv_n + \xi_n(\omega)$ evaluated at v_n.

Proof of claim. This is ultimately an application of the fundamental theorem of calculus, because $t \mapsto f(tv_n + \xi_n(\omega))$ is an antiderivative of

$$t \mapsto Df(tv_n + \xi_n(\omega))(v_n).$$

In order to see this, we can compute the derivative directly, using the limit definition:

$$\frac{d}{dt}\Big[f(tv_n + \xi_n(\omega))\Big] = \lim_{h \to 0} \frac{f((t+h)v_n + \xi_n(\omega)) - f(tv_n + \xi_n(\omega))}{h}$$

$$= \lim_{h \to 0} \frac{f(tv_n + \xi_n(\omega) + hv_n) - f(tv_n + \xi_n(\omega))}{h}$$

$$= Df(tv_n + \xi_n(\omega))(v_n).$$

Therefore,

$$\int_0^1 Df(tv_n + \xi_n(\omega))(v_n)\,dt = f(tv_n + \xi_n(\omega))\Big|_0^1 = f(v_n + \xi_n(\omega)) - f(\xi_n(\omega)).$$

This verifies Claim 2. $\diamond$

Recall that $\mathrm{Lip}(X)$ is the dual space of $\cancel{E}(X)$. Denote the dual action by $\langle \cdot, \cdot \rangle$. For $f \in \mathrm{Lip}(X)$ and $n \in \mathbb{N}$, Claim 2 tells us that

$$\langle \phi_n, f \rangle = \int_{\Omega_n} \Big[f(v_n + \xi_n(\omega)) - f(\xi_n(\omega))\Big]\,\mathbb{P}_n(d\omega)$$

$$= \int_{\Omega_n} \Big[\int_0^1 Df(tv_n + \xi_n(\omega))(v_n)\,dt\Big]\,\mathbb{P}_n(d\omega),$$

provided that f is a Gâteaux differentiable function. Therefore, by Fubini's theorem, we have that
$$\langle \phi_n, f \rangle = \int_\Omega Df(\xi(\omega))(v_n)\, \mathbb{P}(d\omega).$$
It follows that
$$\langle T_V(v), f \rangle = \int_\Omega Df(\xi(\omega))(v)\, \mathbb{P}(d\omega),$$
for all $v \in V$ and $f \in \mathrm{Lip}(X)$ that are Gâteaux differentiable functions, and so we conclude that
$$|\langle T_V(v), f \rangle| \leq \|Df(\xi(\omega))(v)\|_{L_\infty(\Omega)} \leq \mathrm{Lip}(f)\|v\|_X$$
for all such v and f. Since any function $f \in \mathrm{Lip}(X)$ can be uniformly approximated by Lipschitz continuous Gâteaux differentiable functions, by Theorem 6.9, we conclude that $\|T_V(v)\|_{Æ(X)} \leq \|v\|_X$, and consequently $\|T_V\| \leq 1$.

Finally, since $X = \overline{V}$, we may extend $T_V : V \to Æ(X)$ to a bounded linear operator $T : X \to Æ(X)$ such that $\beta T = \mathrm{Id}_X$ and $\|T\| = 1$. $\square$

Theorem 6.13 answers Question 6.1 (by means of Corollary 6.14), giving a positive answer to Question 4.15 when X is a separable Banach space.

THEOREM 6.13. *Let X and Y be Banach spaces and assume that X is separable. If there exists a Lipschitz function $f : X \to Y$ and a quotient map $Q : Y \to X$ such that $Q \circ f = \mathrm{Id}_X$, then there exists a bounded linear function $T_f : X \to Y$ such that $Q \circ T_f = \mathrm{Id}_X$ and $\|T_f\| \leq \mathrm{Lip}(f)$.*

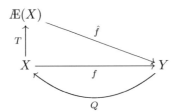

PROOF. Without loss of generality, we may assume that $f(0) = 0$. Therefore, by Corollary 5.10, there exists a unique linear function $\hat{f} : Æ(X) \to Y$ such that $\|\hat{f}\| = \mathrm{Lip}(f)$ and $\hat{f} \circ \delta_X = f$, where $\delta_X : X \to Æ(X)$ is the Dirac lifting defined on X.

The Banach space X is assumed to be separable, and consequently has the isometric lifting property, by Theorem 6.12. Thus, there exists a linear operator $T : X \to Æ(X)$ with $\|T\| = 1$ such that $\beta_X \circ T = \mathrm{Id}_X$, where $\beta_X : Æ(X) \to X$ is the barycenter map defined on $Æ(X)$.

Since $\hat{f} \circ \delta_X = f$, it follows that
$$Q \circ (\hat{f} \circ \delta_X) = Q \circ f = \mathrm{Id}_X,$$
where the second equality is an assumption made in the statement of the theorem. But this means that $(Q \circ \hat{f}) \circ \delta_X = \mathrm{Id}_X$, and so $Q \circ \hat{f} = \beta_X$.

Define the map $T_f : X \to Y$ by $T_f = \hat{f} \circ T$. Then T_f is a bounded linear map and $\|T_f\| \leq \mathrm{Lip}(f)$. It remains only to show that $Q \circ T_f = \mathrm{Id}_X$; however, this follows immediately from the definitions and properties we have already observed:
$$Q \circ T_f = Q \circ (\hat{f} \circ T) = (Q \circ \hat{f}) \circ T = \beta_X \circ T = \mathrm{Id}_X\,.$$

Therefore, the proof is complete. □

COROLLARY 6.14. *Let X and Y be Banach spaces. If X is separable and isometric to a subset of Y, then X is linearly isometric to a subspace of Y.*

PROOF. We may assume without loss of generality that the isometry maps $0 \in X$ to $0 \in Y$. Therefore, the conclusion follows from Theorem 4.17 (Figiel's theorem) and Theorem 6.13. □

REMARK 6.15. Corollary 6.14 was first proved by Kalton and Godefroy in 2003 [**39**] (using the method presented here).

Additional comments

The linear map T we constructed in the proof of Theorem 6.12 is the derivative of a *convolution*. A convolution can be used to "smooth" a function, which (under the right circumstances) can create a function that is differentiable, even though the original was not differentiable.

DEFINITION 6.16. Suppose that $f : X \to Y$ is a function between Banach spaces and suppose that μ is a compactly supported finite regular Borel measure on X. Define the *convolution* of μ and f by

$$(\mu * f)(x) = \int_X f(x - u)\, \mu(du),$$

where the integral is a Bochner integral.[2]

LEMMA 6.17. *Let $f : X \to Y$ be a Lipschitz function between Banach spaces. If μ is a compactly supported finite regular Borel measure on X, then $\mu * f$ is a Lipschitz function and $\mathrm{Lip}(\mu * f) \leq \mathrm{Lip}(f)\,\|\mu\|_M$, where $\|\mu\|_M$ is the total variation norm of μ.*

PROOF. Let x and x' be points in X. Then

$$\|(\mu * f)(x) - (\mu * f)(x')\| \leq \int_X \|f(x - u) - f(x' - u)\|\, |\mu|(du)$$

$$\leq \int_X \mathrm{Lip}(f)\|x - x'\|\, |\mu|(du) \leq \mathrm{Lip}(f)\|\mu\|_M\|x - x'\|.$$

This completes the proof. □

If μ is a probability measure, then μ can be identified with the distribution of an X-valued random variable $\xi : \Omega \to X$, for some probability space $(\Omega, \mathbb{P})$. With this interpretation,

$$(\mu * f)(x) = \mathbb{E}\big[f(x - \xi)\big] = \int_\Omega f\big(x - \xi(\omega)\big)\, \mathbb{P}(d\omega).$$

EXAMPLE 6.18. Let X be a Banach space and let $v \in X$. If $\Omega = [-1, 1]$ and $\mathbb{P}$ is normalized Lebesgue measure on Ω, and if $\xi(t) = tv$, then

$$(\mu * f)(x) = \frac{1}{2} \int_{-1}^{1} f(x + tv)\, dt,$$

where μ is the probability measure on X defined by ξ.

---

[2]See Appendix A for the definition of the Bochner integral.

COMMENT 6.19. It may look as though there is a sign error in the previous displayed equation, but $\int_{-1}^{1} f(x-tv)\,dt = \int_{-1}^{1} f(x+sv)\,ds$, by means of the substitution $s = -t$. (Then replace s with t to get the integral in the example.)

The linear map T in the proof of Theorem 6.12 is the derivative of a convolution between the Dirac lifting $\delta : X \to \text{Æ}(X)$ and something called a *cube measure*.

DEFINITION 6.20. Suppose that X is a Banach space and $(x_k)_{k=1}^{\infty}$ is a sequence of points in X such that $\sum_{k=1}^{\infty} \|x_k\| < \infty$. Let $(\eta_k)_{k=1}^{\infty}$ be a sequence of independent random variables, each of which is uniformly distributed on $[0,1]$. Define a random variable ξ by

$$\xi = \sum_{k=1}^{\infty} \eta_k x_k.$$

The range of ξ is a compact set in X and ξ defines a Borel measure μ_ξ on X by the formula

$$\mu_\xi(B) = \mathbb{P}(\xi \in B),$$

where B is a Borel set in X and $\mathbb{P}$ is a probability measure on $[0,1]^{\mathbb{N}}$. Note that the random variable ξ can be interpreted as a function $\xi : \bigl([0,1]^{\mathbb{N}}, \mathbb{P}\bigr) \to X$, defined by

$$\xi(t_1, t_2, \ldots) = \sum_{k=1}^{\infty} t_k x_k.$$

The measure μ_ξ is called a *cube measure* on X. It is called a *dense cube measure* if additionally the set $\left\{ \frac{x_k}{\|x_k\|} : k \in \mathbb{N} \right\}$ is dense in the unit sphere of X.

EXAMPLE 6.21. Let X be a separable Banach space and let $\delta : X \to \text{Æ}(X)$ be the Dirac lifting. Let $(\Omega, \mathbb{P})$ be the Hilbert cube $\Omega = [0,1]^{\mathbb{N}}$ with product measure $\mathbb{P}$. Let μ_ξ be a dense cube measure with $\xi : \Omega \to X$ the corresponding X-valued random variable on Ω. The convolution of μ_ξ with δ is given by

$$(\mu_\xi * \delta)(x) = \int_\Omega \delta\bigl(x + \xi(\omega)\bigr)\, \mathbb{P}(d\omega),$$

for $x \in X$. Since this gives an element of $\text{Æ}(X)$, we can evaluate it at $f \in \text{Lip}(X)$ to get

$$(\mu_\xi * \delta)_x(f) = \int_\Omega f\bigl(x + \xi(\omega)\bigr)\, \mathbb{P}(d\omega),$$

where we write $(\mu_\xi * \delta)_x$ in place of $(\mu_\xi * \delta)(x)$ for notational simplicity. Then it can be shown that the derivative of $\mu_\xi * \delta$ at 0 is given by

$$D\bigl((\mu_\xi * \delta)_0(v)\bigr)(f) = \int_\Omega Df\bigl(\xi(\omega)\bigr)(v)\, \mathbb{P}(d\omega),$$

for $v = x_k$ for some k (where x_k is the x_k given in Definition 6.20) and $f \in \text{Lip}(X)$. We recognize that this gives us the map T from the proof of Theorem 6.12.

While the convolution $\mu_\xi * \delta$ can be differentiated, the Dirac lifting δ does not have a Gâteaux derivative at any $x \in X$. If the map $t \mapsto \delta(x + tv)$ is differentiable at 0, then the map $t \mapsto f(x + tv)$ is differentiable at 0 for each $f \in \text{Lip}(X)$. Since that is not true, we conclude that δ is not Gâteaux differentiable at x.

CHAPTER 7

Differentiation and Haar-null sets

If $F : \mathbb{R} \to \mathbb{R}$ is a Lipschitz function, then it is known that F is differentiable almost everywhere. Indeed, the same conclusion is true if $F : \mathbb{R} \to X$ is a Lipschitz function and $\dim(X) < \infty$. (See Example 6.5.) The following definition is due to Diestel and Uhl [27].[1]

DEFINITION 7.1. A Banach space X has the *Radon–Nikodym property (RNP)* if any Lipschitz function $F : [0,1] \to X$ is necessarily differentiable almost everywhere (with respect to Lebesgue measure on $\mathbb{R}$).

> COMMENT 7.2. In the case of a function $F : \mathbb{R} \to X$, where the domain is $\mathbb{R}$ (or a closed interval in $\mathbb{R}$), the definitions of Fréchet and Gâteaux differentiability coincide, and so we may unambiguously use the word "differentiable".

EXAMPLE 7.3. Each of the following functions is a Lipschitz function that is not differentiable at any point in its domain. (See Examples 6.6 and 6.7.)

(a) $F : [0,1] \to L_1(0,1)$ defined by $F(t) = \chi_{[0,t)}$.

(b) $F : [0,1] \to c_0$ defined by the formula $F(t) = \left(\dfrac{\sin(nt)}{n} \right)_{n=1}^{\infty}$.

It follows, therefore, that neither $L_1(0,1)$ nor c_0 possesses the Radon–Nikodym property.

PROPOSITION 7.4. *Let X be a separable Banach space. If $F : \mathbb{R} \to X^*$ is a Lipschitz function, then F is weak* differentiable almost everywhere. That is,*

$$w^* F'(t) = \lim_{h \to 0} \frac{F(t+h) - F(t)}{h}$$

exists for almost every $t \in \mathbb{R}$, where convergence is in the weak* topology on X^*.

PROOF. Let $(x_n)_{n=1}^{\infty}$ be a countable dense subset of X. For each $n \in \mathbb{N}$, let

$$A_n = \Big\{ t \in \mathbb{R} \ : \ \langle x_n, F(\cdot) \rangle \text{ is differentiable at } t \Big\}.$$

Observe that $\bigcap_{n=1}^{\infty} A_n$ has full measure. If $t \in \bigcap_{n=1}^{\infty} A_n$, then

$$\lim_{h \to 0} \left\langle x_n, \frac{F(t+h) - F(t)}{h} \right\rangle = c_n$$

---

[1] Diestel and Uhl introduced the notion of the Radon–Nikodym property, but in a different context than this. The original context was that of vector measure theory, where a space X had the Radon–Nikodym property if X-valued measures satisfied a certain Radon–Nikodym type of theorem. (See Appendix B for more on this topic.)

exists for each $n \in \mathbb{N}$. Suppose that $\{a_1, \ldots, a_m\}$ is a finite set of real numbers. Then
$$a_1 c_1 + \cdots + a_m c_m = \lim_{h \to 0} \left\langle a_1 x_1 + \cdots + a_m x_m, \frac{F(t+h) - F(t)}{h} \right\rangle.$$
Consequently,
$$|a_1 c_1 + \cdots + a_m c_m| \le \|a_1 x_1 + \cdots + a_m x_m\|_X \operatorname{Lip}(F).$$
Therefore, there is a bounded linear functional $x^* \in X^*$ such that $x^*(x_n) = c_n$ for all $n \in \mathbb{N}$ and $\|x^*\| \le \operatorname{Lip}(F)$, and
$$\lim_{h \to 0} \left\langle x, \frac{F(t+h) - F(t)}{h} \right\rangle = \langle x, x^* \rangle, \quad x \in X.$$
This completes the proof. □

EXAMPLE 7.5 (Example 7.3 continued). The Lipschitz functions in Example 7.3 are not differentiable, but they are weak* differentiable, by Proposition 7.4. The classical function space $L_1(0,1)$ can be viewed as a subspace of the dual space $M[0,1] = \mathcal{C}[0,1]^*$. Similarly, the classical sequence space c_0 can be viewed as a subspace of the dual space $\ell_\infty = \ell_1^*$. Both $\mathcal{C}[0,1]$ and ℓ_1 are separable Banach spaces.

DEFINITION 7.6. Let X be a Banach space. A function $f : \mathbb{R} \to X$ is called a *Borel function* provided that $f^{-1}(U)$ is a Borel set in $\mathbb{R}$ for every set U that is open in the topology on X.

PROPOSITION 7.7. *Let X be a separable Banach space and assume the function $f : [0,1] \to X$ is a bounded Borel function (with the norm topology on X). If*
$$F(t) = \int_0^t f(s) \, ds,$$
where the integral is taken in the Bochner sense, then $F : [0,1] \to X$ is differentiable almost everywhere and $F'(t) = f(t)$ for almost every t in $[0,1]$.[2]

PROOF. Let $\epsilon > 0$ be given. There exists a function $g : [0,1] \to X$ such that $g = \sum_{n=1}^\infty x_n \chi_{A_n}$, where $(x_n)_{n=1}^\infty$ is a sequence in X and $(A_n)_{n=1}^\infty$ is a sequence of pairwise disjoint Borel sets, such that $\|f(s) - g(s)\| < \epsilon$ for all $s \in [0,1]$.

For each $n \in \mathbb{N}$, let $h_n : [0,1] \to \mathbb{R}$ be the scalar valued function
$$h_n(t) = \int_0^t \chi_{A_n}(s) \, ds,$$
and define $G : [0,1] \to X$ by
$$G(t) = \sum_{n=1}^\infty x_n h_n(t) = \int_0^t g(s) \, ds.$$
Then G is differentiable almost everywhere (see Theorem A.3) and $G'(t) = g(t)$ for almost every t. Therefore, for a given $t \in [0,1]$ that is a point where G is differentiable, choose $\delta > 0$ so that
$$\left\| \frac{G(t+h) - G(t)}{h} - g(t) \right\| < \epsilon,$$

---
[2] See Appendix A for the definition of Bochner integrability.

whenever $0 < |h| < \delta$. Also,
$$\left\| \frac{F(t+h) - F(t)}{h} - \frac{G(t+h) - G(t)}{h} \right\| = \frac{1}{|h|} \left\| \int_t^{t+h} \left(f(s) - g(s) \right) ds \right\| \leq \epsilon.$$
Consequently,
$$\left\| \frac{F(t+h) - F(t)}{h} - f(t) \right\| \leq \left\| \frac{F(t+h) - F(t)}{h} - \frac{G(t+k) - G(t)}{h} \right\|$$
$$+ \left\| \frac{G(t+h) - G(t)}{h} - g(t) \right\| + \|g(t) - f(t)\|$$
$$< 3\epsilon,$$
whenever $0 < |h| < \delta$. The result follows. $\square$

THEOREM 7.8. *If X^* is a separable dual space, then X^* has RNP.*[3]

PROOF. Let $F : [0, 1] \to X^*$ be a Lipschitz function. Without loss of generality, assume $F(0) = 0$. Let f be the weak* derivative of F, which is known to exist by Proposition 7.4. Let
$$G(t) = \int_0^t f(s)\, ds, \quad t \in [0, 1].$$
Then $G'(t) = f(t)$ for almost every t, by Proposition 7.7, provided that f is a bounded Borel function.

Let U be a set open in the norm topology on X^*. The goal is to show that $f^{-1}(U)$ is a Borel set in $[0, 1]$. Since f is known to be a weak* derivative, it is sufficient to show that U is a weak* Borel set in X^*. Because U is an open set in a separable Banach space X^*, it is a countable union of closed balls. Consequently, it suffices to show that B_{X^*} is a weak* Borel set. Observe that
$$B_{X^*} = \{x^* : |x^*(x)| \leq 1 \text{ for all } x \text{ with } \|x\| \leq 1\}.$$
Let the sequence $(x_n)_{n=1}^\infty$ be a countable dense subset of B_X. Then
$$B_{X^*} = \{x^* : |x^*(x_n)| \leq 1 \text{ for all } n \in \mathbb{N}\} = \bigcap_{n=1}^\infty \{x^* : |x^*(x_n)| \leq 1\},$$
and thus B_{X^*} is a weak* Borel set.

Now, for all $x \in X$,
$$\langle x, G(t) \rangle = \left\langle x, \int_0^t f(s)\, ds \right\rangle = \int_0^t \langle x, f(s) \rangle\, ds = \langle x, F(t) \rangle,$$
and so $G = F$. Therefore F is differentiable since G is differentiable. $\square$

COMMENT 7.9. The property that we used at the end of Theorem 7.8 is a property of the Bochner integral (see Corollary A.5 in Appendix A):
$$\left\langle x, \int_0^t f(s)\, ds \right\rangle = \int_0^t \langle x, f(s) \rangle\, ds.$$

---

[3]This result is due to Dunford and Pettis [**29**], although their proof was different.

EXAMPLE 7.10. If X is a separable reflexive Banach space, or if $X = \ell_1$, then X has RNP, by Theorem 7.8.

If Y is a Banach space with RNP and if $F : [0,1] \to Y$ is a Lipschitz function, then F is differentiable outside of a set having Lebesgue measure zero. This result can be extended to Lipschitz functions $F : X \to Y$ when X is finite dimensional. Can this be extended to X an infinite dimensional Banach space? In order to answer this question, it is first necessary to determine how to extend the notion of a "measure zero" set. In an infinite dimensional space, there need not be a unique invariant measure.

So, what is meant by a set of "small" measure? There are three traditional approaches, due to Christensen (1973) [**23**], Mankiewicz (1973) [**78**], and Aronszajn (1976) [**8**]. The approach taken here is that of Christensen.

DEFINITION 7.11. Let X be a separable Banach space and suppose A is a Borel subset of X. The set A is called *Haar-null* if there exists a probability measure μ on X having compact support such that $\mu(A + x) = 0$ for all $x \in X$.

REMARK 7.12. Suppose that X is a separable Banach space and A is a Haar-null set. If μ is a compactly supported probability measure such that $\mu(A + x) = 0$ for all $x \in X$, then so is μ_{x_0}, where $x_0 \in X$ and $\mu_{x_0}(B) = \mu(B + x_0)$ for all Borel subsets B of X. So too is the measure $\frac{\mu|_C}{\mu(C)}$, whenever C is a Borel set for which $\mu(C) > 0$. Consequently, it can be assumed that μ is supported in ϵB_X for any $\epsilon > 0$.

PROPOSITION 7.13. *Let X be a separable Banach space. A Borel subset A of X is Haar-null if and only if there exists an X-valued random variable ξ on a probability space $(\Omega, \mathbb{P})$ with compact range in X such that*

$$\mathbb{P}(\xi + x \in A) = \mathbb{E}[\chi_A(\xi + x)] = 0, \quad x \in X.$$

PROOF. See Remark 7.12. □

PROPOSITION 7.14. *The countable union of Haar-null sets is a Haar-null set.*

PROOF. Let X be a separable Banach space and suppose that $(A_n)_{n=1}^\infty$ is a sequence of Haar-null sets. For each $n \in \mathbb{N}$, there exists an X-valued random variable ξ_n such that $\|\xi_n\| < 2^{-n}$ and such that

$$\mathbb{E}[\chi_{A_n}(\xi_n + x)] = 0, \quad x \in X.$$

Without loss of generality, assume that $(\xi_n)_{n=1}^\infty$ is a sequence of independent random variables. Let $\xi = \sum_{n=1}^\infty \xi_n$ and $A = \bigcup_{n=1}^\infty A_n$. Then,

$$\mathbb{E}[\chi_A(\xi + x)] \leq \sum_{n=1}^\infty \mathbb{E}\Big[\chi_{A_n}\Big(\sum_{j=1}^\infty \xi_j + x\Big)\Big] = \sum_{n=1}^\infty \mathbb{E}\Big[\chi_{A_n}\Big(\xi_n + \sum_{j \neq n}^\infty \xi_j + x\Big)\Big] = 0.$$

Therefore, A is a Haar-null set, as required. □

DEFINITION 7.15. Let $f : X \to Y$ be a Lipschitz function between two Banach spaces. Define

$$\Delta_t f(x, v) = \frac{f(x + tv) - f(x)}{t},$$

and write
$$Df(x,v) = \lim_{t \to 0} \Delta_t f(x,v),$$
if the limit exists (in a specified topology). These are the *directional difference quotient* and the *directional derivative* of f at x in the direction of v, respectively.

REMARK 7.16. Note that
$$\|\Delta_t f(x,v) - \Delta_t f(x,w)\| \leq \text{Lip}(f) \|v - w\|,$$
and so for each t, the map $v \mapsto \Delta_t f(x,v)$ is a Lipschitz function with Lipschitz $\text{Lip}(f)$ for any given x in X. Furthermore, the set $\{v \in X : Df(x,v) \text{ exists}\}$ is a closed set in X and the map $v \mapsto Df(x,v)$ is continuous on this set.

REMARK 7.17. Let X and Y be Banach spaces. A function $f : X \to Y$ is Gâteaux differentiable at a point $x \in X$ provided that $Df(x,v)$ is defined for all $v \in X$ and provided that the map $v \mapsto Df(x,v)$ is linear. (See Definition 6.2.) The map $v \mapsto Df(x,v)$ is denoted $Df(x)$. The linear map $Df(x)$ is the Gâteaux derivative of f at x.

LEMMA 7.18. *Let X and Y be Banach spaces and suppose that $f : X \to Y$ is a Lipschitz function. If $Df(x,v)$ exists for all $x \in X$ and $v \in X$, and if the map $x \mapsto Df(x,v)$ is continuous for each $v \in X$, then f is Gâteaux differentiable.*

PROOF. First consider the case $Y = \mathbb{R}$. That is, assume that f is a real-valued function. Let x and v_1 and v_2 be members of X, and define a multivariable real-valued function $\phi : \mathbb{R} \times \mathbb{R} \to \mathbb{R}$ by
$$\phi(s,t) = f(x + sv_1 + tv_2)$$
for each pair $(s,t) \in \mathbb{R} \times \mathbb{R}$. By assumption, the function ϕ has continuous partial derivatives, and so ϕ is differentiable (in particular) at $(0,0)$. It follows that
$$Df(x, v_1 + v_2) = Df(x, v_1) + Df(x, v_2),$$
and so f is Gâteaux differentiable when f is real-valued.

Now assume that Y is a Banach space and suppose that $f : X \to Y$ is a Lipschitz function that satisfies the hypotheses of the lemma. We want to show that $Df(x)$ is linear for each $x \in X$. To that end, let x be an element of X.

Let $y^* \in Y^*$. The function $y^* \circ f : X \to \mathbb{R}$ is a real-valued Lipschitz function that satisfies the hypotheses of the lemma, and so by what has already been shown (in the first part of this proof), the map $D(y^* \circ f)(x)$ is linear for each $x \in X$. That is, for v_1 and v_2 in X,
$$D(y^* \circ f)(x)(v_1 + v_2) = D(y^* \circ f)(x)(v_1) + D(y^* \circ f)(x)(v_2).$$
Since y^* is an element of Y^*, we have that $D(y^* \circ f)(x) = y^*\big(D(f)(x)\big)$, and so
$$y^*\Big(D(f)(x)(v_1 + v_2)\Big) = y^*\Big(D(f)(x)(v_1) + D(f)(x)(v_2)\Big).$$
This equation is true for each $y^* \in Y^*$, and consequently we have that
$$D(f)(x)(v_1 + v_2) = D(f)(x)(v_1) + D(f)(x)(v_2)$$
for each v_1 and v_2 in X. Therefore, we conclude that $Df(x)$ is linear, and so the function f is Gâteaux differentiable at x. □

COMMENT 7.19. In order to see how the differentiability of ϕ at $(0,0)$ implies linearity of $Df(x)$, compute the directional derivatives:

$$\begin{aligned}
Df(x, v_1 + v_2) &= \lim_{h \to 0} \frac{f(x + hv_1 + hv_2) - f(x)}{h} \\
&= \lim_{h \to 0} \frac{f(x + hv_1 + hv_2) - f(x + hv_1)}{h} + \frac{f(x + hv_1) - f(x)}{h} \\
&= \lim_{h \to 0} \frac{\phi(h, h) - \phi(h, 0)}{h} + \frac{\phi(h, 0) - \phi(0, 0)}{h} \\
&= \frac{\partial \phi}{\partial s}(0, 0) + \frac{\partial \phi}{\partial t}(0, 0) = Df(x, v_1) + Df(x, v_2).
\end{aligned}$$

The second-to-last equality is where we used the differentiability of ϕ, since differentiability of ϕ at $(0,0)$ implies the existence of continuous partial derivatives.

For the next proposition, we recall the definition of a dense cube measure. (See Definition 6.20.) Suppose that X is a Banach space and $(x_k)_{k=1}^\infty$ is a sequence of points in X such that $\sum_{k=1}^\infty \|x_k\| < \infty$. Let $(\eta_k)_{k=1}^\infty$ be a sequence of independent random variables, each of which is uniformly distributed on $[0, 1]$. Define a random variable ξ by

$$\xi = \sum_{k=1}^\infty \eta_k x_k.$$

The range of ξ is a compact set in X and ξ defines a Borel measure μ_ξ on X by the formula

$$\mu_\xi(B) = \mathbb{P}(\xi \in B),$$

where B is a Borel set in X and $\mathbb{P}$ is a probability measure on $[0,1]^\mathbb{N}$. The random variable ξ can be interpreted as a function $\xi : ([0,1]^\mathbb{N}, \mathbb{P}) \to X$, defined by

$$\xi(t_1, t_2, \ldots) = \sum_{k=1}^\infty t_k x_k.$$

The measure μ_ξ is called a *cube measure* on X. It is called a *dense cube measure* if additionally the set $\left\{ \frac{x_k}{\|x_k\|} : k \in \mathbb{N} \right\}$ is dense in the unit sphere of X.

LEMMA 7.20. *Suppose X and Y are separable Banach spaces, where Y has the Radon–Nikodym property. Let $\Omega = [0,1]^\mathbb{N}$ be the Hilbert cube with product measure $\mathbb{P}$. Suppose ξ is a random variable on Ω corresponding to a dense cube measure. If $f : X \to Y$ is a Lipschitz function, then the set*

$$\Omega \setminus \{\omega \in \Omega : Df(x + \xi(\omega), v) \text{ exists for all } v \in X\}$$

has $\mathbb{P}$-measure zero for each $x \in X$.

PROOF. We assume ξ is a random variable on $([0,1]^\mathbb{N}, \mathbb{P})$ such that

$$\xi = \sum_{k=1}^\infty \eta_k v_k,$$

for some sequence $(v_k)_{k=1}^\infty$ of terms in X with $\sum_{k=1}^\infty \|v_k\| < \infty$ and such that $\left\{ \frac{v_k}{\|v_k\|} : k \in \mathbb{N} \right\}$ is a dense set in the unit sphere of X, and $(\eta_k)_{k=1}^\infty$ is a sequence of independent random variables, each of which is uniformly distributed on $[0, 1]$.

Let x be an element in X, and let $(t_k)_{k=1}^\infty$ be a sequence of real numbers taken from the interval $[0,1]$ and for each $n \in \mathbb{N}$ define a function $f_n : [0,1] \to Y$ by

$$f_n(t) = f\Big(x + \sum_{k \neq n} t_k v_k + t v_n\Big).$$

For each $n \in \mathbb{N}$,

$$\|f_n(t) - f_n(s)\|_Y \leq \mathrm{Lip}(f) \|v_n\|_X |t - s|,$$

and so the function f_n is Lipschitz continuous with $\mathrm{Lip}(f_n) \leq \mathrm{Lip}(f)\|v_n\|_X$. In particular, since $f_n : [0,1] \to Y$ is a Lipschitz function and Y has the Radon–Nikodym property, it follows that f_n is differentiable[4] almost everywhere with respect to Lebesgue measure on $[0,1]$.

Consequently, for each $n \in \mathbb{N}$, the directional derivative

$$Df\Big(x + \sum_{k \neq n} t_k v_k + t v_n, v_n\Big)$$

exists for almost every t in $[0,1]$. To see this, observe that

$$Df\Big(x + \sum_{k \neq n} t_k v_k + t v_n, v_n\Big)$$

$$= \lim_{h \to 0} \frac{f\big(x + \sum_{k \neq n} t_k v_k + t v_n + h v_n\big) - f\big(x + \sum_{k \neq n} t_k v_k + t v_n\big)}{h}$$

$$= \lim_{h \to 0} \frac{f_n(t+h) - f_n(t)}{h} = f_n'(t).$$

It follows that $Df(x + \xi(\omega), v_n)$ exists for almost every $\omega \in [0,1]^{\mathbb{N}}$, with respect to the probability measure $\mathbb{P}$. Thus,

$$\mathbb{P}\Big(\{\omega \in [0,1]^{\mathbb{N}} : Df(x + \xi(\omega), v_n) \text{ does not exist}\}\Big) = 0,$$

for all $n \in \mathbb{N}$ (because the countable union of sets having measure zero is a set having measure zero). Using the density of the set $\{v_k : k \in \mathbb{N}\}$, we conclude that $Df(x + \xi, v)$ exists $\mathbb{P}$-almost everywhere. $\square$

Using the previous result, we can show that a Lipschitz function between separable Banach spaces has directional derivatives in all directions outside of a Haar-null set when the range space has the Radon–Nikodym property.

COROLLARY 7.21. *Suppose X and Y are separable Banach spaces, where Y has the Radon–Nikodym property. If $f : X \to Y$ is a Lipschitz function, and*

$$A = X \setminus \{x \in X : Df(x, v) \text{ exists for all } v \in X\},$$

then $\mu(x + A) = 0$ if $x \in X$ whenever μ is a dense cube measure on X. In particular, the set of points where f fails to have directional derivatives in all directions is a Haar-null set.

---

[4]Gâteaux and Fréchet differentiability coincide when the domain is $[0,1]$, and so we can use the term "differentiable" without ambiguity.

PROOF. Since μ is a dense cube measure on X, there is a sequence $(x_k)_{k=1}^\infty$ of points in X such that $\sum_{k=1}^\infty \|x_k\| < \infty$ and a sequence $(\eta_k)_{k=1}^\infty$ of independent random variables, each of which is uniformly distributed on $[0, 1]$, such that

$$\xi = \sum_{k=1}^\infty \eta_k x_k$$

and

$$\mu(x + A) = \mathbb{P}(\xi \in x + A) = \mathbb{P}(-x + \xi \in A).$$

This quantity equals zero, by Lemma 7.20, and so we have completed the proof. □

The previous result tells us that a Lipschitz function $f : X \to Y$, where X and Y are separable Banach spaces and Y has the Radon–Nikodym property and has directional derivatives in all directions outside a set that is Haar-null. This on its own is not enough to determine that f is Gâteaux differentiable outside a Haar-null set, however, because we still need to establish that the maps $v \mapsto Df(x, v)$ are linear for each x outside of that Haar-null set. To that end, we will start by proving a special case of Rademacher's theorem. (We originally stated Rademacher's theorem as Example 6.5.)

LEMMA 7.22 (Rademacher's theorem). *Suppose E is a finite dimensional real normed vector space. If $f : E \to \mathbb{R}$ is a Lipschitz function, then f is Gâteaux differentiable outside a Haar-null set.*

PROOF. Since E is separable and $\mathbb{R}$ has the Radon–Nikodym property, we know from Corollary 7.21 that there is a Haar-null set A_0 so that for all $x \notin A_0$, the directional derivative $Df(x, v)$ exists for each $v \in E$. Without loss of generality, we may assume that $E = \mathbb{R}^n$ for some $n \in \mathbb{N}$ and A_0 has λ-measure zero, where λ is Lebesgue measure on $\mathbb{R}^n$.

In order to show that f is Gâteaux differentiable at $x \notin A_0$, we need to verify that the map $v \mapsto Df(x, v)$ is linear on V for that given x. In light of Lemma 6.10, it suffices to show linearity on a dense subset. To that end, let W be a countable dense additive subgroup of $\mathbb{R}^n$.

We first observe that the directional derivatives $Df(x, v)$ are always homogeneous in v whenever they exist. That is, for each $x \notin A_0$ and each $v \in \mathbb{R}^n$, if $\alpha \in \mathbb{R}$, then

$$Df(x, \alpha v) = \lim_{t \to 0} \frac{f(x + t\alpha v) - f(x)}{t} = \alpha \lim_{t \to 0} \frac{f(x + (t\alpha) v) - f(x)}{t\alpha} = \alpha Df(x, v).$$

Consequently, in order to show that f is Gâteaux differentiable at some x, it suffices to show that $Df(x, w_1 + w_2) = Df(x, w_1) + Df(x, w_2)$ for each w_1 and w_2 in W.

Let ϕ be a nonnegative continuously differentiable real valued function with compact support such that $\int_{\mathbb{R}^n} \phi \, d\lambda = 1$. Furthermore, for each $k \in \mathbb{N}$, define a function $\phi_k : \mathbb{R}^n \to \mathbb{R}$ by $\phi_k(x) = k^n \phi(kx)$ for each $x \in \mathbb{R}^n$. Then $\phi_k * g \to g$ almost everywhere as $k \to \infty$ for any bounded measurable function $g : \mathbb{R}^n \to \mathbb{R}$.

Now let $k \in \mathbb{N}$. Then

$$(f * \phi_k)(x) = \int_{\mathbb{R}^n} f(u) \, \phi_k(x - u) \, \lambda(du)$$

is continuously differentiable, and so $v \mapsto D(f * \phi_k)(x,v)$ is linear for each $x \in \mathbb{R}^n$. Let us compute $D(f * \phi_k)(x,v)$ for a given x and v in $\mathbb{R}^n$:

$$D(f * \phi_k)(x,v) = \lim_{t \to 0} \Delta(f * \phi_k)(x,v) = \lim_{t \to 0} \frac{(f * \phi_k)(x+tv) - (f * \phi_k)(x)}{t}$$

$$= \lim_{t \to 0} \int_{\mathbb{R}^n} \phi_k(u) \left[\frac{f(x+tv-u) - f(x-u)}{t} \right] \lambda(du).$$

Since f is a Lipschitz function, the integrand in the integral above is dominated by the integrable function $\phi_k \operatorname{Lip}(f) \|v\|$, and so by the Lebesgue dominated convergence theorem,

$$D(f * \phi_k)(x,v) = \int_{\mathbb{R}^n} \phi_k(u) \left[\lim_{t \to 0} \frac{f(x+tv-u) - f(x-u)}{t} \right] \lambda(du)$$

$$= \int_{\mathbb{R}^n} \phi_k(u) \, Df(x-u,v) \, \lambda(du) = (\phi_k * Df(\cdot,v))(x).$$

We recall that the integral in the previous line makes sense because the directional derivatives of f are known to exist outside of a set of measure zero.

We know that $D(f * \phi_k)(x,v)$ is linear in v for any $x \in \mathbb{R}^n$, and so for $x \in \mathbb{R}^n$, we now have that

$$(\phi_k * Df(\cdot, w_1 + w_2))(x) = (\phi_k * Df(\cdot, w_1))(x) + (\phi_k * Df(\cdot, w_2))(x),$$

for w_1 and w_2 in W, which means that

$$\phi_k * \Big(Df(\cdot, w_1 + w_2) - Df(\cdot, w_1) - Df(\cdot, w_2) \Big) = 0.$$

But now we recall that $\phi_k * g \to g$ almost everywhere as $k \to \infty$ for any bounded real valued measurable function g on $\mathbb{R}^n$, and so

$$Df(x, w_1 + w_2) - Df(x, w_1) - Df(x, w_2) = 0$$

for almost every $x \in \mathbb{R}^n$. Since the set W is countable, it follows that there is a measure-zero set A containing A_0 so that

$$Df(x, w_1 + w_2) = Df(x, w_1) + Df(x, w_2)$$

for all w_1 and w_2 in W whenever $x \notin A$. This is what we needed to show, and so we conclude that f is Gâteaux differentiable at x for all $x \in \mathbb{R}^n \setminus A$, where A is a measurable set having measure zero. □

We can now extend Lemma 7.22 to domains that are separable Banach spaces.

LEMMA 7.23. *Suppose X is a separable Banach space. If $f : X \to \mathbb{R}$ is a Lipschitz function, then f is Gâteaux differentiable outside a Haar-null set.*

PROOF. We may assume that X is infinite dimensional. Let $(v_i)_{i=1}^\infty$ be a countable dense subset in X. For each $n \in \mathbb{N}$, let V_n be the closed linear span of the finite set $\{v_1, \ldots, v_n\}$. Then $(V_n)_{n=1}^\infty$ is an increasing set of finite dimensional subspaces of X and their union is dense in X.

Now let $n \in \mathbb{N}$ be given and let B_n be the collection of all $x \in X$ for which there exists a bounded linear functional $v_x^* \in V_n^*$ (that in general depends on x) with

$$\lim_{t \to 0} \frac{|f(x+tv) - f(x) - tv_x^*(v)|}{t} = 0$$

for all $v \in V_n$. Let $A_n = X \setminus B_n$. Then A_n is Haar-null, by Rademacher's theorem (Lemma 7.22).

Let $A = \bigcup_{n=1}^{\infty} A_n$. Then A is Haar-null, by Proposition 7.14. The set A is the collection of points in X for which f fails to have a Gâteaux derivative, and so the proof is complete. □

EXAMPLE 7.24. Let X be a separable Banach space and suppose that ξ is a random variable on $([0,1]^{\mathbb{N}}, \mathbb{P})$ associated with a dense cube measure on the separable Banach space X. That is, suppose that

$$\xi = \sum_{k=1}^{\infty} \eta_k v_k,$$

for some sequence $(v_k)_{k=1}^{\infty}$ of terms in X with $\sum_{k=1}^{\infty} \|v_k\| < \infty$ and such that $\{\frac{v_k}{\|v_k\|} : k \in \mathbb{N}\}$ is a dense set in the unit sphere of X, where $(\eta_k)_{k=1}^{\infty}$ is a sequence of independent random variables, each of which is uniformly distributed on $[0,1]$.

For ease of notation, let Ω denote $[0,1]^{\mathbb{N}}$. Define a function $F : X \to L_2(\Omega, \mathbb{P})$ by

$$F(x)(\omega) = f\big(x + \xi(\omega)\big), \quad \omega \in \Omega.$$

The function F is a Lipschitz function with Lipschitz constant $\text{Lip}(f)$. To see this, let x_1 and x_2 be in X. Then

$$\|F(x_1) - F(x_2)\|_{L_2(\Omega, \mathbb{P})} = \bigg(\int_{\Omega} |F(x_1)(\omega) - F(x_2)(\omega)|^2 \, \mathbb{P}(d\omega)\bigg)^{1/2}$$

$$= \bigg(\int_{\Omega} |f(x_1 + \xi(\omega)) - f(x_2 + \xi(\omega))|^2 \, \mathbb{P}(d\omega)\bigg)^{1/2}.$$

Since f is a Lipschitz function, this is bounded by

$$\bigg(\int_{\Omega} \big[\text{Lip}(f)\|x_1 - x_2\|_X\big]^2 \, \mathbb{P}(d\omega)\bigg)^{1/2} = \text{Lip}(f)\|x_1 - x_2\|_X.$$

We claim that F is Gâteaux differentiable everywhere. We first show that for any $x \in X$, the directional derivative $DF(x, v)$ exists for each $v \in X$. To that end, let $v \in X$. Then for each $\omega \in \Omega$, the directional difference quotient of F at x in the direction of v evaluated at ω is

$$\Delta_t F(x, v)(\omega) = \frac{F(x+tv)(\omega) - F(x)(\omega)}{t} = \frac{f(x+tv+\xi(\omega)) - f(x+\xi(\omega))}{t}.$$

Consequently,

$$\Delta_t F(x, v)(\omega) = \Delta_t f\big(x + \xi(\omega), v\big)$$

for every x and v in X and every ω in Ω. From Lemma 7.20, we know that $\Delta_t f\big(x + \xi(\omega), v\big)$ converges to $Df\big(x + \xi(\omega), v\big)$ as $t \to 0$ for almost every $\omega \in \Omega$. Therefore,

$$\lim_{t \to 0} \Delta_t F(x, v)(\omega) = Df\big(x + \xi(\omega), v\big) \quad \text{a.e.}(\omega).$$

Furthermore, the function $\Delta_t F(x, v) : \Omega \to \mathbb{R}$ is bounded as a function on Ω. To see this, observe that for $\omega \in \Omega$,

$$|\Delta_t F(x, v)(\omega)| = \left|\frac{f(x+tv+\xi(\omega)) - f(x+\xi(\omega))}{t}\right| \leq \frac{\text{Lip}(f)\|tv\|}{|t|} = \text{Lip}(f)\|v\|.$$

The bound is uniform in t, and so by the bounded convergence theorem, we are able to conclude that $\Delta_t F(x, v)$ converges to $Df(x + \xi, v)$ in $L_2(\Omega, \mathbb{P})$. Therefore, $DF(x, v) = Df(x + \xi, v)$ for all x and v in X.

The next step is to show that the map $v \mapsto DF(x,v)$ is linear for a given $x \in X$, but this follows from Lemma 7.23: If A is the set of points where f fails to be Gâteaux differentiable, then A is Haar-null, and so $\mathbb{P}(x+\xi \in A) = 0$. Therefore, the map $DF(x,v) = Df(x+\xi,v)$ is linear in v almost everywhere with respect to $\mathbb{P}$. We conclude that if w_1 and w_2 are in X and a and b are in $\mathbb{R}$, then

$$DF(x, aw_1 + bw_2) = aDF(x, w_1) + bDF(x, w_2)$$

as functions in $L_2(\Omega, \mathbb{P})$.

We have shown that for all $x \in X$, the derivational derivative $DF(x,v)$ exists for each $v \in X$ and the map $v \mapsto DF(x,v)$ is linear. It follows that F is Gâteaux differentiable at each $x \in X$.[5]

We now consider the case where Y is a separable Banach space with the Radon–Nikodym property.

THEOREM 7.25. *Suppose that X and Y are separable Banach spaces, and suppose Y has the Radon–Nikodym property. If $f : X \to Y$ is a Lipschitz function, then f is Gâteaux differentiable outside of a Haar-null set.*

PROOF. Since Y is separable, there exists a countable sequence $(y_n^*)_{n=1}^\infty$ in Y^* such that if $y_n(y) = 0$ for all n, then $y = 0$. For each $n \in \mathbb{N}$, the scalar valued function $y_n \circ f : X \to \mathbb{R}$ is a Lipschitz function. Consequently, by Lemma 7.23, the Gâteaux derivative $D(y_n^* \circ f)(x)$ exists away from a Haar-null set. Call this set A_n. Note that

$$D(y_n^* \circ f)(x,v) = y_n^* Df(x,v).$$

Let $A = \bigcup_{n=1}^\infty A_n$. This is the countable union of Haar-null sets, and is therefore also a Haar-null set, and $y_n^* Df(x)$ exists for each $n \in \mathbb{N}$ outside of A. Therefore, the derivative $Df(x)$ exists and is linear outside of a Haar-null set, which completes the proof. □

REMARK 7.26. In the statement of Theorem 7.25, it is not necessary for Y to be separable. Since X is separable and f is Lipschitz continuous, the image of f in Y will be separable, and this is sufficient for the proof. It is, however, necessary for Y to have the Radon–Nikodym property.

If Y does not have the Radon–Nikodym property, but is the dual space of a separable Banach space, then Y has what one might describe as a weak* version of the Radon–Nikodym property, as is shown in Proposition 7.4. That is, if Z is a separable Banach space and $Y = Z^*$, then a Lipschitz function $f : [0,1] \to Z^*$ is weak* differentiable almost everywhere.

THEOREM 7.27. *Suppose X and Z are separable Banach spaces and let $Y = Z^*$. If $f : X \to Y$ is a Lipschitz function, then f is weak* Gâteaux differentiable outside a Haar-null set.*

PROOF. The proof is similar to the proof of Theorem 7.25, but uses Proposition 7.4 in place of the Radon–Nikodym property. □

It is now possible to at least partially answer Question 5.1.

---

[5]In Kalton's original lectures, he used the function F of Example 7.24 to prove Lemma 7.23. However, the details of that proof proved elusive to the current author, and so we have instead opted to follow the proof given in [**62**]. (See Theorem 2.2 in that document.)

THEOREM 7.28. *Let X be a separable Banach space which Lipschitz embeds into a Banach space Y. If Y has the Radon–Nikodym property, then X linearly embeds into Y.*

PROOF. Suppose $f : X \to Y$ is a Lipschitz embedding from X into Y. Then there exist positive constants c and C such that
$$c\|x - y\|_X \leq \|f(x) - f(y)\|_Y \leq C\|x - y\|_X,$$
for all x and y in X. By Theorem 7.25, the function f is Gâteaux differentiable at some point $x_0 \in X$. Therefore, for each $x \in X$,
$$Df(x_0, x) = \lim_{t \to 0} \frac{f(x_0 + tx) - f(x_0)}{t},$$
where the limit exists in norm. For each $t \neq 0$,
$$c\|x\|_X = c\frac{\|tx\|_X}{|t|} \leq \frac{\|f(x_0 - tx) - f(x_0)\|_Y}{|t|} \leq C\frac{\|tx\|_X}{|t|} = C\|x\|_X.$$
Therefore,
$$c\|x\|_X \leq \|Df(x_0, x)\|_Y \leq C\|x\|_X,$$
for all $x \in X$. Consequently, $Df(x_0)$ is a linear embedding of X into Y. □

EXAMPLE 7.29. For any finite $p \geq 1$, the space ℓ_p has RNP. If $p > 1$, then ℓ_p is reflexive, and $\ell_1 = c_0^*$, so that ℓ_1 is a separable dual space. Consequently, if X Lipschitz embeds into ℓ_p, for any $p \in [1, \infty)$, then X linearly embeds into ℓ_p. (In fact, if $p = 2$, it must be that X is linearly isomorphic to ℓ_2 itself.)

The preceding example raises a natural question: If X is a separable Banach space that Lipschitz embeds into ℓ_∞, does it also linearly embed into ℓ_∞? The space ℓ_∞ does not have the Radon–Nikodym property, but it is the dual of the separable space ℓ_1, and so we can use Theorem 7.27 to find a weak* Gâteaux derivative to fill the role of the Gâteaux derivative used in the proof of Theorem 7.28. However, since the convergence of the derivative is not in norm, there is something more to prove.

THEOREM 7.30. *Let X be a separable Banach space which Lipschitz embeds into a Banach space Y. If Y is the dual space of a separable space, then X linearly embeds into Y.*

PROOF. Assume that $f : X \to Y$ is a Lipschitz embedding from the separable Banach space X into the space Y, which is the dual of a separable Banach space Z. There exist positive constants c and C such that
$$c\|x - y\|_X \leq \|f(x) - f(y)\|_Y \leq C\|x - y\|_X,$$
for all x and y in X. By Theorem 7.27, the function f is weak* Gâteaux differentiable except on a Haar-null set. That is, for each $x \in X$ outside of a Haar-null set,
$$\lim_{t \to 0} \frac{f(x + tv) - f(x)}{t} = Df(x, v)$$
exists for all $v \in X$, where the limit converges in the weak* topology. As usual, we let $Df(x)$ be the map defined by $Df(x)(v) = Df(x, v)$ for each $v \in X$.

7. DIFFERENTIATION AND HAAR-NULL SETS

If x is a point for which f is weak* Gâteaux differentiable, then $Df(x)$ is a linear map, and so $\|Df(x)\| < \infty$. In fact, $\|Df(x)\| \leq C$. To see this, let $v \in X$. Then

$$\|Df(x)(v)\|_Y = \sup_{z \in B_Z} |\langle z, Df(x,v)\rangle| = \sup_{z \in B_Z} \left|\lim_{t \to 0} \left\langle z, \frac{f(x+tv) - f(x)}{t}\right\rangle\right|$$

$$\leq \sup_{t \neq 0} \frac{\|f(x+tv) - f(x)\|_Y}{|t|} \leq C\|v\|_X.$$

The question now becomes: Is there an $x \in X$ such that $Df(x,v)$ exists for all $v \in X$ and such that

$$\|Df(x,v)\|_Y \geq c\|v\|_X, \quad v \in X?$$

Since X is separable, it suffices to show that for any fixed $v \in X$, the above inequality holds except on a Haar-null set. We will therefore fix v and show that for any $b < c$, the set

$$A = \{x \in X : \|Df(x,v)\|_Y \leq b\|v\|_X\}$$

is a Haar-null set. (Note that A is a Borel set because Y is the dual space of a separable Banach space Z, and therefore the norm in Y can be achieved with a countable subset of Z.)

Assume to the contrary that A is not Haar-null. Let $\xi = \nu v$, where ν is a uniformly distributed random variable on the interval $[0,1]$, and let λ be Lebesgue measure on $[0,1]$. Because A is not Haar-null, there exists a point $x \in X$ such that the set

$$B = \{t \in [0,1] : Df(x+tv) \text{ exists and } \|Df(x+tv,v)\|_Y \leq b\|v\|_X\}$$

has positive Lebesgue measure.

Given $\epsilon > 0$, there exists an interval $I = [\alpha, \beta]$ such that

$$\lambda(I \cap B) > (1-\epsilon)(\beta - \alpha) \quad \text{and} \quad \lambda(I \cap B^c) \leq \epsilon(\beta - \alpha).$$

By assumption,

$$\|f(x+\beta v) - f(x+\alpha v)\|_Y \geq c(\beta - \alpha)\|v\|_X.$$

Consequently, there exists a $z \in Z$ such that $\|z\|_Z = 1$ and

$$\langle z, f(x+\beta v)\rangle - \langle z, f(x+\alpha v)\rangle \geq (c-\epsilon)(\beta - \alpha)\|v\|_X.$$

Define $\phi : [0,1] \to \mathbb{R}$ by the formula $\phi(t) = \langle z, f(x+tv)\rangle$. Then ϕ is a Lipschitz function with $\text{Lip}(\phi) \leq \text{Lip}(f)$ and

$$\phi(\beta) - \phi(\alpha) \geq (c-\epsilon)(\beta - \alpha)\|v\|_X.$$

The function f was assumed to be weak* Gâteaux differentiable outside of a Haar-null set, and so ϕ is an almost everywhere differentiable scalar valued function, because

$$\phi'(t) = \lim_{h \to 0} \frac{\phi(t+h) - \phi(t)}{h} = \lim_{h \to 0} \frac{\langle z, f(x+tv+hv)\rangle - \langle z, f(x+tv)\rangle}{h}$$

$$= \lim_{h \to 0} \left\langle z, \frac{f(x+tv+hv) - f(x+tv)}{h}\right\rangle = \langle z, Df(x+tv,v)\rangle$$

and the set where $\langle z, Df(x+tv,v)\rangle$ does not exist has Lebesgue measure zero (since the function $\langle z, f\rangle$ is Gâteaux differentiable outside of a Haar-null set.) The function ϕ is Lipschitz continuous, and hence absolutely continuous, and therefore

$$\phi(\beta) - \phi(\alpha) = \int_\alpha^\beta \langle z, Df(x+tv,v)\rangle\, dt.$$

It follows that

$$\phi(\beta) - \phi(\alpha) = \int_{I \cap B} \langle z, Df(x+tv,v)\rangle\, dt + \int_{I \cap B^c} \langle z, Df(x+tv,v)\rangle\, dt$$

$$\leq b\|v\|_X (\beta - \alpha) + C\|v\|_X \epsilon(\beta - \alpha)$$

$$= (b + C\epsilon)\|v\|_X (\beta - \alpha).$$

Putting all of these inequalities together,

$$(c - \epsilon)(\beta - \alpha)\|v\|_X \leq \phi(\beta) - \phi(\alpha) \leq (b + C\epsilon)\|v\|_X (\beta - \alpha),$$

from which it follows that

$$c - \epsilon \leq b + C\epsilon.$$

The choice of $\epsilon > 0$ was arbitrary, and so we conclude that $c \leq b$, which is a contradiction. $\square$

REMARK 7.31. The Banach space ℓ_∞ does not have RNP, but $\ell_\infty = \ell_1^*$, and so ℓ_∞ is the dual space of a separable Banach space. Consequently, if a separable Banach space X Lipschitz embeds into ℓ_∞, then X linearly embeds into ℓ_∞, by Theorem 7.30. It is not necessary to use this theorem, however, since, in fact, every separable Banach space embeds isometrically into ℓ_∞.

PROPOSITION 7.32. *Every separable Banach space is isometrically isomorphic to a subspace of ℓ_∞.*

PROOF. Let $(x_n)_{n=1}^\infty$ be a countable dense subset of the separable Banach space X. For each $n \in \mathbb{N}$, there is a bounded linear functional x_n^* in X^* such that $\|x_n^*\| = 1$ and $x_n^*(x_n) = \|x_n\|$. Define a linear operator $T : X \to \ell_\infty$ by

$$T(x) = \left(x_n^*(x)\right)_{n=1}^\infty, \quad x \in X.$$

The linear operator T is the required isometric isomorphism. $\square$

REMARK 7.33. The Banach space $L_p(0,1)$ of p-integrable functions is separable and reflexive if $p \in (1, \infty)$, and thus has RNP, by Theorem 7.8. The space $L_\infty(0,1)$ of essentially bounded measurable functions does not have RNP, but it is the dual space of the separable Banach space $L_1(0,1)$. Consequently, if the separable space X Lipschitz embeds into $L_p(0,1)$, for any $p \in (1, \infty]$, then X linearly embeds into $L_p(0,1)$, by Theorem 7.28 if $1 < p < \infty$ and Theorem 7.30 if $p = \infty$.

As in the case of sequence spaces, it is not necessary to use Theorem 7.30 to embed separable Banach spaces into $L_\infty(0,1)$. Indeed, since $L_\infty(0,1)$ and ℓ_∞ are isomorphic Banach spaces, it follows from Proposition 7.32 that every separable Banach space is isometrically isomorphic to a subspace of $L_\infty(0,1)$.

In fact, every separable Banach space embeds into $\mathcal{C}[0,1]$, the space of continuous functions on $[0,1]$, and $\mathcal{C}[0,1]$ is a subspace of $L_\infty(0,1)$. This fact is called the Banach–Mazur theorem, which is stated here without proof.[6]

THEOREM 7.34 (Banach–Mazur Theorem). *Every separable Banach space is isometrically isomorphic to a subspace of $\mathcal{C}[0,1]$.*

The case of $L_1(0,1)$ remains to be addressed. In the case of sequence spaces, ℓ_1 has RNP because it is the separable dual space of c_0. This argument will not work in this case, however. The dual space of $\mathcal{C}[0,1]$, the Banach space of continuous functions on $[0,1]$, is $\mathcal{M}[0,1]$, the space of countably additive measures on $[0,1]$ having finite total variation. Since $\mathcal{M}[0,1]$ is the dual space of a separable Banach space, it follows from Threorem 7.30 that a Lipschitz embedding of a separable space into $\mathcal{M}[0,1]$ implies the existence of a linear embedding into $\mathcal{M}[0,1]$.

Even so, it is true that if a separable Banach space Lipschitz embeds into $L_1(0,1)$, then it linearly embeds into $L_1(0,1)$. To see this, observe that $L_1(0,1)$ linearly embeds into $\mathcal{M}[0,1]$. Thus, if X Lipschitz embeds into $L_1(0,1)$, then

$$X \stackrel{\text{Lip}}{\hookrightarrow} L_1(0,1) \hookrightarrow \mathcal{M}[0,1].$$

Consequently, by Theorem 7.30, X linearly embeds into $\mathcal{M}[0,1]$. The conclusion that X then linearly embeds in $L_1(0,1)$ follows from the fact that a separable subset of $\mathcal{M}[0,1]$ must linearly embed into $L_1(0,1)$, which is a fact we will not prove here.[7] We summarize this in the following corollary to Theorem 7.30.

COROLLARY 7.35. *If a separable Banach space Lipschitz embeds into $L_1(0,1)$, it linearly embeds into $L_1(0,1)$.*

COMMENT 7.36. Neither $\mathcal{C}[0,1]$ nor c_0 have RNP, nor are they dual spaces of separable Banach spaces, so we cannot use the derivative techniques developed above to understand the interplay between the Lipschitz and the linear structure of these spaces. The Banach–Mazur theorem tells us that every separable Banach space is a subspace of $\mathcal{C}[0,1]$, but ℓ_1 (for example) is not isomorphic to a subspace of c_0.

In the case of $L_1(0,1)$, which also does not have RNP and is not a dual space of a separable space, we made use of the fact that $L_1(0,1)$ could be viewed as a subspace of $\mathcal{M}[0,1]$, which is the dual space of a separable space. However, this worked only because every separable subspace of $\mathcal{M}[0,1]$ is isomorphic to a subspace of $L_1(0,1)$. This can be viewed in a broader context due to the following theorem of Heinrich and Mankiewicz (which we will prove later, as Theorem 9.7).

THEOREM 7.37 (Heinrich–Mankiewicz, 1982 [**45**]). *If X and Y are separable Banach spaces and if there is a Lipschitz embedding of X into Y, then there exists a linear embedding of X into Y^{**}.*

When X Lipschitz embeds into $Y = L_1(0,1)$, then the Heinrich–Mankiewicz theorem states that X linearly embeds into $Y^{**} = L_\infty(0,1)^*$. The only separable

---

[6] For a proof of the Banach–Mazur theorem, see Theorem 1.4.4 in [**5**].
[7] A separable subset of $\mathcal{M}[0,1]$ linearly embeds into $L_1(0,1)$ because $\mathcal{M}[0,1]$ is an ℓ_1-sum of $L_1(\mu)$-spaces, where each μ is a measure on $[0,1]$. See the proof of Proposition 4.3.8(iii) in [**5**] for more.

subspaces of $L_\infty(0,1)^*$ are isomorphic to subspaces of $L_1(0,1)$, which is why the only separable subspaces of $\mathcal{M}[0,1]$ are isomorphic to subspaces of $L_1(0,1)$. This means that although we have to extend our codomain to the second dual, which is a very large space indeed, the only possible choices for separable subspaces are isomorphic to subspaces of $L_1(0,1)$.

A natural question then, is will this work with c_0? Unfortunately, if a separable space X Lipschitz embeds into c_0, the Heinrich–Mankiewicz theorem states that X linearly embeds into $c_0^{**} = \ell_\infty$, and ℓ_∞ has a lot of separable subspaces. In fact, as Proposition 7.32 makes evident, every separable Banach space can be found as a subspace of ℓ_∞.

Theorem 7.38, due to Aharoni, shows that c_0 has a special role in the Lipschitz theory of Banach spaces.

THEOREM 7.38 (Aharoni, 1974 [**2**]). *Every separable metric space Lipschitz embeds into c_0.*

In particular, Aharoni showed that there is a constant $K > 0$ such that for every separable metric space M, there is a map $f : M \to c_0$ satisfying

$$d_M(x,y) \le \|f(x) - f(y)\|_{c_0} \le K d_M(x,y),$$

for every x and y in M. Aharoni showed that K could be any number greater than 6. To be more precise, for any $\epsilon > 0$ and any metric space M, one can find a Lipschitz map $f : M \to c_0$ for which $K = 6 + \epsilon$ in the above inequality.

In 1978, Assouad was able to replace $6 + \epsilon$ with $3 + \epsilon$ [**9**]. Much later, in 1994, Pelant improved the constant to exactly 3 [**91**]. Pelant actually found embeddings into c_0^+, the set of positive sequences converging to zero, and in this context, 3 is the optimal value of K. (Pelant demonstrated the optimality of 3 in this context by means of an example.)

In the case of c_0, the optimal constant is 2, which was shown by Kalton and Lancien in 2008 [**66**]. The proof of this will be given in Chapter 8. For now, we will provide Aharoni's example that shows $K \ge 2$. (Proposition 3 in [**2**].)

PROPOSITION 7.39. *If $f : \ell_1 \to c_0$ is a Lipschitz function such that*

$$\|x - y\|_{\ell_1} \le \|f(x) - f(y)\|_{c_0} \le \lambda \|x - y\|_{\ell_1},$$

for all x and y in ℓ_1, then $\lambda \ge 2$.

PROOF. Suppose that $f : \ell_1 \to c_0$ is a Lipschitz function such that

$$\|x - y\|_{\ell_1} \le \|f(x) - f(y)\|_{c_0} \le \lambda \|x - y\|_{\ell_1},$$

for all x and y in ℓ_1. Without loss of generality, assume $f(0) = 0$. The codomain of f is c_0, and consequently, we may write

$$f(x) = \left(f_j(x)\right)_{j=1}^\infty, \quad x \in \ell_1,$$

where $(f_j)_{j=1}^\infty$ is a sequence of real-valued functions on ℓ_1 such that $\lim_{j \to \infty} f_j(x) = 0$ for each $x \in \ell_1$. Observe that for each $j \in \mathbb{N}$ the function $f_j : \ell_1 \to \mathbb{R}$ is Lipschitz continuous with $\mathrm{Lip}(f_j) \le \lambda$.

Let $(e_n)_{n=1}^{\infty}$ be the standard basis for ℓ_1. For $j \in \mathbb{N}$, consider the sequences

$$\left\{f_j(e_1 + e_n)\right\}_{n=1}^{\infty} \quad \text{and} \quad \left\{f_j(-e_1 + e_n)\right\}_{n=1}^{\infty}.$$

Both of these sequences are bounded by 2λ, and so (passing to subsequences, as necessary), we may assume without loss of generality that

$$\lim_{n \to \infty} f_j(e_1 + e_n) = \alpha_j \quad \text{and} \quad \lim_{n \to \infty} f_j(-e_1 + e_n) = \beta_j,$$

for each $j \in \mathbb{N}$.

By assumption, $\lim_{j \to \infty} f_j(e_1) = 0$ and $\lim_{j \to \infty} f_j(-e_1) = 0$. Therefore, for $\epsilon > 0$, there exists a $N \in \mathbb{N}$ such that $|f_j(e_1)| < \epsilon$ and $|f_j(-e_1)| < \epsilon$ for all $j > N$. Now let n and m be arbitrary positive integers. For each $j \in \mathbb{N}$,

$$|f_j(e_1 + e_n) - f_j(e_1)| \leq \lambda \|(e_1 + e_n) - e_1\|_{\ell_1} = \lambda \|e_n\|_{\ell_1} = \lambda.$$

Thus, for each $j > N$,

$$|f_j(e_1 + e_n)| \leq \lambda + |f_j(e_1)| < \lambda + \epsilon.$$

By a similar argument (replacing e_1 with $-e_1$ and replacing e_n with e_m), we see that $|f_j(-e_1 + e_m)| < \lambda + \epsilon$, for each $j > N$. It follows that

$$|f_j(e_1 + e_n) - f_j(-e_1 + e_m)| < 2\lambda + 2\epsilon, \quad j > N.$$

Suppose, contrary to the desired conclusion, that $\lambda < 2$. Under this assumption, we may choose ϵ so that $0 < \epsilon < 2 - \lambda$. This means that

$$|f_j(e_1 + e_n) - f_j(-e_1 + e_m)| < 4$$

for any $j > N$.

On the other hand, if $n \neq m$,

$$\|f(e_1 + e_n) - f(-e_1 + e_m)\|_{c_0} \geq \|2e_1 + e_n - e_m\|_{c_0} = 4.$$

This means the maximum value of $|f_j(e_1 + e_n) - f_j(-e_1 + e_m)|$ is at least 4, and so must occur at a value of j in the set $\{1, 2, \ldots, N\}$. That is,

$$\max_{1 \leq j \leq N} |f_j(e_1 + e_n) - f_j(-e_1 + e_m)| \geq 4,$$

when $m \neq n$. Since the choice of m and n were arbitrary (subject to the constraint that they are unequal), it follows that

$$\max_{1 \leq j \leq N} |\alpha_j - \beta_j| \geq 4.$$

If we now consider the case where $n = m$, however, we see that

$$|f_j(e_1 + e_n) - f_j(-e_j + e_n)| \leq \lambda \|2e_1\|_{\ell_1} = 2\lambda,$$

for all $j \in \mathbb{N}$. This is true for all $n \in \mathbb{N}$, and so it follows that $|\alpha_j - \beta_j| \leq 2\lambda$ for (in particular) all j in the set $\{1, 2, \ldots, N\}$. The assumption that $\lambda < 2$ now implies that $|\alpha_j - \beta_j| < 4$, which contradicts what has already been shown. Therefore, it must be the case that $\lambda \geq 2$. $\square$

COMMENT 7.40. The existence of α_j and β_j in the proof of Proposition 7.39 follow from the Bolzano–Weierstrass theorem, which states that a bounded sequence of real numbers has a convergent subsequence. To add some of the details, consider the sequence $\{f_j(e_1 + e_n)\}_{n=1}^{\infty}$. This is a bounded sequence, because for every $n \in \mathbb{N}$,
$$\|f_j(e_1 + e_n)\|_{c_0} = \|f_j(e_1 + e_n) - f_j(0)\|_{c_0} \leq \lambda \|e_1 + e_n\|_{\ell_1} = 2\lambda.$$
It follows that there is a convergent subsequence, say $\{f_j(e_1 + e_{n_k})\}_{k=1}^{\infty}$.

Now consider the sequence $\{f_j(-e_1 + e_n)\}_{n=1}^{\infty}$. This sequence may not converge, and indeed it may be that the subsequence $\{f_j(-e_1 + e_{n_k})\}_{k=1}^{\infty}$ does not converge. However, this sequence, too, is bounded (the argument is essentially the same), and so there exists a convergent subsequence, say $\{f_j(-e_1 + e_{n_{k_\ell}})\}_{\ell=1}^{\infty}$. Now let
$$\alpha_j = \lim_{\ell \to \infty} f_j(e_1 + e_{n_{k_\ell}}) \quad \text{and} \quad \beta_j = \lim_{\ell \to \infty} f_j(-e_1 + e_{n_{k_\ell}}).$$

CHAPTER 8

Property $\Pi(\lambda)$ and embeddings into c_0

The primary objective of this chapter is to prove Aharoni's theorem (Theorem 7.38) following the method of Kalton and Lancien in [**66**]. To that end, we introduce property $\Pi(\lambda)$. But first, we prove a technical lemma.

LEMMA 8.1. *Let M be a metric space. Let δ give the distance between subsets of M. That is, for two subsets A and B of M, let*
$$\delta(A,B) = \inf \{d(a,b) : a \in A, b \in B\}.$$
If A, B, and C are nonempty subsets of M, and if $\epsilon > 0$, then there exists a Lipschitz function $f : M \to \mathbb{R}$ with $\mathrm{Lip}(f) \leq 1$ such that
  (1) $|f(x)| \leq \epsilon$ *for $x \in C$,*
  (2) $|f(x) - f(y)| = \min\{\delta(A,B), \delta(A,C) + \delta(B,C) + 2\epsilon\}$ *for $x \in A$, $y \in B$.*

PROOF. First, we add a point 0 to the metric space M. Let $M^* = M \cup \{0\}$, and define a metric d^* on M^* by
$$d^*(x,y) = \begin{cases} \min\{d(x,y), d(x,C) + d(y,C) + 2\epsilon\} & \text{if } x \in M, y \in M, \\ d(x,C) + \epsilon & \text{if } x \in M, y = 0, \\ d(y,C) + \epsilon & \text{if } y \in M, x = 0, \\ 0 & \text{if } x = 0, y = 0. \end{cases}$$
It can be shown that d^* does indeed define a metric on M^*. Now pick real numbers s and t so that
$$-[\delta(B,C) + \epsilon] \leq s \leq 0 \leq t \leq \delta(A,C) + \epsilon,$$
and
$$t - s = \min\{\delta(A,B), \delta(A,C) + \delta(B,C) + 2\epsilon\}.$$
Define a function $g : A \cup B \cup \{0\} \to \mathbb{R}$ by the rule
$$g(x) = \begin{cases} t & \text{if } x \in A, \\ s & \text{if } x \in B, \\ 0 & \text{if } x = 0. \end{cases}$$
The function g is Lipschitz on $(A \cup B \cup \{0\}, d^*)$ with $\mathrm{Lip}(g) \leq 1$, and so can be extended to a Lipschitz function g^* on (M^*, d^*) with $\mathrm{Lip}(g^*) \leq 1$, by Theorem 1.8. Let f be the restriction of g^* to M. The function f is the required function. □

DEFINITION 8.2. *A metric space (M,d) has property $\Pi(\lambda)$ for a real number $\lambda > 1$ if given any $\mu > \lambda$, there exists a $\nu > \mu$ such that whenever B_1 and B_2 are metric balls of positive radii r_1 and r_2, respectively, there exist finitely many sets $(U_j)_{j=1}^N$ and $(V_j)_{j=1}^N$ such that*
$$\lambda \delta(U_j, V_j) \geq (r_1 + r_2)\nu, \quad 1 \leq j \leq N,$$

and
$$\left\{(x,y) \in B_1 \times B_2 : d(x,y) > (r_1+r_2)\mu\right\} \subseteq \bigcup_{j=1}^{N}(U_j \times V_j).$$

PROPOSITION 8.3. *Every metric space has property* $\Pi(2)$.

PROOF. Let $\mu > 2$ be given. Set $\nu = 2\mu - 2 = 2(\mu - 1)$. Let B_1 and B_2 be metric balls with radii r_1 and r_2, respectively. Without loss of generality, assume that $r_1 \leq r_2$. Define
$$U = \left\{x \in B_1 : \exists y \in B_2 \text{ such that } d(x,y) > (r_1+r_2)\mu\right\},$$
and
$$V = \left\{y \in B_2 : \exists x \in B_1 \text{ such that } d(x,y) > (r_1+r_2)\mu\right\}.$$

Suppose that $x \in U$ and $y \in V$. Because $y \in V$, there exists a point $x' \in U$ such that $d(x',y) > (r_1+r_2)\mu$. Thus,
$$(r_1+r_2)\mu < d(x',x) + d(x,y),$$
and so
$$d(x,y) > (r_1+r_2)\mu - d(x',x) > (r_1+r_2)\mu - 2r_1.$$
It was assumed that $r_1 \leq r_2$, and so
$$d(x,y) > (r_1+r_2)\mu - (r_1+r_2) = (r_1+r_2)(\mu - 1).$$
Therefore,
$$2\delta(U,V) \geq 2(r_1+r_2)(\mu-1) = (r_1+r_2)\nu,$$
as required. □

DEFINITION 8.4. A metric space (M,d) is said to λ-*Lipschitz embed* into a metric space (M',d') if there exists a Lipschitz function $f : M \to M'$ such that
$$d(x,y) \leq d'\bigl(f(x),f(y)\bigr) \leq \lambda d(x,y)$$
for all x and y in M. A function f satisfying this condition is called a λ-*Lipschitz embedding* of metric spaces.

THEOREM 8.5. *Suppose the metric space M λ_0-embeds into c_0. Then M has property $\Pi(\lambda)$ for any $\lambda > \lambda_0$.*

PROOF. By assumption, there is a function $f : M \to c_0$ such that
$$d(x,y) \leq \|f(x) - f(y)\|_{c_0} \leq \lambda_0 d(x,y).$$

Suppose that λ is any real number such that $\lambda > \lambda_0$. We will show that M has $\Pi(\lambda)$. To that end, let $\mu > \lambda$ and let B_1 and B_2 be two metric balls in M. Let B_1 have center a_1 and radius r_1, and let B_2 have center a_2 and radius r_2. Furthermore, let
$$\Delta = \left\{(x,y) \in B_1 \times B_2 : d(x,y) > \mu(r_1+r_2)\right\}.$$
The objective is to cover Δ in a finite collection of sets of the form $U_j \times V_j$ having the property
$$\lambda \delta(U_j, V_j) \geq \nu(r_1+r_2),$$
for some choice of $\nu > \mu$.

8. PROPERTY $\Pi(\lambda)$ AND EMBEDDINGS INTO c_0

Pick an $\epsilon > 0$ for which $\lambda(\mu - \epsilon) > \lambda_0 \mu$ and then choose

$$\nu = \frac{\lambda}{\lambda_0}(\mu - \epsilon) > \mu.$$

We will construct the sets U_j and V_j that work for this ν.

Write $f(x) = \bigl(f_i(x)\bigr)_{i=1}^{\infty}$. Observe that each function f_i is Lipschitz continuous with Lipschitz constant λ_0, because

$$|f_i(x) - f_i(y)| \leq \|f(x) - f(y)\|_{c_0} \leq \lambda_0 d(x, y).$$

Observe also that $f(a_1) - f(a_2)$ is an element of c_0, and so $\lim_{i \to \infty} |f_i(a_1) - f_i(a_2)| = 0$. Consequently, there exists a positive integer n such that

$$|f_i(a_1) - f_i(a_2)| < (\mu - \lambda)(r_1 + r_2), \quad i \geq n + 1.$$

Then, for $(x, y) \in \Delta$ and $i \geq n + 1$,

$$|f_i(x) - f_i(y)| \leq |f_i(x) - f_i(a_1)| + |f_i(a_1) - f_i(a_2)| + |f_i(a_2) - f_i(y)|$$

$$\leq \lambda_0 d(x, a_1) + (\mu - \lambda)(r_1 + r_2) + \lambda_0 d(a_2, y)$$

$$\leq \lambda_0 r_1 + (\mu - \lambda)(r_1 + r_2) + \lambda_0 r_2 = (\mu - \lambda + \lambda_0)(r_1 + r_2).$$

Since $(x, y) \in \Delta$ and $\lambda > \lambda_0$, it follows that $|f_i(x) - f_i(y)| < d(x, y)$ for $i \geq n + 1$. However, by assumption, $d(x, y) \leq \|f(x) - f(y)\|_{c_0}$, and so it must be that

$$d(x, y) \leq \max_{1 \leq i \leq n} |f_i(x) - f_i(y)|, \quad (x, y) \in \Delta.$$

By compactness, there is a covering $\bigl(W_k^{(1)}\bigr)_{k=1}^{m_1}$ of B_1 such that

$$|f_i(x) - f_i(x')| \leq \frac{\epsilon}{2}(r_1 + r_2), \quad x \in W_k^{(1)}, \quad x' \in W_k^{(1)}, \quad 1 \leq i \leq n,$$

for each k in $\{1, \ldots, m_1\}$. Similarly, there is a covering $\bigl(W_k^{(2)}\bigr)_{k=1}^{m_2}$ of B_2 such that

$$|f_i(y) - f_i(y')| \leq \frac{\epsilon}{2}(r_1 + r_2), \quad y \in W_k^{(2)}, \quad y' \in W_k^{(2)}, \quad 1 \leq i \leq n,$$

for each k in $\{1, \ldots, m_2\}$. Now define

$$S = \left\{(k_1, k_2) : W_{k_1}^{(1)} \times W_{k_2}^{(2)} \cap \Delta \neq \emptyset\right\}.$$

Let $(U_j \times V_j)_{j=1}^{N}$ be an enumeration of the sets in $\bigl\{W_{k_1}^{(1)} \times W_{k_2}^{(2)} : (k_1, k_2) \in S\bigr\}$.

Now let $x \in U_j$ and $y \in V_j$. By construction, $(U_j \times V_j) \cap \Delta \neq \emptyset$, and so there exists some $x' \in U_j$ and some $y' \in V_j$ such that $d(x', y') > \mu(r_1 + r_2)$. Then there exists some $i \in \{1, \ldots, n\}$ for which $|f_i(x') - f_i(y')| > \mu(r_1 + r_2)$. On the other hand,

$$|f_i(x') - f_i(y')| \leq |f_i(x') - f_i(x)| + |f_i(x) - f_i(y)| + |f_i(y) - f_i(y')|$$

$$\leq \frac{\epsilon}{2}(r_1 + r_2) + |f_i(x) - f_i(y)| + \frac{\epsilon}{2}(r_1 + r_2)$$

$$= |f_i(x) - f_i(y)| + \epsilon(r_1 + r_2),$$

and so

$$|f_i(x) - f_i(y)| \geq |f_i(x') - f_i(y')| - \epsilon(r_1 + r_2) > (\mu - \epsilon)(r_1 + r_2).$$

Consequently,

$$d(x, y) > \frac{\mu - \epsilon}{\lambda_0}(r_1 + r_2).$$

Therefore,
$$\delta(U_j, V_j) > \frac{\mu - \epsilon}{\lambda_0}(r_1 + r_2),$$
and so
$$\lambda \delta(U_j, V_j) > \frac{\lambda}{\lambda_0}(\mu - \epsilon)(r_1 + r_2) = \nu(r_1 + r_2).$$
We conclude that M has property $\Pi(\lambda)$ for any $\lambda > \lambda_0$, as required. □

EXAMPLE 8.6. The classical sequence space ℓ_p has $\Pi(2^{1/p})$ for $p \in [1, \infty)$.[1]

The next proposition provides a lower bound condition for embedding Banach spaces into c_0.

PROPOSITION 8.7. *Suppose X is a Banach space that λ_0-embeds into c_0. If u is an element of X for which $\|u\| = 1$ and if Y is an infinite dimensional subspace of X, then*
$$\inf_{y \in \partial B_Y} \|u + y\| \leq \lambda_0.$$

In the above proposition, we are using the notation ∂B_Y to denote the collection of all $y \in Y$ for which $\|y\| = 1$. In order to prove Proposition 8.7, we will make use of the Borsuk–Ulam theorem.

THEOREM 8.8 (Borsuk–Ulam Theorem). *Let S^n be the unit sphere in $\mathbb{R}^n$. There is no continuous antipodal mapping $f : S^n \to S^k$ for $k < n$. That is, there is no function $f : S^n \to S^k$ for which $f(-x) = -f(x)$ for all $x \in S^n$.*

The proof of Theorem 8.8 will not be given here. (See [79] for a detailed proof and discussion.) However, an example of the application of the Borsuk–Ulam theorem can be found in the following proof of Riesz's lemma.[2]

THEOREM 8.9 (Riesz's Lemma). *Let E be a finite dimensional proper subspace of a Banach space X. There exists an $x \in X \setminus E$ for which $\|x + e\| \geq \|x\|$ for all $e \in E$.*

PROOF. Let $F \subseteq X$ be a finite dimensional subspace with $\dim(F) > \dim(E)$. First, suppose X is strictly convex. Then, for each $x \in \partial B_F = \{x \in F : \|x\| = 1\}$, there exists a unique element $\phi(x) \in E$ such that
$$d(x, E) = \|x - \phi(x)\|.$$
(See Corollary D.8.) This defines a continuous map $\phi : \partial B_F \to E$ with the property $\phi(-x) = -\phi(x)$. Define a new map $\psi : \partial B_F \to \partial B_E$ by
$$\psi(x) = \frac{\phi(x)}{\|\phi(x)\|}, \quad x \in \partial B_F.$$
If $\|\phi(x)\| \neq 0$ for all $x \in \partial B_F$, then ψ is a continuous antipodal mapping from ∂B_F to ∂B_E. This contradicts the Borsuk–Ulam theorem (Theorem 8.8), and so there must be some $x_0 \in \partial B_F$ for which $\|\phi(x_0)\| = 0$. It follows that
$$d(x_0, E) = \|x_0\|,$$
where $\|x_0\| = 1$. Then choosing $x = -x_0$ gives us the desired point.

---

[1] A detailed proof can be found in Theorem 3.2 of [66].
[2] Adapted from Lemma 2.c.8 in [76].

Now suppose that X is not strictly convex. Since E and F are both finite dimensional, we can without loss of generality, assume that X is finite dimensional. We can therefore find an equivalent norm $|||\cdot|||$ that is strictly convex. For each $n \in \mathbb{N}$, let $\|\cdot\|_n$ be the strictly convex norm on X given by

$$\|x\|_n = \|x\| + \frac{|||x|||}{n}, \quad x \in X.$$

Applying the first half of the proof to F using the strictly convex norm $\|\cdot\|_n$, we can find a point $x_n \in F$ with $\|x_n\|_n = 1$ and $d(x_n, E) = 1$. Let x be any limit point of the sequence $(x_n)_{n=1}^{\infty}$. Then $\|x\| = 1$ and $d(x, E) = 1$, and the proof is complete. $\square$

COMMENT 8.10. In the previous proof, for the case where the norm $\|\cdot\|$ on X is assumed to be not strictly convex, we took X to be a finite dimensional Banach space and equipped it with an equivalent norm $|||\cdot|||$ that is strictly convex. We were able to assert the existence of such an equivalent norm because X is assumed to be finite dimensional, although we only needed X to be separable. This follows from a result of Clarkson (Theorem 9 in [**24**]).

THEOREM 8.11 (Clarkson). *Any separable Banach space may be given a strictly convex norm that is equivalent to the original norm.*

Furthermore, we took it for granted that $\|\cdot\|_n$ is a strictly convex norm on X for each $n \in \mathbb{N}$. This may not be obvious, but can be seen as a special case of the following theorem (given here without proof).

THEOREM 8.12. *Assume that $(X, \|\cdot\|_X)$ and $(Y, \|\cdot\|_Y)$ are Banach spaces and suppose that $(Y, \|\cdot\|_Y)$ is strictly convex. If $T : X \to Y$ is an injective bounded linear operator, then*

$$\|x\|_T = \|x\|_X + \|T(x)\|_Y, \quad x \in X$$

defines a strictly convex norm on X that is equivalent to $\|\cdot\|_X$.

(For a proof, see Proposition 152 in [**44**].) In our case, the Banach spaces are $(X, \|\cdot\|_X)$ and $(X, |||\cdot|||)$, and the bounded linear injection is the map $T : (X, \|\cdot\|_X) \to (X, |||\cdot|||)$ given by $T(x) = \frac{x}{n}$ for $x \in X$.

For our purposes, we will not use the Borsuk–Ulam theorem directly, but rather a classical corollary that is due to Lyusternik and Šnirel′man.

LEMMA 8.13 (Lyusternik–Šnirel′man). *Let E be a finite dimensional subspace of a Banach space and let $\partial B_E = \{x \in E : \|x\| = 1\}$. If $A_1, \ldots, A_N$ is a cover of ∂B_E by closed sets and $\dim(E) > N$, then one of the sets contains a pair of antipodal points. That is, there is an index $j \in \{1, \ldots, N\}$ and a point $e \in \partial B_E$ such that $\{e, -e\} \subseteq A_j$.*

Lemma 8.13 is viewed as a corollary to the Borsuk–Ulam theorem, but the work of Lyusternik and Šnirel′man in [**77**] actually predates that of Borsuk in [**17**]. (For a detailed discussion, see pages 23–25 of [**79**].)

We now return to Proposition 8.7 and provide a proof using Lemma 8.13 (the Lyusternik–Šnirel′man lemma).

PROOF OF PROPOSITION 8.7. Assume that X is a Banach space and suppose that $f : X \to c_0$ is a Lipschitz embedding with Lipschitz constant λ_0. Let $u \in X$ be given such that $\|u\| = 1$. The objective is to show that if Y is an infinite dimensional subspace of X, then

$$\inf_{y \in \partial B_Y} \|u + y\| \leq \lambda_0,$$

where $\partial B_Y = \{y \in Y : \|y\| = 1\}$. By Theorem 8.5, the Banach space X has property $\Pi(\lambda)$ for any $\lambda > \lambda_0$. Let μ and λ be real numbers such that $\mu > \lambda > \lambda_0$. Let $B_1 = -u + B_X$ and $B_2 = u + B_X$ be two metric balls in X, each having radius 1. By the $\Pi(\lambda)$ property, there exist some number $\nu > \mu$ and finite collections of closed sets $(U_j)_{j=1}^N$ and $(V_j)_{j=1}^N$ such that

$$\lambda \delta(U_j, V_j) \geq 2\nu, \quad j \in \{1, \ldots, N\}$$

and

$$\{(x, y) \in B_1 \times B_2 : \|x - y\| > 2\mu\} \subseteq \bigcup_{j=1}^N (U_j \times V_j).$$

Let E be a finite dimensional subspace of Y for which $\dim(E) > N$. For each j in the set $\{1, \ldots, N\}$, let

$$A_j = \{e \in E : \|e\| = 1, \, (-u + e, u - e) \in U_j \times V_j\}.$$

Assume, contrary to that which is to be shown, that $\|u - e\| > \mu$ for all $e \in \partial B_E$. If this is the case, then

$$\partial B_E \subseteq \bigcup_{j=1}^N A_j.$$

By the Lyusternik–Šnirel'man lemma (Lemma 8.13), there exists some positive integer $j \in \{1, \ldots, N\}$ and some $e \in \partial B_E$ such that $e \in A_j$ and $-e \in A_j$. Since $e \in A_j$, it follows that

$$(-u + e, u - e) \in U_j \times V_j,$$

and since $-e \in A_j$, it follows that

$$(-u - e, u + e) \in U_j \times V_j.$$

From this we conclude that $\delta(U_j, V_j) \leq 2$. This is a contradiction however, because it was assumed that $\lambda \delta(U_j, V_j) \geq 2\nu$, and this implies that

$$\delta(U_j, V_j) \geq \frac{2\nu}{\lambda} > 2,$$

because $\nu > \lambda$. This contradiction leads to the conclusion that $\|u - e\| \leq \mu$ for some $e \in \partial B_E$. Since E is a subspace of Y, and since μ is an arbitrary real number with $\mu > \lambda$, we arrive at the desired conclusion. □

COMMENT 8.14. Let us say just a little bit more about why assuming that $\|u - e\| > \mu$ for all $e \in \partial B_E$ implies that

$$\partial B_E \subseteq \bigcup_{j=1}^N A_j,$$

in the proof of Proposition 8.7, in case it is not clear. Let $e \in \partial B_E$, so that $\|e\| = 1$. In order for this e to be in A_j for some j, we need to show that $(-u+e, u-e)$ is in $U_j \times V_j$ for some j. To demonstrate that this is true, let $x = -u+e$ and $y = u-e$. Then
$$\|x - y\| = \|(-u+e) - (u-e)\| = \|-2u + 2e\| = 2\|u - e\| > 2\mu.$$
Consequently, it follows that
$$(x,y) \in \bigcup_{j=1}^{N}(U_j \times V_j)$$
and so there exists some $j \in \{1, \ldots, N\}$ for which
$$(-u+e, u-e) = (x,y) \in U_j \times V_j$$
which is what we needed to show.

We now show that property $\Pi(\lambda)$ implies a formally stronger property.

LEMMA 8.15. *Assume (M,d) has property $\Pi(\lambda)$ for some real number $\lambda > 1$. Given any $\mu > \lambda$, there exists a $\nu > \mu$ such that whenever B_1 and B_2 are metric balls of positive radii r_1 and r_2, respectively, there exist finitely many sets $(U_j)_{j=1}^{N}$ and $(V_j)_{j=1}^{N}$ such that if $(x,y) \in B_1 \times B_2$ and $d(x,y) > (r_1+r_2)\mu$, then there exists a $j \in \{1, \ldots, N\}$ such that $x \in U_j$ and $y \in V_j$ and*
$$\lambda \delta(U_j, V_j) \geq \frac{\nu}{\mu} d(x,y).$$

PROOF. Let μ be a real number such that $\mu > \lambda$. Since M has property $\Pi(\lambda)$, there exists a $\nu' > \mu$ such that whenever B_1 and B_2 are metric balls of positive radii r_1 and r_2, respectively, there exist finitely many sets $(U_j)_{j=1}^{N}$ and $(V_j)_{j=1}^{N}$ such that
$$\lambda \delta(U_j, V_j) \geq (r_1 + r_2)\nu', \quad 1 \leq j \leq N,$$
and
$$\left\{(x,y) \in B_1 \times B_2 : d(x,y) > (r_1+r_2)\mu\right\} \subseteq \bigcup_{j=1}^{N}(U_j \times V_j).$$

Pick ν so that $\mu < \nu < \nu'$ and pick $\epsilon > 0$ so that $(1+\epsilon)\nu = \nu'$. Let B_1 and B_2 be metric balls of positive radii r_1 and r_2, respectively, and let
$$D = \sup\left\{d(x,y) : x \in B_1, y \in B_2\right\}.$$
Let m be the greatest integer such that $(1+\epsilon)^m(r_1+r_2)\mu \leq D$.

For each $k \in \{0, 1, \ldots, m\}$, let $B_1^{(k)}$ be the metric ball having the same center as B_1, but radius $(1+\epsilon)^k r_1$. Similarly, let $B_2^{(k)}$ be the metric ball having the same center as B_2, but radius $(1+\epsilon)^k r_2$. By the $\Pi(\lambda)$ property, for each $k \in \{0, 1, \ldots, m\}$, there exist sets $(U_{k,\ell})_{\ell=1}^{N_k}$ and $(V_{k,\ell})_{\ell=1}^{N_k}$ such that
$$\lambda \delta(U_{k,\ell}, V_{k,\ell}) \geq (1+\epsilon)^k (r_1 + r_2)\nu', \quad 1 \leq \ell \leq N_k,$$
and
$$\left\{(x,y) \in B_1^{(k)} \times B_2^{(k)} : d(x,y) > (1+\epsilon)^k(r_1+r_2)\mu\right\} \subseteq \bigcup_{\ell=1}^{N_k}(U_{k,\ell} \times V_{k,\ell}).$$

If $x \in B_1$ and $y \in B_2$ are such that $d(x,y) > (r_1+r_2)\mu$, then there is a $k \in \{0,1,\ldots,m\}$ for which
$$(1+\epsilon)^k (r_1+r_2)\mu < d(x,y) \leq (1+\epsilon)^{k+1}(r_1+r_2)\mu.$$
Therefore, there exists an $\ell \in \{1,\ldots,N_k\}$ such that $x \in U_{k,\ell}$ and $y \in V_{k,\ell}$ and
$$\lambda \delta(U_{k,\ell}, V_{k,\ell}) \geq (1+\epsilon)^k (r_1+r_2)\nu' = (1+\epsilon)^{k+1}(r_1+r_2)\nu$$
$$= \frac{\nu}{\mu}(1+\epsilon)^{k+1}(r_1+r_2)\mu \geq \frac{\nu}{\mu} d(x,y).$$
This is the desired result. □

LEMMA 8.16. *Let M be a metric space that has property $\Pi(\lambda)$ for $\lambda > 1$. Suppose that α and β are real numbers such that $0 < \alpha < \beta$, and suppose that F and G are finite subsets of M. Let $\Delta(F,G,\alpha,\beta)$ be the set of points $(x,y) \in M \times M$ such that*
$$\lambda\bigl[d(x,G)+d(y,G)\bigr] + \alpha \leq d(x,y) < \lambda\bigl[d(x,F)+d(y,F)\bigr] + \beta.$$
There exists a finite set $\mathcal{F}$ of Lipschitz continuous functions $f : M \to \mathbb{R}$ with $\mathrm{Lip}(f) \leq \lambda$ for which
$$|f(x)| \leq \lambda\beta, \quad x \in F,$$
and
$$d(x,y) < \max_{f \in \mathcal{F}} |f(x) - f(y)|, \quad (x,y) \in \Delta(F,G,\alpha,\beta).$$

PROOF. Let R be the diameter of the set G. For $(x,y) \in \Delta(F,G,\alpha,\beta)$,
$$\lambda\bigl[d(x,y) - R\bigr] < d(x,y).$$

COMMENT 8.17. To see why this inequality is true, note that G is a finite set, and so there exists a $g_x \in G$ such that $d(x,G) = d(x,g_x)$ and a $g_y \in G$ such that $d(y,G) = d(y,g_y)$. Thus,
$$d(x,y) \leq d(x,g_x) + d(g_x,g_y) + d(g_y,y) \leq d(x,G) + R + d(y,G)$$
and so
$$\lambda\bigl[d(x,y) - R\bigr] \leq \lambda\bigl[d(x,G) + d(y,G)\bigr] < \lambda\bigl[d(x,G) + d(y,G)\bigr] + \alpha.$$
By assumption, this last quantity is bounded above by $d(x,y)$, and so
$$\lambda\bigl[d(x,y) - R\bigr] < d(x,y),$$
as claimed.

Rewriting the inequality, we have that
$$(\lambda - 1)\, d(x,y) < \lambda R,$$
and so,
$$d(x,y) \leq \frac{\lambda R}{\lambda - 1}.$$
Consequently,
$$\lambda\bigl[d(x,G)+d(y,G)\bigr] < \lambda\bigl[d(x,G)+d(y,G)\bigr] + \alpha \leq d(x,y) \leq \frac{\lambda R}{\lambda - 1}.$$

Therefore,
$$d(x,G) + d(y,G) < \frac{R}{\lambda - 1}$$
for $(x,y) \in \Delta(F,G,\alpha,\beta)$.

Now we let

(8.1) $$\mu = \lambda + \frac{(\lambda - 1)\alpha}{2R}$$

and apply Lemma 8.15 for this choice of μ. According to Lemma 8.15, there exists a $\nu > \mu$ such that whenever B_1 and B_2 are metric balls of positive radii r_1 and r_2, respectively, there exist finitely many sets $(U_j)_{j=1}^N$ and $(V_j)_{j=1}^N$ such that if $(x,y) \in B_1 \times B_2$ and $d(x,y) > (r_1 + r_2)\mu$, then there exists a $j \in \{1, \ldots, N\}$ such that $x \in U_j$ and $y \in V_j$ and

$$\lambda \mu \, \delta(U_j, V_j) \geq \nu \, d(x,y).$$

Choose $\epsilon > 0$ so that $4\mu\epsilon < \alpha$ and let

$$E = \left\{ x \in M \; : \; d(x,G) < \frac{R}{\lambda - 1} \right\}.$$

Because F and G are finite sets, we can find a finite collection of pairwise-disjoint sets $E_1, \ldots, E_m$ such that $E = E_1 \cup \cdots \cup E_m$ and such that

$$|d(x,z) - d(x',z)| \leq \epsilon, \quad \{x, x'\} \subseteq E_j, \quad z \in F \cup G,$$

for each $j \in \{1, \ldots, m\}$. (Note that the value of the positive integer m may depend on the choice of ϵ.) Since G is finite, for each $j \in \{1, \ldots, m\}$ there exists a $z_j \in G$ such that

$$\inf_{x \in E_j} d(x,G) = \inf_{x \in E_j} d(x, z_j).$$

Let r_j be this value. That is, for each $j \in \{1, \ldots, m\}$, let $r_j = \inf_{x \in E_j} d(x, z_j)$, which is a nonnegative real number. For each $x \in E_j$, it follows from the defining property of the set E_j that

$$d(x, z_j) \leq \epsilon + d(x', z_j)$$

for all $x' \in E_j$. Taking the infimum over all x' in E_j, we have $d(x, z_j) \leq \epsilon + r_j$, and so

$$E_j \subseteq B(z_j, r_j + \epsilon)$$

for each $j \in \{1, \ldots, m\}$.

By Lemma 8.15, for each (j,k), we can find finitely many sets $(U_{j,k,\ell})_{\ell=1}^{N_{j,k}}$ and $(V_{j,k,\ell})_{\ell=1}^{N_{j,k}}$ such that if $(x,y) \in E_j \times E_k$ and $d(x,y) > (r_j + r_k + 2\epsilon)\mu$, then there exists an index $\ell \in \{1, \ldots, N_{j,k}\}$ such that $x \in U_{j,k,\ell}$ and $y \in V_{j,k,\ell}$ and

$$\lambda \mu \, \delta(U_{j,k,\ell}, V_{j,k,\ell}) \geq \nu \, d(x,y).$$

Furthermore, we can choose $U_{j,k,\ell}$ and $V_{j,k,\ell}$ so that $U_{j,k,\ell} \subseteq E_j$ and $V_{j,k,\ell} \subseteq E_k$.

We now apply the technical lemma from the start of this chapter (Lemma 8.1). For each (j,k,ℓ), there exists a Lipschitz continuous function $f_{j,k,\ell} : M \to \mathbb{R}$ with $\mathrm{Lip}(f_{j,k,\ell}) \leq \lambda$ such that $|f_{j,k,\ell}(x)| \leq \lambda\beta$ for all $x \in F$ and

$$|f_{j,k,\ell}(x) - f_{j,k,\ell}(y)| \geq \lambda \theta_{j,k,\ell}, \quad (x,y) \in U_{j,k,\ell} \times V_{j,k,\ell},$$

where

$$\theta_{j,k,\ell} = \min\left[\delta(U_{j,k,\ell}, V_{j,k,\ell}), \delta(U_{j,k,\ell}, F) + \delta(V_{j,k,\ell}, F) + 2\beta\right].$$

We claim the collection of functions $f_{j,k,\ell}$ satisfies the conclusions of the lemma. We already know that $\operatorname{Lip}(f_{j,k,\ell}) \leq \lambda$ and $|f_{j,k,\ell}(x)| \leq \lambda\beta$ for all $x \in F$ for all choices of j, k, and ℓ. Consequently, it remains only to show that we can find some choice of indices (j,k,ℓ) for which $|f_{j,k,\ell}(x) - f_{j,k,\ell}(y)| > d(x,y)$ whenever (x,y) is a pair in $\Delta(F,G,\alpha,\beta)$.

Let $(x,y) \in \Delta(F,G,\alpha,\beta)$. Then $x \in E_j$ and $y \in E_k$ for some index pair (j,k). It follows that
$$d(x,y) \geq \lambda\bigl[d(x,G) + d(y,G)\bigr] + \alpha \geq \lambda(r_j + r_k) + \alpha,$$
because $r_j = \inf_{x \in E_j} d(x,G)$ and $r_k = \inf_{y \in E_k} d(y,G)$. Now rewrite the term on the right of the above inequality as
$$\lambda(r_j + r_k) + \alpha = \mu(r_j + r_k + 2\epsilon) + \alpha - 2\mu\epsilon - (\mu - \lambda)(r_j + r_k).$$
Next, we observe that
$$r_j + r_k \leq d(x,G) + d(y,G) < \frac{R}{\lambda - 1},$$
and so (putting all of these together), we have that
$$d(x,y) \geq \mu(r_j + r_k + 2\epsilon) + \alpha - 2\mu\epsilon - (\mu - \lambda)(r_j + r_k)$$
$$> \mu(r_j + r_k + 2\epsilon) + \alpha - 2\mu\epsilon - (\mu - \lambda)\frac{R}{\lambda - 1}.$$
Finally, we recall that $4\mu\epsilon < \alpha$ and use the fact that $\mu - \lambda = \frac{(\lambda-1)\alpha}{2R}$ (see (8.1)) to conclude that
$$d(x,y) > \mu(r_j + r_k + 2\epsilon).$$
But this condition on x and y implies (see where we used Lemma 8.15 above) that there exists an index ℓ such that $x \in U_{j,k,\ell}$ and $y \in V_{j,k,\ell}$ and
$$\lambda\delta(U_{j,k,\ell}, V_{j,k,\ell}) \geq \frac{\nu}{\mu}d(x,y) > d(x,y),$$
since $\nu > \mu$.

We claim that for this choice of j, k, and ℓ, the function $f_{j,k,\ell}$ will give us the desired inequality. We know that
$$|f_{j,k,\ell}(x) - f_{j,k,\ell}(y)| \geq \lambda\theta_{j,k,\ell}$$
because $(x,y) \in U_{j,k,\ell} \times V_{j,k,\ell}$, and so it will suffice to show that $\lambda\theta_{j,k,\ell} > d(x,y)$. Recall that
$$\theta_{j,k,\ell} = \min\bigl[\delta(U_{j,k,\ell}, V_{j,k,\ell}), \delta(U_{j,k,\ell}, F) + \delta(V_{j,k,\ell}, F) + 2\beta\bigr].$$
We just showed that $\lambda\delta(U_{j,k,\ell}, V_{j,k,\ell}) > d(x,y)$. Consequently, we will have the desired result if we show that
$$\lambda\Bigl[\delta(U_{j,k,\ell}, F) + \delta(V_{j,k,\ell}, F) + 2\beta\Bigr] > d(x,y).$$

Recall that $|d(x,z) - d(x',z)| \leq \epsilon$ for each $x' \in E_j$ and $z \in F$. Consequently, $d(x,z) \leq d(x',z) + \epsilon$ for each $x' \in E_j$ and $z \in F$. Computing the infimum over x' in $U_{j,k,\ell} \subseteq E_j$ and $z \in F$, we conclude that
$$d(x,F) \leq \delta(U_{j,k,\ell}, F) + \epsilon.$$
Similarly, we also have that
$$d(y,F) \leq \delta(V_{j,k,\ell}, F) + \epsilon.$$

(This argument is similar to an earlier argument involving G.) It follows that
$$\lambda\Big[\delta(U_{j,k,\ell}, F) + \delta(V_{j,k,\ell}, F) + 2\beta\Big] \geq \lambda\Big[d(x,F) + d(y,F) + 2\beta - 2\epsilon\Big]$$
$$> \lambda\Big[d(x,F) + d(y,F) + \beta\Big].$$

The last inequality follows from the fact that we chose ϵ small enough to satisfy $4\mu\epsilon < \alpha$, and so $\epsilon < \frac{\alpha}{4\mu} < \frac{\beta}{2}$. The original hypotheses of the lemma include the assumption that
$$\lambda\big[d(x,F) + d(y,F)\big] + \beta > d(x,y),$$
and so we have that
$$\lambda\big[d(x,F) + d(y,F) + \beta\big] > d(x,y) + (\lambda - 1)\beta > d(x,y).$$

Therefore, if (x,y) is chosen in the set $\Delta(F, G, \alpha, \beta)$, there is a function $f_{j,k,\ell}$ for which
$$|f_{j,k,\ell}(x) - f_{j,k,\ell}(y)| \geq \lambda\theta_{j,k,\ell} > d(x,y),$$
and so we can take $\mathcal{F}$ to be the collection of functions $f_{j,k,\ell}$ for j and k in $\{1, \ldots, m\}$ and ℓ in $\{1, \ldots, N_{j,k}\}$. This completes the proof. $\square$

The following result will be the primary tool needed to prove Aharoni's theorem (Theorem 7.38).

THEOREM 8.18. *If (M, d) is a separable metric space that has property $\Pi(\lambda)$ for some $\lambda > 1$, then there exists a function $f : M \to c_0$ such that*
$$d(x,y) < \|f(x) - f(y)\| \leq \lambda d(x,y), \quad (x,y) \in M \times M, \quad x \neq y.$$

PROOF. Let $(u_n)_{n=1}^\infty$ be a countable dense subset of distinct points in M. For each $k \in \mathbb{N}$, let $F_k = \{u_1, \ldots, u_k\}$. Let $(\epsilon_n)_{n=1}^\infty$ be a strictly decreasing sequence with $\epsilon_n \to 0$ as $n \to \infty$.

Using Lemma 8.16, construct a strictly increasing sequence $(n_k)_{k=0}^\infty$ of integers, with $n_0 = 0$, and a sequence $(f_j)_{j=1}^\infty$ of Lipschitz functions from M into $\mathbb{R}$, each with Lipschitz constant λ, such that
$$|f_j(x)| \leq \lambda\epsilon_k, \quad x \in F_k, \quad n_{k-1} < j \leq n_k,$$
and if
$$\lambda\big[d(x, F_{k+1}) + d(y, F_{k+1})\big] + \epsilon_{k+1} \leq d(x,y) < \lambda\big[d(x, F_k) + d(y, F_k)\big] + \epsilon_k$$
then
$$\max_{n_{k-1} < j \leq n_k} |f_j(x) - f_j(y)| > d(x,y).$$

Define a map $f : M \to c_0$ by $f(x) = \big(f_j(x)\big)_{j=1}^\infty$. Then f is a Lipschitz embedding of M into c_0 with Lipschitz constant λ, as required. $\square$

We are now able to prove Aharoni's theorem with the optimal constant 2.

THEOREM 8.19 (Aharoni-Kalton-Lancien, 2008). *Every separable metric space 2-Lipschitz embeds into c_0, and 2 is the best possible constant.*

PROOF. Every separable metric space has property $\Pi(2)$, by Proposition 8.3, and so 2-Lipschitz embeds into c_0, by Theorem 8.18. The optimality of 2 is given by Proposition 7.39. $\square$

At the end of Chapter 7, we remarked that Aharoni showed in 1974 (in [**2**]) that every separable metric space K-Lipschitz embeds into c_0 for some positive real number K, but did not provide the optimal value for K. Aharoni showed that K could be any number greater than 6. That is, $K = 6 + \epsilon$ for any $\epsilon > 0$. In 1978, Assouad was able to replace $6 + \epsilon$ with $3 + \epsilon$ [**9**]. This, too, was improved to $K = 3$ by Pelant in 1994 [**91**]. Pelant was actually working within the context of embeddings into c_0^+, the set of positive sequences converging to zero. In fact, for embeddings into c_0^+, the optimal value is in fact 3. In this case, a property similar to property $\Pi(\lambda)$ is used.

DEFINITION 8.20. A metric space (M, d) has *property* $\Pi_+(\lambda)$ for a real number $\lambda > 1$ if given any $\mu > \lambda$, there exists a $\nu > \mu$ such that whenever B_1 and B_2 are metric balls both having radius r, there exist finitely many sets $(U_j)_{j=1}^N$ and $(V_j)_{j=1}^N$ such that
$$\lambda \delta(U_j, V_j) \geq r\nu, \quad 1 \leq j \leq N,$$
and
$$\left\{ (x, y) \in B_1 \times B_2 \ : \ d(x, y) > r\mu \right\} \subseteq \bigcup_{j=1}^N (U_j \times V_j).$$

The key result, given here without proof, is the following proposition. (A proof can be found in [**66**] under Lemma 4.2.)

PROPOSITION 8.21. *Every metric space has property* $\Pi_+(3)$.

In Chapter 7, we showed that if X is a separable Banach space which Lipschitz embeds into a Banach space Y that has the Radon–Nikodym property, then X linearly embeds into Y. (See Theorem 7.28.) This statement does not hold true for $Y = c_0$ because c_0 does not have the Radon–Nikodym property. For example, the space $\mathcal{C}[0, 1]$ of continuous functions on $[0, 1]$ Lipschitz embeds into c_0 (since every separable metric space does Lipschitz embed into c_0), but $\mathcal{C}[0, 1]$ does not *linearly* embed into c_0.

Since every separable metric space embeds into c_0, a natural question arises which does not yet have an answer.

QUESTION 8.22. *If every separable metric space Lipschitz embeds into X, does c_0 linearly embed into X? Or, equivalently, if c_0 Lipschitz embeds into X, does c_0 linearly embed into X?*

If a Banach space X Lipschitz embeds into a Banach space Y, then we cannot conclude that X necessarily linearly embeds into Y. However, we can conclude that X is what we call *crudely finitely representable* in Y.

DEFINITION 8.23. A Banach space X is said to be λ-*finitely representable* in the Banach space Y if given any finite dimensional subspace E of X, there is a finite dimensional subspace F of Y and a linear isomorphism $T : E \to F$ such that $\|T\|\|T^{-1}\| \leq \lambda$. If X is λ-finitely representable in Y for some $\lambda > 1$, then X is said to be *crudely finitely representable* in Y. If X is λ-finitely representable in Y for every $\lambda > 1$, then X is said to be *finitely representable* in Y.

We note that a Banach space X is finitely representable in Y if for any finite dimensional subspace E of X and given any $\epsilon > 0$, there exists a finite dimensional subspace F of Y with $\dim(F) = \dim(E)$ and a linear isomorphism $T : E \to F$

such that $\|T\|\|T^{-1}\| < 1+\epsilon$. It will sometimes be convenient to formulate finite representability in this way.

EXAMPLE 8.24. Every Banach space is finitely representable in c_0. To see why this is true, let E be a finite-dimensional subspace of X and let $\epsilon > 0$. We choose ν to be a positive number such that $\frac{1}{1-\nu} < 1+\epsilon$. It is possible then to find a finite collection of elements $\{e_1^*, \ldots, e_N^*\}$ in B_{E^*} such that every element in B_{E^*} is within ν of e_k^* for at least one value of $k \in \{1, \ldots, N\}$. Let $T : E \to \ell_\infty^N$ be defined by $T(e) = (e_1^*(e), \ldots, e_N^*(e))$ for any $e \in E$. If we let $F = T(E)$, then it can be shown that $\|T\|\|T^{-1}\| < 1+\epsilon$.

COMMENT 8.25. The set $\{e_1^*, \ldots, e_N^*\} \subset B_{E^*}$ in Example 8.24 is known as a ν-net. In general, if (X, d) is a metric space and $\epsilon > 0$, then a set N of elements in X is called an ϵ-net if $d(n_1, n_2) \geq \epsilon$ for all n_1 and n_2 in N such that $n_1 \neq n_2$ and
$$\sup\left\{d(x, N) : x \in X\right\} \leq \epsilon.$$
More precisely, this is known as an ϵ-separated ϵ-net, because the elements of N are separated by at least ϵ, but every element in X is at most ϵ away from some element of N. We know that we can find a ν-net for B_{E^*} in Example 8.24 because of the following proposition.

PROPOSITION 8.26. *If X is a metric space and $\epsilon > 0$, then there is an ϵ-separated ϵ-net in X.*

We will not prove Proposition 8.26 here, other than to say it follows from Zorn's lemma. A proof can be found in [**5**]. (See Lemma 14.1.17.)

EXAMPLE 8.27. For any $p \in [1, \infty)$, the space $L_p(0,1)$ is λ-finitely representable in ℓ_p for every $\lambda > 1$. That is, $L_p(0,1)$ is finitely representable in ℓ_p.

Verifying Example 8.27 would take us too far afield, so we will not include the details here. (See Proposition 12.1.8 in [**5**] for the details.) The relationship between Lipschitz embeddings and λ-finite representability is given in the following theorem (due to Ribe [**100**]).

THEOREM 8.28 (Ribe). *If a Banach space X κ-Lipschitz embeds into a Banach space Y, then X is λ-finitely representable in Y for every $\lambda > \kappa$.*

We do not currently have the necessary tools to prove Theorem 8.28, but it follows from Corollary 11.38, which we will prove later. (See Theorem 14.2.27 in [**5**] for an explicit proof.)

Theorem 8.19 states that any separable Banach space 2-Lipschitz embeds into c_0. Consequently, Theorem 8.28 tells us that every separable Banach space is λ-finitely representable in c_0 for every $\lambda > 2$. However, we saw in Example 8.24 that every Banach space is finitely representable in c_0, so Theorem 8.28 does not give us the best possible result in this case.

For our next example, we give a significant, nontrivial result.

EXAMPLE 8.29 (Principle of Local Reflexivity). If X is a Banach space, then the second dual X^{**} is finitely representable in X.

The Principle of Local Reflexivity is an important result of Lindenstrauss and Rosenthal [74] that tells us that every Banach space is reflexive in a local sense. (For a proof of the Principle of Local Reflexivity, see Theorem 12.2.4 in [5].)

CHAPTER 9

Local complementation and the Heinrich–Mankiewicz theorem

Let X be a subspace of a Banach space Y. Recall that X is said to be *complemented* in Y if there exists a bounded linear map $P : Y \to X$, called a *projection*, such that $P(x) = x$ for all $x \in X$.

In order to define what it means for X to be locally complemented in Y, we will use the language of short exact sequences. First, suppose that X is a subspace of Y and let $Q : Y \to Y/X$ be the associated quotient map. The map Q determines a short exact sequence

$$0 \longrightarrow X \longrightarrow Y \xrightarrow{Q} Y/X \longrightarrow 0 \ .$$

If X is a complemented subspace of Y, then this short exact sequence splits. If X is not a complemented subspace of Y, the sequence does not split. Local complementation concerns the dual short exact sequence

$$0 \longrightarrow X^\perp \longrightarrow Y^* \longrightarrow X^* \longrightarrow 0 \ .$$

DEFINITION 9.1. Suppose that X is a subspace of the Banach space Y. We say that X is *locally complemented* in Y if the dual short exact sequence splits.

COMMENT 9.2. It might be useful at this point to recall the terminology of short exact sequences. A sequence of Banach spaces and continuous linear maps

$$A \xrightarrow{f} B \xrightarrow{g} C$$

is called *exact* at B if $\operatorname{ran}(f) = \ker(g)$. A *short exact sequence* is a sequence

$$0 \longrightarrow A \xrightarrow{f} B \xrightarrow{g} C \longrightarrow 0$$

that is exact at A, B, and C. In particular, that means f is injective, g is surjective, and $\operatorname{ran}(f) = \ker(g)$. The maps from 0 to A and C to 0 are unique, and so are not labeled. The short exact sequence above is said to *split* if there exists a continuous map $h : C \to B$ such that $g \circ h$ is the identity on C:

$$0 \longrightarrow A \xrightarrow{f} B \xrightarrow{g} C \longrightarrow 0 \ .$$
$$\underset{h}{\longleftarrow}$$

Equivalently, this short exact sequence splits if there is a bounded linear projection from B to A, and so when the short exact sequence splits, it follows that B is isomorphic to the direct sum $A \oplus C$.

The dual short exact sequence is the short exact sequence between dual spaces induced by the adjoint maps

$$0 \longrightarrow C^* \xrightarrow{g^*} B^* \xrightarrow{f^*} A^* \longrightarrow 0 .$$

where f^* and g^* are the adjoint maps of f and g, respectively.

In the case we are considering, we suppose that X is a subspace of Y, and so the map f from before is the inclusion mapping (and is often omitted from the diagram), and the map g from before is the quotient map $Q: Y \to Y/X$. In this case, the diagram for the short exact sequence is

$$0 \longrightarrow X \longrightarrow Y \xrightarrow{Q} Y/X \longrightarrow 0 .$$
$$\underset{L}{\curvearrowleft}$$

We say that this sequence splits if there is a continuous linear lifting (or section) $L: Y/X \to Y$ such that $QL = \mathrm{Id}_{Y/X}$. Equivalently, the sequence splits if there is a projection map $P: Y \to X$, in which case X is complemented in Y.

The dual short exact sequence is the short exact sequence that is induced by the adjoint maps and involves the duals of the spaces in the original short exact sequence, as we mentioned before. Consequently, in this context, the dual short exact sequence has the diagram

$$0 \longrightarrow X^\perp \longrightarrow Y^* \longrightarrow X^* \longrightarrow 0 .$$

Here, the set $X^\perp$ is the annihilator of X in Y^*. That is,

$$X^\perp = \{y^* \in Y^* : y^*(x) = 0 \text{ for all } x \in X\}.$$

Note that $X^\perp$ can be identified with the dual of Y/X (Proposition 3.51 in [**21**]), which is why it appears in the dual short exact sequence.

If the dual short exact sequence splits, then $X^\perp$ is a complemented subspace of Y^*, and indeed this is sometimes given as the definition of a locally complemented subspace. That is, a subspace X of Y is called locally complemented in Y if $X^\perp$ is complemented in Y^*. This is the definition given in [**59**], for example. When X is locally complemented in Y and the projection $P: Y^* \to X^\perp$ has norm $\|P\| \leq \lambda$, we say that X is λ-*locally complemented* in Y.

The name "locally complemented" suggests that there may be a local formulation of the property, and indeed there is a local formulation.

DEFINITION 9.3. A closed subspace X of a Banach space Y is λ-*locally complemented* in Y if there is a constant $\lambda > 0$ such that whenever F is a finite-dimensional subspace of Y and $\epsilon > 0$, there is a linear operator $T: F \to X$ (that in general depends on F and ϵ) such that $\|T\| \leq \lambda$ and $\|T(f) - f\| \leq \epsilon \|f\|$ for all $f \in X \cap F$.

This is the definition of λ-locally complemented that is given in [**57**], for example. In Theorem 3.5 of [**57**], the Principle of Local Reflexivity (Example 8.29) is used to show the equivalence of the formulations given in Definition 9.1 and Definition 9.3. (For more on the Principle of Local Reflexivity, including the proof, see Theorem 12.2.4 in [**5**].)

REMARK 9.4. Suppose that X is locally complemented in a Banach space Y, so that the dual short exact sequence

$$0 \longrightarrow X^\perp \longrightarrow Y^* \underset{L}{\longrightarrow} X^* \longrightarrow 0$$

splits. Then there is a linear lifting $L : X^* \to Y^*$. Therefore, there is a linear projection $L^* : Y^{**} \to X^{**}$. Consequently, if X is locally complemented in Y, then X^{**} is complemented in Y^{**}. In fact, the converse of this statement is also true, which can be shown by using the Principle of Local Reflexivity (Example 8.29). (See Theorem 3.5 in [57] for the proof.)

REMARK 9.5. A Banach space X need not be complemented in its bidual X^{**}. The classic example, which we have encountered before, is $X = c_0$. We know from Phillips' lemma (Theorem 1.27) that there is no linear projection $P : \ell_\infty \to c_0$. Consequently, the short exact sequence

$$0 \longrightarrow c_0 \longrightarrow \ell_\infty \longrightarrow \ell_\infty/c_0 \longrightarrow 0$$

does not split. On the other hand, X is always locally complemented in its bidual. To see this, let j be the inclusion map that takes X into its bidual X^{**}. In this case, we have the short exact sequence

$$0 \longrightarrow X \xrightarrow{j} X^{**} \xrightarrow{Q} X^{**}/X \longrightarrow 0$$

where Q is the quotient map from X^{**} to X^{**}/X. Since X need not be complemented in X^{**}, there is no reason to assume that this short exact sequence splits. However, the dual short exact sequence

$$0 \longrightarrow X^\perp \longrightarrow X^{***} \underset{\mathrm{Id}-j^*}{\xrightarrow{j^*}} X^* \longrightarrow 0$$

does split, because the adjoint map $j^* : X^{***} \to X^*$ is a projection. Furthermore, the kernel of j^* can be identified with the annihilator $X^\perp$, and so we have the linear isomorphism $X^{***} \approx X^* \oplus X^\perp$. Since the dual short exact sequence splits, we conclude that the Banach space X is locally complemented in X^{**}. In fact, since the projection $\mathrm{Id} - j^*$ of X^{***} onto $X^\perp$ has norm 1, we have that X is 1-locally complemented in X^{**}.

We summarize our findings in the following proposition.

PROPOSITION 9.6. *If X is a Banach space, then X is locally complemented in X^{**} and X^* is complemented in X^{***}. In particular, any dual space is complemented in its bidual.*

We can now prove the following classic theorem of Heinrich and Mankiewicz [45] (which was previously stated as Theorem 7.37 herein).

THEOREM 9.7 (Heinrich–Mankiewicz). *If X and Y are separable Banach spaces and if there is a Lipschitz embedding of X into Y, then there exists a linear embedding of X into Y^{**}.*

Heinrich and Mankiewicz proved Theorem 9.7 as a corollary to the following theorem, which we previously stated (and proved) as Theorem 7.30.

THEOREM 9.8. *If X and Z are separable Banach spaces and X Lipschitz embeds into Z^*, then X linearly embeds into Z^*.*

If X Lipschitz embeds into Y, then X Lipschitz embeds into Y^{**}, and Y^{**} is a dual space (of the Banach space Y^*). However, it need not be true that Y^* is separable even if Y is separable. For example, if $Y = \ell_1$, then Y is separable, but $Y^* = \ell_\infty$ is not separable. Thus, we cannot apply Theorem 9.8 directly to prove Theorem 9.7. Instead, we use the following result, due to Heinrich and Mankiewicz (Proposition 3.4 in [45])[1].

LEMMA 9.9. *Let E be a Banach space. If F is a separable subspace of E, then there exists a separable subspace G such that $F \subset G \subset E$ and such that G^* is a norm-one complemented subspace of E^*.*

To be more technically precise, we should say that G^* is isometric to a complemented subspace of E^*, but we will identify G^* with the subspace of E^* to which it is isometric. To see how this relates to the primary theme of this chapter, we note that if G^* is complemented in E^*, then G is locally complemented in E, and this is the fact we will use. The proof of Lemma 9.9 given in [45] makes use of ultrafilters, which we will not formally introduce until Chapter 11, and so we will not prove it here.

PROOF OF THEOREM 9.7. We assume that X and Y are separable Banach spaces, and let $f : X \to Y$ be a Lipschitz embedding of X into Y. We wish to show that there exists a linear embedding of X into Y^{**}. Since Y is separable, there exists a separable subspace F of Y^* that norms Y. According to Lemma 9.9 (with Y^* filling the role of E), there exists a subspace G such that $F \subset G \subset Y^*$ and such that G^* is a complemented subspace of Y^{**}.

Since G^* is complemented in Y^{**}, the short exact sequence

$$0 \longrightarrow G \longrightarrow Y^* \longrightarrow Y^*/G \longrightarrow 0$$

has a dual sequence

$$0 \longrightarrow G^\perp \longrightarrow Y^{**} \xrightarrow{P} G^* \longrightarrow 0$$
$$\underset{L}{\curvearrowleft}$$

that splits, where $P : Y^{**} \to G^*$ is the norm-one continuous projection map that we know exists by Lemma 9.9. Since the dual sequence splits, G is locally complemented in Y^*, and so there exists a continuous linear lifting $L : G^* \to Y^{**}$ such that $PL = \text{Id}_{G^*}$.

If we let $i : Y \to Y^{**}$ be the inclusion map from Y into it second dual, then the map $P \circ i \circ f : X \to G^*$ is a Lipschitz embedding from X into G^*. (See the diagram below.)

$$X \xrightarrow{f} Y \xrightarrow{i} Y^{**} \xrightarrow{P} G^*$$

That is, we have shown that X Lipschitz embeds into G^*, where G is separable. Therefore, we conclude that X linearly embeds into G^*, by Theorem 9.8. But there is a continuous linear lifting L from G^* to Y^{**}, and so X linearly embeds into Y^{**}. □

---

[1]While the original result comes from [45], we have reproduced the formulation given by Benyamini and Lindenstrauss as Proposition F.8 in [13].

COMMENT 9.10. When we proved Theorem 7.30 (which we restated here as Theorem 9.8), we showed that if $f : X \to Y$ is a Lipschitz embedding from the separable Banach space X into the space Y (which is the dual space of a separable Banach space), then there also exists a linear embedding from X into Y. More specifically, we showed that if there exist positive constants c and C such that

(9.1) $$c\|x-y\|_X \leq \|f(x) - f(y)\|_Y \leq C\|x-y\|_X,$$

for all x and y in X, then there exists a linear embedding $T : X \to Y$ that satisfied

(9.2) $$c\|x\| \leq \|T(x)\| \leq C\|x\|,$$

for all $x \in X$.

If $f : X \to Y$ is a Lipschitz embedding, then f is a Lipschitz isomorphism of X onto $f(X)$. That is, the function f is a bijection such that both $f : X \to f(X)$ and $f^{-1} : f(X) \to X$ are Lipschitz functions. With this in mind, we see that the best possible constants in (9.1) are $C = \mathrm{Lip}(f)$ and $c = 1/\mathrm{Lip}(f^{-1})$. In this case, we define the *distortion constant* of the embedding f to be $\mathrm{dist}(f) = \mathrm{Lip}(f)\mathrm{Lip}(f^{-1})$.

There is a linear analogue of the distortion constant for Lipschitz embeddings. If $T : X \to Y$ is a linear embedding, then the *condition number* of the operator T is defined as the number $\|T\|\|T^{-1}\|$. The fact that constants C and c in (9.2) are the same as those in (9.1) means that the condition number for the T we found in Theorem 7.30 is no larger than the distortion constant for the f we started with. That is, for the linear map T that we found, we have that

$$\|T\|\|T^{-1}\| \leq \mathrm{Lip}(f)\mathrm{Lip}(f^{-1}).$$

In the proof of Theorem 9.7, since our assumptions on the Banach space Y are different, we actually applied Theorem 7.30 not to f, but to the composition $P \circ i \circ f : X \to G^*$. However, the projection P and the inclusion i are both norm-one linear operators, and so the distortion constant remains $\mathrm{Lip}(f)\mathrm{Lip}(f^{-1})$, and consequently, we end up with a linear embedding $T : X \to G^*$ satisfying the same inequality

$$\|T\|\|T^{-1}\| \leq \mathrm{Lip}(f)\mathrm{Lip}(f^{-1}).$$

We recall that the *Banach–Mazur distance* between two Banach spaces X and Y is defined to be

$$d_{\mathrm{BM}}(X,Y) = \inf\left\{\|T\|\|T^{-1}\| \;\middle|\; T : X \to Y \text{ is an isomorphism}\right\}.$$

We can recast the conclusion of Theorem 9.7 in terms of the Banach–Mazur distance. Specifically, according to Theorem 9.7, if X and Y are separable Banach spaces and $f : X \to Y$ is a Lipschitz embedding, then there exists a linear subspace W of Y^{**} such that $d_{\mathrm{BM}}(X,W) \leq \mathrm{Lip}(f)\mathrm{Lip}(f^{-1})$.

EXAMPLE 9.11. Let X be a Banach space with Arens–Eells space $\text{Æ}(X)$. Recall that the associated barycenter map $\beta : \text{Æ}(X) \to X$ is a bounded linear quotient map (Lemma 5.9). Then we have the following short exact sequence

$$0 \longrightarrow \ker(\beta) \longrightarrow \text{Æ}(X) \xrightarrow{\beta} X \longrightarrow 0 \;.$$

This sequence splits if X is separable. (This is shown in Theorem 3.1 of [**39**].) On the other hand, the dual short exact sequence

$$0 \longrightarrow X^* \longrightarrow \mathrm{Lip}(X) \longrightarrow \ker(\beta)^* \longrightarrow 0$$
$$\underset{P}{\underbrace{}}$$

always splits, because there exists a bounded linear projection $P : \mathrm{Lip}(X) \to X^*$, which is shown in Theorem 9.13.

In order to prove Theorem 9.13, we first introduce the notion of an invariant mean.

DEFINITION 9.12. Let G be an abelian group and let $\ell_\infty(G)$ be the set of bounded real-valued functions on G. A linear function $\mathcal{M} : \ell_\infty(G) \to \mathbb{R}$ is called a *mean* on G if (a) $\mathcal{M}(1) = 1$ and (b) $\mathcal{M}(f) \geq 0$ whenever $f \geq 0$. ($\mathcal{M}$ is said to be *nonnegative*.) Additionally, we call $\mathcal{M}$ an *invariant mean* on G if it is invariant under translations. That is, if $\mathcal{M}(\tau_g f) = \mathcal{M}(f)$, where

$$\tau_g f(x) = f(x - g), \quad x \in G$$

is translation by $g \in G$.

We remark that if G is an abelian group, then there exists an invariant mean on G by the Axiom of Choice.[2]

THEOREM 9.13. *If X is a Banach space, then there exists a norm one linear projection $P : \mathrm{Lip}(X) \to X^*$.*

PROOF. Consider the space $\ell_\infty(X)$ of bounded functions on X. Viewing X as a commutative group, there is an invariant mean $\mathcal{M}$ on X. (As we remarked before the statement of the theorem, invariant means exist as a consequence of accepting the Axiom of Choice.) Let $f \in \mathrm{Lip}(X)$. For a fixed $y \in X$, define a function $F_y \in \ell_\infty(X)$ by

$$F_y(x) = f(x + y) - f(x), \quad x \in X.$$

Observe that $|F_y(x)| \leq \mathrm{Lip}(f)\|y\|$ for each $x \in X$, and so F_y is bounded, as claimed. Notice also that the map $y \mapsto F_y$ is a Lipschitz function from X to $\ell_\infty(X)$.

Define a scalar valued function $\widetilde{f}$ on X by

$$\widetilde{f}(y) = \mathcal{M}(F_y)$$

for each $y \in X$. Then $\widetilde{f} \in X^*$ and $\|\widetilde{f}\| \leq \mathrm{Lip}(f)$. If we define P by $P(f) = \widetilde{f}$, then P is the desired projection. □

Theorem 9.13 is due to Lindenstrauss [**75**], although the proof given here is due to Pełczyński [**93**]. (This proof can also be found in [**13**] as Proposition 7.5.)

REMARK 9.14. If X is a Banach space and E is a closed subspace of X, then the projection P from Theorem 9.13 can be chosen so that $P(f)|_E = f|_E$ if $f|_E$ is linear.

THEOREM 9.15. *If X is a Banach space and E is a closed subspace such that there exists a Lipschitz retraction $r : X \to E$, then E is locally complemented in X.*

---

[2]When giving the lecture, Kalton said we cannot construct invariant means, but "we just believe in them."

PROOF. Consider the short exact sequence

$$0 \longrightarrow E \xrightarrow{j} X \longrightarrow X/E \longrightarrow 0 .$$

with a retraction r curving back from X to E.

By assumption, there is a Lipschitz retraction $r : X \to E$. We wish to show that the dual short exact sequence

$$0 \longrightarrow E^\perp \longrightarrow X^* \xrightarrow{j^*} E^* \longrightarrow 0$$

with L curving back from E^* to X^*.

splits. That is, we wish to show that there exists a linear map $L : E^* \to X^*$ such that $j^* \circ L = \mathrm{Id}_{E^*}$, where $j : E \to X$ is the inclusion map.

Let $P_E : \mathrm{Lip}(X) \to X^*$ be the norm-one projection from Remark 9.14 chosen so that $P_E(f)|_E = f|_E$ if $f|_E$ is a linear function. Define a map $T : E^* \to \mathrm{Lip}(X)$ according to the rule $T(e^*) = e^* \circ r$ for all $e^* \in E^*$. Then T is a linear function and $\|T\| = \mathrm{Lip}(r)$.

Now let $L = P_E \circ T$. The map L is linear with $\|L\| \leq \mathrm{Lip}(r)$ and for any $e \in E$, we have

$$(j^* \circ L)(e^*)(e) = L(e^*)(j(e)) = P_E(T(e^*))(e) = T(e^*)(e) = (e^* \circ r)(e) = e^*(e).$$

Therefore, $j^* \circ L = \mathrm{Id}_{E^*}$, as required. □

COMMENT 9.16. Theorem 9.15 is due to Lindenstrauss. (See Theorem 3 in [**75**].) A different formulation of the theorem can be found in Godefroy and Kalton's significant paper from 2003 [**39**], where it appears as Proposition 2.6 and is stated in terms of short exact sequences. We say that the short exact sequence

$$0 \longrightarrow A \xrightarrow{f} B \xrightarrow{g} C \longrightarrow 0$$

with L curving back from C to B.

Lipschitz splits if there exists a Lipschitz function (that is not necessarily linear) $L : C \to B$ such that $g \circ L = \mathrm{Id}_C$. (If the function L is linear, then we say the short exact sequence *splits*, or linearly splits, if we wish to be more precise.) With this terminology, we can rephrase Theorem 9.15 as follows.

THEOREM 9.17. *If a short exact sequence Lipschitz splits, then the dual sequence linearly splits.*

COROLLARY 9.18. *Suppose X is a Banach space with closed subspace E. If E is reflexive or if E is complemented in E^{**}, then the existence of a Lipschitz retraction $r : X \to E$ implies the existence of a linear projection $P : X^{**} \to E$ having norm $\|P\| \leq \mathrm{Lip}(r)\|Q\|$, where Q is the projection of E^{**} onto E.*

PROOF. Let $L : E^* \to X^*$ be the linear map from Theorem 9.15. The adjoint map $L^* : X^{**} \to E^{**}$ is a projection of X^{**} onto E^{**}. To see this, let $x^{**} \in E^{**}$ and let $e^* \in E^*$. Then,

$$L^* x^{**}(e^*) = x^{**}(Le^*) = x^{**}(e^*).$$

Consequently, if E is reflexive, then L^* is the desired projection, and if E is complemented in E^{**}, then $Q \circ L^*$ is the desired projection. □

We know from Theorem 9.15 that if there exists a Lipschitz retraction from a Banach space onto a closed subspace, then that closed subspace is locally complemented in the Banach space. We also know from Proposition 9.6 that a Banach space X is locally complemented in X^{**}. This leads us to a natural question, which remains a significant open question in the nonlinear theory of Banach spaces.[3]

QUESTION 9.19. *For any separable Banach space X, is it true that there is a Lipschitz retraction $r : X^{**} \to X$?*

A positive answer to Question 9.19 will have implications for theory of $\mathcal{L}_\infty$-spaces. Let us pause here to recall the definition of an $\mathcal{L}_\infty$-space.

DEFINITION 9.20. A Banach space X is an *$\mathcal{L}_\infty$-space* if there exists a $\lambda \geq 1$ such that given any finite dimensional subspace E of X, there is a finite dimensional subspace F such that $E \subset F$ and a linear isomorphism $T : F \to \ell_\infty^n$ for some $n \in \mathbb{N}$ with $\|T\|\|T^{-1}\| \leq \lambda$.

REMARK 9.21. If the answer to Question 9.19 is "yes," then every separable $\mathcal{L}_\infty$-space is an absolute Lipschitz retract. This follows from the fact that if a Banach space is an $\mathcal{L}_\infty$-space, then its second dual is injective (see Proposition 9.22), and an injective space is an absolute Lipschitz retract. It follows that if X is a separable $\mathcal{L}_\infty$-space space, then X^{**} is an absolute Lipschitz retract. Consequently, if there exists a Lipschitz retraction $r : X^{**} \to X$ then X is an absolute Lipschitz retract, as well.

In the preceding remark, we used the fact that an $\mathcal{L}_\infty$-space has an injective second dual. We will not prove that fact here, but a proof can be found in the classic 1969 paper by Lindenstrauss and Rosenthal [**74**] in the corollary to Theorem 3.2. In fact, the corollary to Theorem 3.2 in that paper says much more. As that corollary is interesting in its own right, and since we will have occasion to use it again soon, we state it here as a proposition (without proof).

PROPOSITION 9.22. *Every injective space is an $\mathcal{L}_\infty$-space. A Banach space is an $\mathcal{L}_\infty$-space if and only if its second dual is injective.*

In Remark 9.21, we noted that if there is a positive answer to Question 9.19, then every separable $\mathcal{L}_\infty$-space is an absolute Lipschitz retract. In particular, we do not know if every separable $\mathcal{L}_\infty$-space is an absolute Lipschitz retract. On the other hand, we do know that any space $\mathcal{C}(K)$ of continuous functions on a compact metric space K is an absolute Lipschitz retract. (See Theorem 1.30.)

While we do not yet know if every separable $\mathcal{L}_\infty$-space is an absolute Lipschitz retract, we do know that every separable Banach space that is an absolute Lipschitz retract is an $\mathcal{L}_\infty$-space. As we will see, this also follows from Proposition 9.22

THEOREM 9.23. *If X is a separable Banach space that is an absolute Lipschitz retract, then X is an $\mathcal{L}_\infty$-space.*

PROOF. Let X be a separable Banach space that is an absolute Lipschitz retract. It follows that there is a Lipschitz retraction $r : \ell_\infty \to X$ (by Corollary 1.14),

---

[3]Kalton phrased this question about Banach spaces in general (not just separable ones). Kalton himself showed that the answer to this question is "no" for general Banach spaces by demonstrating the existence of a nonseparable Banach space X for which there is no Lipschitz retraction $r : X^{**} \to X$ [**63**]. The answer is still unknown for separable Banach spaces.

and hence X is locally complemented in ℓ_∞ (by Theorem 9.15). We claim that if a Banach space X is locally complemented in ℓ_∞, then it is an $\mathcal{L}_\infty$-space.

Proposition 9.22 states that a Banach space is an $\mathcal{L}_\infty$-space if and only if its second dual is injective. This tells us two things: (1) it is enough to show that X^{**} is injective, and (2) the space ℓ_∞^{**} is injective. We know that X is locally complemented in ℓ_∞, which implies that X^{**} is complemented in ℓ_∞^{**}, by Remark 9.4. But ℓ_∞^{**} is injective, and so the complemented subspace X^{**} is injective, too, and this means that X is an $\mathcal{L}_\infty$-space. □

In the proof of Theorem 9.23, we saw that a Banach space that is locally complemented in ℓ_∞ is an $\mathcal{L}_\infty$-space. Since this fact will prove useful to us again later, we state it here as a separate proposition.

PROPOSITION 9.24. *If a Banach space X is locally complemented in ℓ_∞, then it is an $\mathcal{L}_\infty$-space.*

In fact, a stronger statement can be made: A locally complemented subspace of an $\mathcal{L}_\infty$-space is also an $\mathcal{L}_\infty$-space. This argument is more subtle and so we direct the interested reader to Lemma A.12 in [**10**].

Proposition 9.22 states that every injective space is an $\mathcal{L}_\infty$-space. The converse is not true, because not every $\mathcal{L}_\infty$-space is injective. The space c_0, for example, is an $\mathcal{L}_\infty$-space, but it is not injective because it is not complemented in ℓ_∞. However, every $\mathcal{L}_\infty$-space does satisfy a *local* type of injectivity.

PROPOSITION 9.25. *An $\mathcal{L}_\infty$-space is locally complemented in any space that has it as a subspace.*

PROOF. Let X be an $\mathcal{L}_\infty$-space and assume that X is a subspace of a Banach space Y. Since X is an $\mathcal{L}_\infty$-space, its bidual X^{**} is injective, by Proposition 9.22. This means that X^{**} is complemented in Y^{**}, because X^{**} is an injective Banach space that is a subspace of Y^{**}. But this implies that X is locally complemented in Y. (See Remark 9.4.) □

COMMENT 9.26. In our discussion so far, we have been interested in $\mathcal{L}_\infty$-spaces, but we can define $\mathcal{L}_p$-spaces for any $p \geq 1$. For any $p \geq 1$ (including $p = \infty$), a Banach space X is called an $\mathcal{L}_p$-*space* if there exists some $\lambda \geq 1$ such that for every finite dimensional subspace F of X, there is a finite dimensional subspace G containing F such that $d_{\mathrm{BM}}(G, \ell_p^n) \leq \lambda$, where $n = \dim(G)$ and $d_{\mathrm{BM}}(G, \ell_p^n)$ is the Banach–Mazur distance between G and ℓ_p^n.

$\mathcal{L}_p$-spaces were introduced by Lindenstrauss and Pełczyński in their landmark 1968 paper [**73**] where many of the properties of $\mathcal{L}_p$-spaces were catalogued. The following year, in 1969, Lindenstrauss and Rosenthal published [**74**], which further developed the ideas of $\mathcal{L}_p$-spaces.

Whether or not a Banach space is an $\mathcal{L}_p$-space depends on the finite-dimensional subspaces of the space. For this reason, it is considered a "local" property. A theorem of Ribe shows that this local structure is a uniform invariant. Specifically, Theorem 2 in Ribe's 1978 paper [**100**] states that if a real Banach space is uniformly homeomorphic to an $\mathcal{L}_p$-space, then it is also an $\mathcal{L}_p$-space. In Ribe's original paper, the theorem is stated for $1 < p < \infty$; however, the theorem is also true for $p = 1$ and $p = \infty$. The case $p = \infty$ was shown to be true by Heinrich and Mankiewicz in

[45] in 1982. (See Theorem 4.9 in that paper.) The case $p = 1$ took much longer to show. It was not until 2009 that Johnson, Maurey and Schechtman proved that the theorem was true for the case $p = 1$. (See Corollary 4 in [51].)

Since the local structure of a Banach space is a uniform invariant, it is in particular a Lipschitz invariant, which makes it a valuable tool for the Lipschitz classification of Banach spaces.

We now return to Question 5.29, which is one of the central questions in the nonlinear theory of Banach spaces. For completeness, we restate it here.

QUESTION 9.27 (Lipschitz isomorphism problem). *If X and Y are separable Lipschitz isomorphic Banach spaces, are they linearly isomorphic?*

We already saw that it is possible to have two nonseparable Banach spaces that are Lipschitz isomorphic, but not linearly isomorphic (see Theorem 5.22), which is why in Question 5.29 we restrict our attention to separable Banach spaces.

In the separable case, we know that, because ℓ_2 is a Hilbert space, if X is Lipschitz isomorphic to ℓ_2, then X is linearly isomorphic to ℓ_2. For other values of $p \in [1, \infty)$, we know that if X Lipschitz embeds into ℓ_p, then X linearly embeds into ℓ_p. This is true because ℓ_p is reflexive for $p \neq 1$ and because ℓ_1 has the Radon–Nikodym property. For separable reflexive Banach spaces in general, we have the following result.

THEOREM 9.28. *If X and Y are reflexive Lipschitz isomorphic separable Banach spaces, then X is isomorphic to a complemented subspace of Y and Y is isomorphic to a complemented subspace of X.*

By the Heinrich–Mankiewicz theorem (Theorem 9.7), we know that if X and Y are Lipschitz isomorphic separable Banach spaces, then X is isomorphic to a subspace of Y^{**} and Y is isomorphic to a subspace of X^{**}. In the current context, the Banach spaces X and Y are reflexive, and so we already know that X is isomorphic to a subspace of Y and Y is isomorphic to a subspace of X. However, it still remains to show that the subspaces are complemented.

PROOF OF THEOREM 9.28. Let X and Y be separable reflexive Banach spaces and let $f : X \to Y$ be a Lipschitz isomorphism such that $f(0) = 0$. By Theorem 7.25, f has a point of Gâteaux differentiability. Without loss of generality, assume that the function f is Gâteaux differentiable at 0. Let $T = Df(0)$. Then $T : X \to Y$ is a linear embedding such that

$$c\|x\|_X \leq \|T(x)\|_Y \leq C\|x\|_X, \quad x \in X,$$

where c and C are positive constants. For later, we observe that

(9.3) $$T(x) = \lim_{n \to \infty} nf\left(\frac{x}{n}\right), \quad x \in X,$$

where the limit is in the norm topology.

For each $n \in \mathbb{N}$, let

$$g_n(y) = nf^{-1}\left(\frac{y}{n}\right), \quad y \in Y.$$

Then we have that $g_n : Y \to X$ is a Lipschitz continuous function for each $n \in \mathbb{N}$ and $\text{Lip}(g_n) \leq \text{Lip}(f^{-1})$. Let $\mathcal{U}$ be a non-principal ultrafilter[4] on $\mathbb{N}$ and let

$$g(y) = \lim_{\mathcal{U}} g_n(y), \quad y \in Y,$$

where the limit is taken in the weak topology on X. By that, we mean that $g(y)$ is the element in X^{**} for which

$$\langle g(y), x^* \rangle = \lim_{\mathcal{U}} \langle g_n(y), x^* \rangle, \quad (y, x^*) \in Y \times X^*.$$

Then $g(y) \in X^{**}$, but since X is assumed to be reflexive, we have that $g(y) \in X$. Since for each $x^* \in X^*$, we have

$$\langle g(x) - g(y), x^* \rangle = \lim_{\mathcal{U}} \langle g_n(x) - g_n(y), x^* \rangle,$$

it follows that g is a Lipschitz continuous function and $\text{Lip}(g) \leq \text{Lip}(f^{-1})$.

Now we start putting all of the pieces together. First, we note that since g_n is Lipschitz continuous with Lipschitz constant no more than $\text{Lip}(f^{-1})$, we have

$$\left\| g_n\left(nf\left(\tfrac{x}{n}\right)\right) - g_n(T(x)) \right\| \leq \text{Lip}(f^{-1}) \left\| nf\left(\tfrac{x}{n}\right) - T(x) \right\|.$$

The right hand side of this inequality converges to zero, by (9.3). Thus, for each $x \in X$ and each $x^* \in X^*$,

$$\langle g(T(x)), x^* \rangle = \lim_{\mathcal{U}} \langle g_n(T(x)), x^* \rangle = \lim_{\mathcal{U}} \left\langle g_n\left(nf\left(\tfrac{x}{n}\right)\right), x^* \right\rangle.$$

Recalling the definition of g_n, we see that for each $x \in X$, we have

$$g_n\left(nf\left(\tfrac{x}{n}\right)\right) = nf^{-1}\left(f\left(\tfrac{x}{n}\right)\right) = x,$$

for each $n \in \mathbb{N}$. (We are using the formula for $g_n(y)$ with $y = nf(\tfrac{x}{n})$.) Consequently, for each $x^* \in X^*$, we have that

$$\langle g(T(x)), x^* \rangle = \lim_{\mathcal{U}} \langle x, x^* \rangle.$$

Thus, we conclude that $g(T(x)) = x$ for each $x \in X$.

The map $T \circ g : Y \to T(X)$ is a Lipschitz mapping. Furthermore, by what we just showed in the previous paragraph, for any $T(x) \in T(X)$, we have that

$$(T \circ g)(T(x)) = T(g(T(x))) = T(x),$$

and so $T \circ g$ is a Lipschitz retraction of Y onto $T(X)$. Since X is reflexive, it is also true that $T(X)$ is reflexive, and so it follows that $T(X)$ is complemented in the second dual Y^{**}, by Corollary 9.18. But Y is also assumed to be reflexive, and so we have that $T(X)$ is complemented in Y. Therefore, since T is an isomorphism of X onto its image, we conclude that X is isomorphic to a complemented subspace of Y, as required.

A similar argument shows that Y is isomorphic to a complemented subspace of X, so we leave the details to the reader. $\square$

COROLLARY 9.29 (Pełczyński [**92**]). *If X is Lipschitz isomorphic to ℓ_p for $1 < p < \infty$, then X is linearly isomorphic to ℓ_p.*

---

[4]We will formally define an ultrafilter in Definition 11.1 in Chapter 11. The interested reader can skip ahead to find the definition of an ultrafilter and a non-principal ultrafilter, but for the time being it may be enough to know that using a non-principal ultrafilter allows for a more general method for computing a limit.

PROOF. Since ℓ_p is reflexive, X is isomorphic to a complemented subspace of ℓ_p. However, any complemented infinite dimensional subspace of ℓ_p is isomorphic to ℓ_p. (That is, ℓ_p is a *prime* space.) □

COROLLARY 9.30 (Pełczyński [**92**]). *If X is Lipschitz isomorphic to $L_p(0,1)$ for $1 < p < \infty$, then X is linearly isomorphic to $L_p(0,1)$.*

We will not give the proof, other than to say it uses a technique now known as the *Pełczyński decomposition trick*. These results are related to a problem in the linear theory known as the *Schroeder–Bernstein Problem*, due to its similarity to the Schroeder–Bernstein theorem.

QUESTION 9.31 (Schroeder–Bernstein Problem). *If X is complemented in Y and Y is complemented in X, is it true that X is linearly isomorphic to Y?*

We know there is a positive answer to the Schroeder–Bernstein Problem in certain cases; for example, when one of the spaces is c_0 or ℓ_p or $L_p(0,1)$ for any p in the interval $[1,\infty)$. The answer is also "yes" when X is isomorphic to $X \oplus X$ and Y is isomorphic to $Y \oplus Y$. These positive results go back to Pełczyński's landmark paper of 1960 [**92**]. It was not until 1996 that Gowers showed that the answer is not "yes" in general [**42**]. Gowers (with the aid of B. Maurey) gave an example of a Banach space Z that is *not* isomorphic to $Z \oplus Z$ but *is* isomorphic to $Z \oplus Z \oplus Z$. The negative result is then obtained by choosing $X = Z$ and $Y = Z \oplus Z$.

Theorem 9.28 is a statement about reflexive Banach spaces. For non-reflexive Banach spaces, we have a similar result, provided that X has RNP and is complemented in its second dual.

THEOREM 9.32. *Suppose X and Y are separable Banach spaces with RNP and suppose that X is complemented in X^{**}. If X and Y are Lipschitz isomorphic, then X is isomorphic to a complemented subspace of Y.*

The proof is similar to the proof of Theorem 9.28, and so we will not include the details here. In the proof, we define g in the same way as in the proof of Theorem 9.28, but the best we can do is conclude that $g(y) \in X^{**}$ for each $y \in Y$. Assuming that X is complemented in X^{**} allows us to project back into X.

$$Y \xrightarrow{g} X^{**} \xrightarrow{P} X$$

In the above diagram, $P : X^{**} \to X$ is the projection that is assumed to exist in the hypothesis of the theorem.

EXAMPLE 9.33. Note that ℓ_1 is not reflexive, but has RNP (since separable dual spaces always have RNP) and is complemented in its bidual (since a dual space is always complemented in its bidual). Therefore, if ℓ_1 is Lipschitz isomorphic to a Banach space Y, then ℓ_1 is linearly isomorphic to a complemented subspace of Y.

If Y is also a separable dual space, so that Y has RNP and is complemented in its bidual, and if Y is Lipschitz isomorphic to ℓ_1, then we have that Y is linearly isomorphic to a complemented subspace of ℓ_1. Since ℓ_1 is prime this implies that Y is linearly isomorphic to ℓ_1.

This last example leads us to an open question.

QUESTION 9.34. *Does ℓ_1 have unique Lipschitz structure? That is, if a Banach space is Lipschitz isomorphic to ℓ_1, is it necessarily linearly isomorphic to ℓ_1?*

The same question can be asked about c_0. That question has been answered, and is the topic of the next chapter.

CHAPTER 10

The Lipschitz structure of c_0

In this chapter, we answer the following question.

QUESTION 10.1. *Does c_0 have unique Lipschitz structure? That is, if a Banach space is Lipschitz isomorphic to c_0, is it necessary linearly isomorphic to c_0?*

In Theorem 1.25 we showed that c_0 is an absolute Lipschitz retract (ALR). If X is Lipschitz isomorphic to c_0, then X is also an ALR. This implies that X is an $\mathcal{L}_\infty$-space (by Theorem 9.23). We now recall a theorem of Johnson and Zippin from [54].

THEOREM 10.2. *An $\mathcal{L}_\infty$-space that linearly embeds in c_0 is isomorphic to c_0.*

We state the theorem of Johnson and Zippin here without proof, but we will provide a proof of the result later. (See Theorem 10.19.)

As a consequence of Theorem 10.2, if a Banach space X that is Lipschitz isomorphic to c_0 can be shown to be linearly isomorphic to a subspace of c_0, then it will follow from what we said above that X is isomorphic to c_0 itself. Godefroy, Kalton, and Lancien succeeded in showing that a space Lipschitz isomorphic to c_0 is indeed linearly isomorphic to a subspace of c_0 in 2000 [38], thus answering Question 10.1 affirmatively.

THEOREM 10.3 (Godefroy–Kalton–Lancien). *If X is Lipschitz isomorphic to a subspace of c_0, then X is linearly isomorphic to a subspace of c_0.*

Our objective now is to prove Theorem 10.3. We will follow the approach in [38]. The key tool in the proof is a result known as the *Gorelik principle*, which loosely says that if $f : X \to Y$ is a Lipschitz isomorphism between Banach spaces, and if E has finite codimension in X, then $f(B_E)$ will be "big" in some sense (which will be made more precise shortly). First, we provide a technical result that is also due to Gorelik [41].

PROPOSITION 10.4. *Let X be a Banach space and let E be a closed subspace of finite codimension in X. Suppose α is a real number such that $0 < \alpha < 1$. There exists a compact subset K of B_X with the following property: If $f : K \to X$ is continuous and $\|f(x) - x\| \leq \alpha$ for all $x \in K$, then there exists some $x_0 \in K$ such that $f(x_0) \in E$.*

Proposition 10.4 is an application of the Brouwer fixed-point theorem. For that reason, we state (without proof) the Brouwer fixed-point theorem here.

THEOREM 10.5 (Brouwer Fixed-Point Theorem). *If Z is a finite dimensional normed space and $g : B_Z \to B_Z$ is a continuous function, then there exists a $z \in B_Z$ such that $g(z) = z$.*

PROOF OF PROPOSITION 10.4. Let $Z = X/E$ and let $q : X \to Z$ be the quotient map. Given $z \in \alpha B_Z$, the map
$$z \mapsto q^{-1}(z) \cap [\text{Int}(B_X)]$$
is lower semi-continuous. Thus, by the Michael selection theorem (Theorem 3.22), there exists a continuous selection $\psi : \alpha B_Z \to B_X$ such that $q \circ \psi = \text{Id}_{\alpha B_Z}$. Let $K = \psi(\alpha B_Z)$. Then K is compact because ψ is continuous. We will show that K is the compact set claimed in the statement of the proposition.

Suppose that $f : K \to X$ is a continuous function and $\|f(x) - x\| \leq \alpha$ for all $x \in K$. Define a function $g : \alpha B_Z \to \alpha B_Z$ by $g(z) = z - qf(\psi(z))$ for each $z \in \alpha B_Z$. To show that g is well-defined, observe that
$$g(z) = z - qf(\psi(z)) = q\psi(z) - qf(\psi(z)) = q\big(\psi(z) - f(\psi(z))\big).$$
Since $\psi(z) \in K$, it follows that
$$\|g(z)\| \leq \|q\| \|\psi(z) - f(\psi(z))\| \leq \alpha,$$
and so $g(z) \in \alpha B_Z$. Consequently, by the Brouwer fixed-point theorem, there is some $z_0 \in \alpha B_Z$ such that $g(z_0) = z_0$. Therefore,
$$g(z_0) = z_0 - qf(\psi(z_0)) = z_0,$$
which means that $qf(\psi(z_0)) = 0$, and hence $f(\psi(z_0)) \in E$. The point $x = \psi(z_0)$ is the point claimed to exist in the statement of the proposition, which completes the proof. $\square$

We are now prepared to state and prove the Gorelik principle, which will be the key tool in proving that c_0 has unique Lipschitz structure.

THEOREM 10.6 (Gorelik Principle [41]). *Suppose that X and Y are Banach spaces and $E \subseteq X$ is a subspace having finite codimension. Suppose $f : X \to Y$ is a homeomorphism such that f^{-1} is uniformly continuous. Let r be a fixed positive number. There exists a compact subset $K \subseteq rB_X$ such that if $\omega_{f^{-1}}(\|y\|) < \frac{r}{2}$, then there is a point $x \in K$ and a point $e \in E$ such that $\|e\| < \frac{3r}{2}$ and $y = f(e) - f(x)$. That is, there exists a compact subset C in Y such that*
$$\left\{ y : \omega_{f^{-1}}(\|y\|) < \tfrac{r}{2} \right\} \subseteq f\left(\tfrac{3r}{2} B_X \cap E\right) + C.$$

PROOF. By Proposition 10.4 (with $\alpha = \frac{1}{2}$), there exists a compact set K_1 in B_X with the property that whenever $g : K_1 \to X$ is a continuous function that satisfies the inequality $\|g(x) - x\| \leq \frac{1}{2}$ for all $x \in K_1$, then there exists a point $x_1 \in K_1$ such that $g(x_1) \in E$. Let $K = rK_1$, which is a compact set in rB_X.

Suppose that $y \in Y$ is a point such that $\omega_{f^{-1}}(\|y\|) < \frac{r}{2}$. Define a function $h : K \to X$ by
$$h(x) = f^{-1}\big(f(x) + y\big), \quad x \in K.$$
Then h is a continuous function on K, and for all $x \in K$,
$$\|h(x) - x\| = \left\| f^{-1}\big(f(x) + y\big) - f^{-1}\big(f(x)\big) \right\| \leq \omega_{f^{-1}}(\|y\|) < \frac{r}{2}.$$
Dividing by the positive number r, it follows that
$$\left\| \frac{h(x)}{r} - \frac{x}{r} \right\| < \frac{1}{2}, \tag{10.1}$$
for all $x \in K$.

Now define a function $g : K_1 \to X$ according to the rule $g(x) = \frac{h(rx)}{r}$ for each $x \in K_1$. Note that this is well-defined, because $K = rK_1$, and so if $x \in K_1$, then $rx \in K$. It follows that g is a continuous function on K_1, and for every $x \in K_1$, we have that
$$\|g(x) - x\| = \left\|\frac{h(rx)}{r} - \frac{rx}{r}\right\| < \frac{1}{2},$$
by (10.1), keeping in mind that $rx \in K$. Therefore, by the defining property of the compact set K_1, there exists a point $x_1 \in K_1$ such that $g(x_1) \in E$. That is, there is a point $x_1 \in K_1$ such that $\frac{h(rx_1)}{r} \in E$, and so $h(rx_1) \in rE$. Since E is a subspace of X, we have that $rE = E$. Furthermore, since $K = rK_1$, we conclude that there is a point $x_0 = rx_1 \in K$ for which $h(x_0) \in E$. Consequently, there exists a point $x_0 \in K$ such that
$$h(x_0) = f^{-1}\Big(f(x_0) + y\Big) \in E.$$
This means that there exists a point $e \in E$ such that $f(x_0) + y = f(e)$, and such that
$$\|e - x_0\| = \|h(x_0) - x_0\| < \frac{r}{2},$$
for which we note that $e = h(x_0)$. From this, we conclude that
$$\|e\| \leq \|e - x_0\| + \|x_0\| < \frac{r}{2} + r = \frac{3r}{2}.$$
Observe now that $y = f(e) - f(x_0)$ and so $y \in f(\frac{3r}{2} B_X \cap E) - f(K)$. It remains only to let $C = -f(K)$, which is a compact set in Y, and the proof is complete. □

We can now prove Theorem 10.3. In order to identify a space as being linearly isomorphic to a subspace of c_0, we will make use of the following lemma from [**38**]. (See Comment 10.8.)

LEMMA 10.7. *Let X be a separable Banach space. If there is a $c > 0$ such that for all $x^* \in X^*$,*

(10.2) $$\lim_{n \to \infty} \|x^* + x_n^*\| \geq \|x^*\| + c \lim_{n \to \infty} \|x_n^*\|,$$

whenever $x_n^* \to 0$ in the weak*-topology and the limits exist, then X linearly embeds into c_0.

COMMENT 10.8. If a norm on a separable Banach space X satisfies the hypothesis of Lemma 10.7 for some c in the interval $(0, 1]$, then the norm is said to be *Lipschitz weak-star Kadec–Klee* (or LKK*, for short) with constant c. Lemma 10.7 says that a separable Banach space with a Lipschitz weak-star Kadec–Klee norm (for some c) is linearly isomorphic to a subspace of c_0.

In the paper [**38**], a stronger and more quantitative version of Lemma 10.7 is given (as Theorem 2.4): If X is a separable Banach space having a norm that is LKK* with constant c, then for any $\epsilon > 0$, there is a subspace E of c_0 such that
$$d_{\mathrm{BM}}(X, E) \leq \frac{1}{c^2} + \epsilon,$$
where
$$d_{\mathrm{BM}}(X, E) = \inf\Big\{\|T\|\|T^{-1}\| \,\Big|\, T : X \to E \text{ is an isomorphism}\Big\}$$
is the *Banach–Mazur distance* between X and E.

PROOF OF THEOREM 10.3. Assume the Banach space X is Lipschitz isomorphic to a subspace Y of c_0. We will show that X is linearly isomorphic to a subspace of c_0 (that is not necessarily Y). Let $f : Y \to X$ be a Lipschitz isomorphism. We will use f to define a norm $|||\cdot|||$ on X^* by

$$|||x^*||| = \mathrm{Lip}(x^* \circ f) = \sup\left\{ \frac{|x^*(f(y_1) - f(y_2))|}{\|y_1 - y_2\|} \; : \; y_1 \in Y, \, y_2 \in Y, \, y_1 \neq y_2 \right\},$$

for all $x^* \in X^*$.

We claim that this new norm on X^* is equivalent to the original norm on X^*. To see this, first note that $|||x^*||| \leq \|x^*\| \mathrm{Lip}(f)$, because f is a Lipschitz continuous function. Next, suppose that $x \in X$. Since f has an inverse, for any $y \in Y$, we have that

$$x = f\big(f^{-1}[f(y) + x]\big) - f(y) = f\big(f^{-1}[f(y) + x]\big) - f\big(f^{-1}[f(y)]\big).$$

Thus, because $|||\cdot|||$ is defined as a supremum, we see that

$$|||x^*||| \geq \frac{|x^*(x)|}{\left\| f^{-1}(f(y) + x) - f^{-1}(f(y)) \right\|},$$

where we are choosing $y_1 = f^{-1}(f(y) + x)$ and $y_2 = f^{-1}(f(y))$ in the definition of the norm. Thus,

$$|x^*(x)| \leq |||x^*|||\left\| f^{-1}(f(y) + x) - f^{-1}(f(y)) \right\| \leq |||x^*||| \mathrm{Lip}(f^{-1})\|x\|.$$

It follows that $\|x^*\| \leq |||x^*||| \mathrm{Lip}(f^{-1})$, and so we obtain $\|x^*\|/\mathrm{Lip}(f^{-1}) \leq |||x^*|||$, which gives us a lower bound on $|||x^*|||$. Combining this with our earlier upper bound on $|||x^*|||$, we conclude that

$$\frac{1}{\mathrm{Lip}(f^{-1})}\|x^*\| \leq |||x^*||| \leq \|x^*\| \mathrm{Lip}(f).$$

Therefore, the two norms $|||\cdot|||$ and $\|\cdot\|$ on X^* are equivalent.

We now observe that the norm $|||\cdot|||$ is weak* lower-semicontinuous. That is, the set $\{x^* : |||x^*||| \leq 1\}$ is closed in the weak* topology on X^*. Consequently, there exists a predual norm $|||\cdot|||_*$ on X such that $(X, |||\cdot|||_*)^* = (X^*, |||\cdot|||)$. But our two norms on X^* are equivalent, and so the two norms $|||\cdot|||_*$ and $\|\cdot\|$ on the predual X are equivalent, as well.

In light of Lemma 10.7, we will show there exists a $c > 0$ such that if $x^* \in X^*$ and $x_n^* \to 0$ in the weak* topology, then

$$\lim_{n \to \infty} |||x^* + x_n^*||| \geq |||x^*||| + c \lim_{n \to \infty} |||x_n^*|||,$$

whenever these limits exist. To that end, we let x^* be an element of X^*. For simplicity, and without loss of generality, we assume $|||x^*||| = 1$.

Suppose that $\epsilon > 0$. There exist points y_1 and y_2 in Y such that

$$1 < \frac{x^*(f(y_1) - f(y_2))}{\|y_1 - y_2\|} + \epsilon,$$

and so (rewriting) we have

$$x^* f(y_1) - x^* f(y_2) > (1 - \epsilon)\|y_1 - y_2\|.$$

We are assuming that X is Lipschitz isomorphic to a subspace of c_0, and we chose f to be a Lipschitz isomorphism from X onto a subspace of c_0; however, we can choose f to be any Lipschitz isomorphism onto some subspace of c_0. Consequently, we may replace our chosen f with a translated version of f, so that $y_1 = y$ and $y_2 = -y$ for some $y \in Y$. Then our inequality above becomes

(10.3) $$x^* f(y) - x^* f(-y) > 2(1-\epsilon)\|y\|.$$

Rewrite (10.3) as
$$x^* f(-y) < x^* f(y) - 2(1-\epsilon)\|y\|.$$

Using the fact that f is a Lipschitz continuous function with $f(0) = 0$, we find an upper bound for $x^* f(y)$ as follows:
$$x^* f(y) = x^* f(y) - x^* f(0) \leq \underbrace{\text{Lip}(x^* f)}_{\||x^*\|| = 1} \|y - 0\| = \|y\|.$$

Consequently,

(10.4) $$x^* f(-y) < \|y\| - 2(1-\epsilon)\|y\| = -(1-2\epsilon)\|y\|.$$

This gives us an upper bound for $x^* f(-y)$ that we will use momentarily. First however, we will use a similar procedure to find a lower bound for $x^* f(y)$. Start by rewriting (10.3) as
$$x^* f(y) > 2(1-\epsilon)\|y\| + x^* f(-y).$$

We once again use the fact that f is a Lipschitz continuous function with $f(0) = 0$, but this time to find a lower bound for $x^* f(-y)$. Observe first that
$$-x^* f(-y) = x^* f(0) - x^* f(-y) \leq \underbrace{\text{Lip}(x^* f)}_{\||x^*\|| = 1} \|0 - y\| = \|y\|.$$

It follows that $x^* f(-y) \geq -\|y\|$, and so

(10.5) $$x^* f(y) > 2(1-\epsilon)\|y\| - \|y\| = (1-2\epsilon)\|y\|.$$

This gives us the second bound we will need momentarily.

By assumption, Y is a subspace of c_0, and consequently $y \in c_0$. It is possible therefore, to find a positive integer N such that if z is any sequence in c_0 for which
$$z_1 = z_2 = \cdots = z_N = 0 \quad \text{and} \quad \|z\| \leq \frac{\|y\|}{2},$$
then $\|y \pm z\| \leq \|y\|$. Consequently, for such a $z \in Y$, we see that
$$x^* f(z) - x^* f(-y) \leq \underbrace{\text{Lip}(x^* f)}_{\||x^*\|| = 1} \|z + y\| \leq \|y\|,$$
and so, using (10.4),
$$x^* f(z) \leq x^* f(-y) + \|y\| < -(1-2\epsilon)\|y\| + \|y\| = 2\epsilon\|y\|.$$

Similarly, we also see that
$$x^* f(y) - x^* f(z) \leq \underbrace{\text{Lip}(x^* f)}_{\||x^*\|| = 1} \|y - z\| \leq \|y\|,$$
and thus it follows from (10.5) that
$$x^* f(z) \geq x^* f(y) - \|y\| > (1-2\epsilon)\|y\| - \|y\| = -2\epsilon\|y\|.$$

Therefore, we have established that

(10.6) $$|x^* f(z)| < 2\epsilon\|y\|$$

for any $z \in Y$ for which $z_1 = z_2 = \cdots = z_N = 0$ and $\|z\| \leq \frac{1}{2}\|y\|$.

Now let $E = \{z \in Y : z_1 = z_2 = \cdots = z_N = 0\}$. (Note that E depends on the choice of y.) The subset E has finite codimension in Y and we have established that if $z \in E$ such that $\|z\| \leq \frac{1}{2}\|y\|$, then $|x^* f(z)| < 2\epsilon \|y\|$.

Apply the Gorelik principle (Theorem 10.6) to the set E with $r = \frac{\|y\|}{3}$. Then there is a compact set K in X such that

$$\left\{x \in X : \omega_{f^{-1}}(\|x\|) < \frac{\|y\|}{6}\right\} \subseteq f\left(\frac{\|y\|}{2} B_Y \cap E\right) + K.$$

That is, there is a compact set K in X with the property that if $x \in X$ is such that $\|x\| \leq \frac{\|y\|}{6 \operatorname{Lip}(f^{-1})}$, then

$$x \in f\left(\frac{\|y\|}{2} B_Y \cap E\right) + K.$$

Suppose that $(x_n^*)_{n=1}^\infty$ is a sequence of terms in X^* that converges to 0 in the weak* topology on X^*. For each $n \in \mathbb{N}$, pick $x_n \in X$ such that

$$\|x_n\| \leq \frac{\|y\|}{6 \operatorname{Lip}(f^{-1})} \quad \text{and} \quad x_n^*(x_n) \geq \frac{\|x_n^*\| \|y\|}{10 \operatorname{Lip}(f^{-1})}.$$

The first condition on x_n ensures that

$$x_n \in f\left(\frac{\|y\|}{2} B_Y \cap E\right) + K,$$

and so there exists some $z_n \in \frac{\|y\|}{2} B_Y \cap E$ such that $x_n - f(z_n) \in K$. The second condition can be met because it is possible to estimate $\|x_n^*\|$ using elements in B_X. (The number 10 in the denominator is somewhat arbitrary. It is only necessary that the constant is larger than 4.)

By assumption, $x_n^* \to 0$ in the weak* topology on X^*. It follows that

$$\lim_{n \to \infty} \sup_{x \in K} |x_n^*(x)| = 0,$$

because K is a compact set and the convergence is uniform on compact sets. Thus, there exists a positive integer n_0 such that if $n \geq n_0$, then $|x_n^*(x)| < \epsilon \|y\|$ for all $x \in K$.

We established that $x_n - f(z_n) \in K$ for all $n \in \mathbb{N}$. Consequently, if $n \geq n_0$, it follows that $x_n^*(x_n - f(z_n)) < \epsilon \|y\|$. Rewrite the inequality as follows:

$$x_n^*(x_n) - x_n^* f(z_n) < \epsilon \|y\|.$$

From this, we have

$$x_n^* f(z_n) > x_n^*(x_n) - \epsilon \|y\| \geq \frac{\|x_n^*\| \|y\|}{10 \operatorname{Lip}(f^{-1})} - \epsilon \|y\|.$$

Since $z_n \in r B_Y \cap E$, we also know from (10.6) that

$$x^* f(z_n) > -2\epsilon \|y\|,$$

and so we see that

$$(x^* + x_n^*) f(z_n) > \frac{\|x_n^*\| \|y\|}{10 \operatorname{Lip}(f^{-1})} - 3\epsilon \|y\|.$$

Furthermore, recalling the bound on $x^* f(-y)$ established in (10.4), we also have the inequality

$$(x^* + x_n^*) f(-y) < -(1 - 2\epsilon) \|y\| + x_n^* f(-y).$$

Thus,
$$\mathrm{Lip}\left((x^* + x_n^*) \circ f\right) \geq \frac{(x^* + x_n^*)f(z_n) - (x^* + x_n^*)f(-y)}{\|z_n + y\|}$$
$$\geq \left[(1 - 5\epsilon)\|y\| + \frac{\|x_n^*\|\,\|y\|}{10\,\mathrm{Lip}(f^{-1})} - x_n^* f(-y)\right] \frac{1}{\|y\|}$$
$$= (1 - 5\epsilon) + \frac{\|x_n^*\|}{10\,\mathrm{Lip}(f^{-1})} - \frac{x_n^* f(-y)}{\|y\|}.$$

Therefore,
$$\liminf_{n \to \infty}\|\|x^* + x_n^*\|\| \geq (1 - 5\epsilon) + \frac{1}{10\,\mathrm{Lip}(f^{-1})} \liminf_{n \to \infty} \|x_n^*\|.$$

Since $\epsilon > 0$ is arbitrary and since $\|x_n^*\| \geq \frac{1}{\mathrm{Lip}(f)}\|\|x_n^*\|\|$,
$$\liminf_{n \to \infty}\|\|x^* + x_n^*\|\| \geq 1 + c\liminf_{n \to \infty}\|\|x_n^*\|\|,$$
where
$$c = \frac{1}{10\,\mathrm{Lip}(f)\,\mathrm{Lip}(f^{-1})}.$$

It follows that X with the norm $\|\|\cdot\|\|_*$ is isomorphic to a subspace of c_0, by Lemma 10.7. Since this norm on X is equivalent to the original norm on X, the proof is complete. $\square$

We can now state and prove the theorem that asserts that the Lipschitz structure of c_0 is unique.

THEOREM 10.9 (Godefroy–Kalton–Lancien[38]). *If a Banach space is Lipschitz isomorphic to c_0, then it is linearly isomorphic to c_0.*

PROOF. The proof was outlined at the start of the chapter. Let X be a Banach space that is Lipschitz isomorphic to c_0. We know that c_0 is an absolute Lipschitz retract (Theorem 1.25), and consequently X, being Lipschitz isomorphic to c_0, is also an absolute Lipschitz retract. This implies that X is locally complemented in ℓ_∞, and hence X is an $\mathcal{L}_\infty$-space. Therefore, since X linearly embeds in c_0 (Theorem 10.3), X is linearly isomorphic to c_0 by the theorem of Johnson and Zippin (Theorem 10.2). $\square$

In order to complete this circle of ideas, we will prove Theorem 10.2, the theorem of Johnson and Zippin that states that an $\mathcal{L}_\infty$-space that linearly embeds into c_0 is isomorphic to c_0. The basic idea behind Theorem 10.2 is that if X is an $\mathcal{L}_\infty$-space that is isomorphic to a subspace of c_0, then X is locally complemented in c_0 (by Proposition 9.25), which in turn means that X^* is complemented in ℓ_1. The only complemented subspaces of ℓ_1 are linearly isomorphic to ℓ_1 (because ℓ_1 is prime). This fact, together with the fact that X linearly embeds into c_0 is sufficient to prove that X is linearly isomorphic to c_0.

The fact that X^* is linearly isomorphic to ℓ_1 is not in general sufficient to show that X is linearly isomorphic to c_0. If K is a compact set, then $\mathcal{M}(K) = \mathcal{C}(K)^*$. If K is a countable set, then $\mathcal{M}(K) \approx \ell_1$. However, $\mathcal{C}(K)$ is not linearly isomorphic to c_0 if K is "too large". As an example, $\mathcal{C}(K)$ is not linearly isomorphic to c_0 if $K = \omega^\omega$, even though ω^ω is a countable set.

COMMENT 10.10. The example $C(\omega^\omega)$ given above is due to Bessaga and Pełczyński [**15**]. In that work from 1960, Bessaga and Pełczyński classified all $C(K)$ spaces that have ℓ_1 as a dual space. In so doing, they showed there are an infinite number of ℓ_1 preduals that are not linearly isomorphic to each other.

In 1972, Benyamini and Lindenstrauss found an ℓ_1 predual that is not even isomorphic to a $C(K)$ space [**12**]. Consequently, knowing that X^* is linearly isomorphic to ℓ_1 is not enough to conclude that X is linearly isomorphic to c_0.

As Theorem 10.2 is a result in the linear theory of Banach spaces, we now recall some definitions from the linear theory of Banach spaces. (See [**5**] for more details.)

DEFINITION 10.11. A *Schauder basis* (or simply a *basis*) of a Banach space X is a sequence $(e_n)_{n=1}^\infty$ such that each x in X can be uniquely represented in the form
$$x = \sum_{n=1}^\infty a_n e_n.$$

REMARK 10.12. Suppose X is a Banach space with a Schauder basis $(e_k)_{k=1}^\infty$. Then each $x \in X$ can be written as $x = \sum_{k=1}^\infty a_k e_k$ for some sequence of real scalars $(a_k)_{k=1}^\infty$. For each $n \in \mathbb{N}$, define a functional e_n^* on the Banach space X by
$$e_n^*\Big(\sum_{k=1}^\infty a_k e_k\Big) = a_n.$$
Since every $x \in X$ can be written in that form, the map $e_n^* : X \to \mathbb{R}$ is well-defined. Furthermore, for each $n \in \mathbb{N}$, the map e_n^* is bounded and linear, and so $e_n^* \in X^*$.

DEFINITION 10.13. The elements in the sequence $(e_n^*)_{n=1}^\infty$ in Remark 10.12 are called the *biorthogonal functionals* (or the *coordinate functionals*) associated to the Schauder basis $(e_n)_{n=1}^\infty$. If $(e_n^*)_{n=1}^\infty$ is a basis for X^*, then $(e_n)_{n=1}^\infty$ is called a *shrinking basis*.

EXAMPLE 10.14. Every basis of a reflexive Banach space is a shrinking basis. The canonical basis of c_0 is shrinking, but the canonical basis of ℓ_1 is *not* shrinking.

DEFINITION 10.15. Suppose that X is a Banach space with a Schauder basis $(e_n)_{n=1}^\infty$, and suppose that $(e_n^*)_{n=1}^\infty$ is the associated sequence of coordinate functionals. The finite rank bounded linear map $S_n : X \to X$ defined by
$$S_n(x) = \sum_{k=1}^n e_k^*(x) e_k$$
is called the *nth partial sum projection* for $n \in \mathbb{N}$. We also set $S_0 = 0$.

REMARK 10.16. If X has a basis and S_n is the nth partial sum projection, then $\|S_n(x) - x\| \to 0$ as $n \to \infty$ for all $x \in X$. If the basis is a shrinking basis, then it is also true that $\|S_n^*(x^*) - x^*\| \to 0$ as $n \to \infty$ for all $x^* \in X^*$.

REMARK 10.17. If X has a basis and S_n is the nth partial sum projection, then the sequence $(S_n)_{n=1}^\infty$ is uniformly bounded in the norm on X. That is, there exists some $M \geq 0$ such that $\|S_n(x)\| \leq M$ for all $n \in \mathbb{N}$. (See Proposition 1.1.4 in [**5**].)

Finally, we will make use of the following theorem of Johnson, Rosenthal, and Zippin from 1971 [**53**], given here without proof.

THEOREM 10.18 (Johnson–Rosenthal–Zippin). *Let X be a Banach space. If X^* has a basis, then X has a shrinking basis.*

We will now prove Theorem 10.2, which we restate here as Theorem 10.19, using an argument based on that given by Johnson and Zippin in [54].

THEOREM 10.19. *An $\mathcal{L}_\infty$-space that linearly embeds in c_0 is isomorphic to c_0.*

PROOF. Let X be an $\mathcal{L}_\infty$-space that linearly embeds into c_0. Then X is a subspace of c_0 that is locally complemented in c_0, by Proposition 9.25. It follows that X^* is complemented in ℓ_1, and thus X^* is linearly isomorphic to ℓ_1 (because ℓ_1 is prime). Therefore, X^* has a basis, and so X has a shrinking basis, by Theorem 10.18. For that shrinking basis of X, let $(S_n)_{n=1}^\infty$ be the corresponding partial sum projections. Let $(P'_n)_{n=1}^\infty$ be the partial sum projections corresponding to the standard basis of c_0 and for each $n \in \mathbb{N}$, let $P_n : X \to c_0$ be the restriction of P'_n to X. That is, let $P_n = P'_n|_X$.

CLAIM 1. The finite rank operator $P_n - S_n$ converges to 0 weakly in the space $\mathcal{K}(X, c_0)$ of compact operators from X to c_0.

Proof of claim. For $T \in \mathcal{K}(X, c_0)$, define a function $f_T : B_{X^{**}} \times B_{c_0^*} \to \mathbb{R}$ by

$$f_T(x^{**}, \xi^*) = x^{**}(T^*\xi^*),$$

where $x^{**} \in B_{X^{**}}$ and $\xi^* \in B_{c_0^*}$, and $T^* : c_0^* \to X^*$ is the adjoint of T. Here we give X^{**} and c_0^* their respective weak* topologies so that the unit ball $B_{X^{**}}$ is compact in X^{**} and the unit ball $B_{c_0^*}$ is compact in c_0^*. If $K = B_{X^{**}} \times B_{c_0^*}$, then $f_T \in \mathcal{C}(K)$, and the map $T \mapsto f_T$ is an isometry from $\mathcal{K}(X, c_0)$ to $C(K)$.

If $(T_n)_{n=1}^\infty$ is a sequence of compact operators in $\mathcal{K}(X, c_0)$ such that $f_{T_n} \to 0$ pointwise and $\sup_{n \in \mathbb{N}} \|T_n\| < \infty$, then $T_n \to 0$ weakly. To see that this is true, observe that under these conditions, for any measure $\mu \in \mathcal{M}(K)$, we have

$$\lim_{n \to \infty} \int_K f_{T_n} \, d\mu = 0,$$

by the bounded convergence theorem. Thus, $f_{T_n} \to 0$ weakly. Therefore, for all $x \in B_X \subseteq B_{X^{**}}$ and $\xi^* \in B_{c_0^*}$,

$$f_{T_n}(x, \xi^*) = x(T_n^*\xi^*) = \xi^*(T_n(x)) \to 0$$

as $n \to \infty$, and so we conclude that $T_n \to 0$ weakly as $n \to \infty$.

Since the elements of the sequence $(P_n - S_n)_{n=1}^\infty$ are uniformly bounded (Remark 10.17), it suffices to show that $f_{P_n - S_n}$ converges to 0 pointwise as $n \to \infty$. Let $x^{**} \in B_{X^{**}}$ and $\xi^* \in B_{c_0^*}$. Then

$$f_{P_n - S_n}(x^{**}, \xi^*) = x^{**}\Big((P_n^* - S_n^*)\xi^*\Big).$$

However, we know that

$$P_n^* \xi^* \to \xi^*|_X \quad \text{and} \quad S_n^* \xi^* \to \xi^*|_X$$

in norm as $n \to \infty$ (Remark 10.16), and so

$$\lim_{n \to \infty} x^{**}\Big((P_n^* - S_n^*)\xi^*\Big) = 0,$$

and so $P_n - S_n \to 0$ weakly as $n \to \infty$, as claimed. ◇

CLAIM 2. There exists a sequence of commuting finite rank operators $(\Delta_n)_{n=1}^\infty$, where $\Delta_n : X \to X$ for each $n \in \mathbb{N}$, such that $\Delta_m \Delta_n = 0$ if $|m - n| \geq 2$ and such that for each $x \in X$, we have that $\|\Delta_n(x)\| \to 0$ as $n \to \infty$ and

$$(10.7) \qquad \frac{1}{2} \max_{n \in \mathbb{N}} \|\Delta_n(x)\| \leq \|x\| \leq 3 \max_{n \in \mathbb{N}} \|\Delta_n(x)\|.$$

Proof of claim. Given a sequence $(\epsilon_n)_{n=1}^\infty$ that decreases to 0, we can find an increasing sequence of integers $(p_k)_{k=0}^\infty$ with $p_0 = 0$ and a sequence of positive real numbers $(a_j)_{j=1}^\infty$ such that

$$\sum_{j=p_{k-1}+1}^{p_k} a_j = 1 \quad \text{and} \quad \left\| \sum_{j=p_{k-1}+1}^{p_k} a_j (P_j - S_j) \right\| < \epsilon_k$$

for all $k \in \mathbb{N}$. We assume that $\epsilon_1 < \frac{1}{12}$. For each $k \in \mathbb{N}$, let

$$T_k = \sum_{j=p_{k-1}+1}^{p_k} a_j S_j \quad \text{and} \quad R_k = \sum_{j=p_{k-1}+1}^{p_k} a_j P_j.$$

Observe that $\|T_k - R_k\| < \epsilon_k$ for each k in $\mathbb{N}$, and $\sup_{k \in \mathbb{N}} \|T_k\| < \infty$. Let $T_0 = 0$ and $R_0 = 0$.

Now let $x \in X$. Denote the standard basis of c_0 by $(e_j)_{j=1}^\infty$ and write

$$x = \sum_{j=1}^\infty x_j e_j$$

for the appropriate sequence $(x_j)_{j=1}^\infty$ of real numbers. Then for each positive integer k, we have that

$$R_k(x) = \left(a_{p_{k-1}+1} P_{p_{k-1}+1} + a_{p_{k-1}+2} P_{p_{k-1}+2} + \cdots + a_{p_k} P_{p_k} \right)(x)$$

$$= a_{p_{k-1}+1} \left[\sum_{j=1}^{p_{k-1}+1} x_j e_j \right] + a_{p_{k-1}+2} \left[\sum_{j=1}^{p_{k-1}+2} x_j e_j \right] + \cdots + a_{p_k} \left[\sum_{j=1}^{p_k} x_j e_j \right].$$

Regrouping the terms, we get

$$R_k(x) = \left(\sum_{j=p_{k-1}+1}^{p_k} a_j \right) \left[\sum_{i=1}^{p_{k-1}+1} x_i e_i \right] + \left(\sum_{j=p_{k-1}+2}^{p_k} a_j \right) \left[x_{p_{k-1}+2} e_{p_{k-1}+2} \right]$$

$$+ \left(\sum_{j=p_{k-1}+3}^{p_k} a_j \right) \left[x_{p_{k-1}+3} e_{p_{k-1}+3} \right]$$

$$+ \cdots + a_{p_k} \left[x_{p_k} e_{p_k} \right].$$

Recalling that $\sum_{j=p_{k-1}+1}^{p_k} a_j = 1$, the above expression can be written in the following way:

$$R_k(x) = \left(x_1, \ldots, x_{p_{k-1}}, x_{p_{k-1}+1}, c_{k,2} x_{p_{k-1}+2}, c_{k,3} x_{p_{k-1}+3}, \ldots, c_{k,\delta_k} x_{p_k}, 0, \ldots \right),$$

where
$$c_{k,i} = \sum_{j=p_{k-1}+i}^{p_k} a_j,$$
for each $i \in \{2,\ldots,p_k - p_{k-1}\}$, and we have written $\delta_k = p_k - p_{k-1}$, so that $c_{k,\delta_k} = a_{p_k}$ (and so that we could fit all of $R_k(x)$ on one line). Note that $0 < c_{k,i} < 1$ for all choices of (k,i).

Now let k be a positive integer such that $k \geq 2$. Then $R_k(x) - R_{k-1}(x) =$
$$\Big(0,\ldots,0, d_{k,2}x_{p_{k-2}+2},\ldots, d_{k,\delta_{k-1}}x_{p_{k-1}}, x_{p_{k-1}+1}, c_{k,2}x_{p_{k-1}+2},\ldots, c_{k,\delta_k}x_{p_k}, 0,\ldots\Big),$$
where
$$d_{k,i} = 1 - c_{k-1,i}$$
for each (k,i).

Consequently, it is necessarily true that $\|(R_k - R_{k-1})(x)\| \leq \|x\|$ for all $k \in \mathbb{N}$. Also, since $d_{k,i} + c_{k-1,i} = 1$, it follows that
$$\|x\| \leq 2\max_{k\in\mathbb{N}}\|(R_k - R_{k-1})(x)\|.$$

For each $k \in \mathbb{N}$, let $D_k = R_k - R_{k-1}$. (Recall that $R_0 = 0$ by definition.) Then we have that
$$\max_{k\in\mathbb{N}}\|D_k(x)\| \leq \|x\| \leq 2\max_{k\in\mathbb{N}}\|D_k(x)\|.$$

Next, for each $k \in \mathbb{N}$, let $\Delta_k = T_k - T_{k-1}$. (Recall that $T_0 = 0$ by definition.) Observe that $\|\Delta_1 - D_1\| \leq \epsilon_1$ and, for each positive integer k such that $k \geq 2$,
$$\|\Delta_k - D_k\| \leq \epsilon_{k-1} + \epsilon_k \leq 2\epsilon_1$$
because the sequence $(\epsilon_k)_{k=1}^\infty$ is a decreasing sequence of positive numbers. Thus, recalling that $\epsilon_1 < \frac{1}{12}$, for each $x \in X$ we have that

(10.8) $$\left|\max_{k\in\mathbb{N}}\|D_k(x)\| - \max_{k\in\mathbb{N}}\|\Delta_k(x)\|\right| \leq 2\epsilon_1\|x\| < \frac{1}{6}\|x\|.$$

Using (10.8) to bound the size of the $\Delta_k(x)$ terms, we get

(10.9) $$\max_{k\in\mathbb{N}}\|\Delta_k(x)\| < \max_{k\in\mathbb{N}}\|D_k(x)\| + \frac{1}{6}\|x\| < 2\|x\|.$$

Next, we use (10.8) again, but this time to bound the size of the $D_k(x)$ terms, and so we obtain the inequality
$$\|x\| \leq 2\max_{k\in\mathbb{N}}\|D_k(x)\| < 2\Big(\max_{k\in\mathbb{N}}\|\Delta_k(x)\| + \frac{1}{6}\|x\|\Big) = 2\max_{k\in\mathbb{N}}\|\Delta_k(x)\| + \frac{1}{3}\|x\|,$$
and hence

(10.10) $$\|x\| < 3\max_{k\in\mathbb{N}}\|\Delta_k(x)\|.$$

Putting together the bounds from (10.9) and (10.10), we obtain the following double inequality
$$\frac{1}{2}\max_{k\in\mathbb{N}}\|\Delta_k(x)\| < \|x\| < 3\max_{k\in\mathbb{N}}\|\Delta_k(x)\|,$$
which provides the claimed inequality in (10.7).

Finally, in order to complete the proof of this claim, we note that
$$T_k T_i = T_{\min(k,i)} = T_i T_k$$

for all $i \neq k$ and
$$\Delta_k \Delta_i = (T_k - T_{k-1})(T_i - T_{i-1}).$$
Therefore, $\Delta_k \Delta_i = \Delta_i \Delta_k$ for all $i \neq k$ and $\Delta_k \Delta_i \neq 0$ only if $|k - i| \leq 1$. This completes the proof of the claim. $\diamond$

CLAIM 3. *There is a $c > 0$ such that $\sum_{n=1}^{\infty} \|\Delta_n^* x^*\| \leq c\|x^*\|$ for all $x^* \in X^*$.*

Proof of claim. For each $n \in \mathbb{N}$, there is a $u_n \in X$ such that $\|u_n\| \leq 1$ and
$$(\Delta_n^* x^*)(u_n) \geq \frac{1}{2}\|\Delta_n^* x^*\|.$$
By (10.7),
$$\Big\|\sum_{n=1}^{N} \Delta_n(u_n)\Big\| \leq 3 \max_{m \in \mathbb{N}} \Big\|\Delta_m\Big(\sum_{n=1}^{N} \Delta_n(u_n)\Big)\Big\|.$$
For any m, there are at most three values of n in the set $\{1, \ldots, N\}$ that are within 1 of m (that is, we could have $n = m - 1$, $n = m$, or $n = m + 1$, depending on the value of m compared to N), and so the above norm (on the right side of the inequality) is at most $3M^2$, where $M = \sup_n \|\Delta_n\| \leq 2$. (The bound on M comes from (10.9).) Consequently,
$$\Big\|\sum_{n=1}^{N} \Delta_n(u_n)\Big\| \leq 9M^2.$$
Therefore,
$$\Big|\sum_{n=1}^{N} (\Delta_n^* x^*)(u_n)\Big| = \Big|\sum_{n=1}^{N} x^*\Big(\Delta_n(u_n)\Big)\Big| \leq 9M^2 \|x^*\|,$$
and so
$$\sum_{n=1}^{\infty} \|\Delta_n^* x^*\| \leq 18M^2 \|x^*\|,$$
which proves the claim with $c = 18M^2$. $\diamond$

Let $k \in \mathbb{N}$ and consider (k, ℓ) for $|k - \ell| \leq 1$. Let $E_{k\ell}$ be a finite dimensional subspace of X^* such that
$$\operatorname{span}\Big\{\Delta_j^*(X^*) \,:\, 1 \leq j \leq \max(k, \ell)\Big\} \subseteq E_{k\ell}.$$
Let $N_{k\ell} = \dim(E_{k\ell})$, which depends on k and ℓ. The idea is that $E_{k\ell} \approx \ell_1^{N_{k\ell}}$ and $E_{k\ell}^* \approx \ell_\infty^{N_{k\ell}}$, and we will show that
$$c_0\Big((E_{k\ell}^*)_{k,\ell}\Big) \subseteq c_0\Big((\ell_\infty^{N_{k\ell}})_{k,l}\Big) \approx c_0,$$
where the subspace is complemented in c_0. The plan, therefore, is to first show that X is linearly isomorphic to $c_0((E_{k\ell}^*)_{k,\ell})$, and then show this space is linearly isomorphic to c_0 by showing it is a complemented subspace of c_0, since c_0 is prime.

COMMENT 10.20. We are using the notation c_0 to denote the set of sequences of real numbers that tend to zero. More generally, if X is a Banach space, then $c_0(X)$ is the set of sequences $(x_k)_{k=1}^\infty$ having terms in X such that $\lim_{k\to\infty}\|x_k\|=0$. In this particular case however, we are using the notation $c_0((E_{k\ell}^*)_{k,\ell})$ to denote the set of arrays $(\xi_{k\ell})_{k,\ell}$ such that $\xi_{k\ell}\in E_{k\ell}^*$ for each pair (k,ℓ) of positive integers, where the terms decay to zero as either k or ℓ increases without bound. Note that in our particular example, the integers k and ℓ must also satisfy the additional condition that $|k-\ell|\le 1$. That means that for a given $k\in\mathbb{N}$, the only choices for ℓ are $k-1$, k, or $k+1$. (Note that if $k=1$, we do not allow $\ell=0$.)

Define a map $L: X \to c_0((E_{k\ell}^*)_{k,\ell})$ by

$$L(x) = \left(\Delta_k(x)\big|_{E_{k\ell}}\right)_{k,\ell}$$

for $x \in X$. Observe that $\|L\| \le 2$ by (10.7). We will now construct a linear map $K: c_0((E_{k\ell}^*)_{k,\ell}) \to X$ for which $KL(x) = x$ for all $x \in X$.

Suppose that we have chosen a point $\xi_{k\ell} \in E_{k\ell}^*$ for every pair of positive integers (k,ℓ) with $|k-\ell|\le 1$ so that the terms in $(\xi_{k\ell})_{k,\ell}$ decay to zero as either k or ℓ increases without bound. That is, so that $(\xi_{k\ell})_{k,\ell} \in c_0((E_{k\ell}^*)_{k,\ell})$. Let $\xi = (\xi_{k\ell})_{k,\ell}$. Since we chose $\xi_{k\ell}$ in $E_{k\ell}^*$, it follows that $\xi_{k\ell} \circ \Delta_\ell^* \in X^{**}$ for each of the pairs (k,ℓ). We wish to show that the sum $\sum_{k,\ell}\xi_{k\ell}\circ\Delta_\ell^*$ converges, where the sum is over all pairs (k,ℓ) with $|k-\ell|\le 1$. Suppose that F is a finite collection of such pairs. Then, for every x^* in X^* with $\|x^*\|\le 1$, we have

$$x^*\left(\sum_{(k,\ell)\in F}\xi_{k\ell}\circ\Delta_\ell^*\right) = \sum_{(k,\ell)\in F}\xi_{k\ell}\left(\Delta_\ell^*x^*\right) \le \|\xi\|_{c_0}\sum_{(k,\ell)\in F}\|\Delta_\ell^*x^*\|,$$

where $\|\xi\|_{c_0}$ is shorthand for the norm of ξ in $c_0((E_{k\ell}^*)_{k,\ell})$. If we now let I be the collection of all pairs of positive integers (k,ℓ) with $|k-\ell|\le 1$, then we certainly have that

$$\sum_{(k,\ell)\in F}\|\Delta_\ell^*x^*\| \le \sum_{(k,\ell)\in I}\|\Delta_\ell^*x^*\|,$$

provided that we can verify that the infinite sum on the right converges. Note that, since the sum on the right is the sum over all pairs (k,ℓ) such that $|k-\ell|\le 1$, we have that for each k, the only possible choices for the index ℓ are those in the set $\{k-1,k,k+1\}$. Consequently, in the sum $\sum_{(k,\ell)\in I}\|\Delta_\ell^*x^*\|$, each summand is counted (at most) three times, and so

$$\sum_{(k,\ell)\in I}\|\Delta_\ell^*x^*\| \le 3\sum_{k=1}^\infty \|\Delta_k^*x^*\|.$$

(The summand is counted only twice if $k=1$.) By Claim 3, this last sum is bounded by $c\|x^*\|$, and thus

$$x^*\left(\sum_{(k,\ell)\in F}\xi_{k\ell}\circ\Delta_\ell^*\right) \le 3c\|\xi\|_{c_0}\|x^*\|,$$

for all $x^* \in X^*$. It follows that
$$\left\|\sum_{(k,\ell)\in F} \xi_{k\ell} \circ \Delta_\ell^*\right\| \leq 3c\|\xi\|_{c_0},$$
and so the collection of terms $\sum_{(k,\ell)\in F} \xi_{k\ell} \circ \Delta_\ell^*$ in X^{**} for all choices of F is bounded uniformly in norm by $3c\|\xi\|_{c_0}$. Dual spaces of Banach spaces with the weak* topology satisfy the Heine–Borel property, and so we conclude that the sum $\sum_{(k,\ell)\in I} \xi_{k\ell} \circ \Delta_\ell^*$ converges to a point in X^{**}. Finally, since this element of X^{**} is weak* continuous, we conclude that $\sum_{(k,\ell)\in I} \xi_{k\ell} \circ \Delta_\ell^*$ is actually an element of X.

Now define $K : c_0((E_{k\ell}^*)_{k,\ell}) \to X$ by
$$K\Big((\xi_{k\ell})_{k,\ell}\Big) = \sum_{(k,\ell)\in I} \xi_{k\ell} \circ \Delta_\ell^*,$$
for all $(\xi_{k\ell})_{k,\ell} \in c_0((E_{k\ell}^*)_{k,\ell})$. Then K is a linear operator and $\|K\| \leq 3c$. If $x \in X$, then
$$KL(x) = K\Big((\Delta_k(x)\big|_{E_{k\ell}})_{k,\ell}\Big) = \sum_{(k,\ell)\in I} \Delta_k(x)\big|_{E_{k\ell}} \circ \Delta_\ell^*.$$
If $x^* \in X^*$, then
$$x^*\Big(KL(x)\Big) = \sum_{(k,\ell)\in I} \Delta_k(x)\Delta_\ell^* x^* = \sum_{(k,\ell)\in I} \Delta_\ell \Delta_k(x) x^*.$$
Consequently,
$$KL(x) = \sum_{(k,\ell)\in I} \Delta_\ell \Delta_k(x) = x.$$
Therefore, it follows that X is linearly isomorphic to $c_0((E_{k\ell}^*)_{k,\ell})$.

COMMENT 10.21. It may be worthwhile to fill in some details explaining the final equality $\sum_{k,\ell} \Delta_\ell \Delta_k(x) = x$. To see this, we first recall that
$$\Delta_k \Delta_\ell = (T_k - T_{k-1})(T_\ell - T_{\ell-1})$$
and
$$T_k T_\ell = T_\ell T_k = T_{\min(k,\ell)}.$$
For each k, the possible choices of ℓ are from the set $\{k-1, k, k+1\}$. We claim that
$$\Delta_{k-1}\Delta_k + \Delta_k \Delta_k + \Delta_{k+1}\Delta_k = \Delta_k.$$
We will compute each of these summands. For the first one, we see that
$$\Delta_{k-1}\Delta_k = (T_{k-1} - T_{k-2})(T_k - T_{k-1})$$
$$= T_{k-1}T_k - T_{k-1}T_{k-1} - T_{k-2}T_k + T_{k-2}T_{k-1}$$
$$= T_{k-1} - T_{k-1}^2 - T_{k-2} + T_{k-2}$$
$$= T_{k-1} - T_{k-1}^2.$$
Similarly, we have
$$\Delta_k \Delta_k = (T_k - T_{k-1})(T_k - T_{k-1})$$
$$= T_k^2 - 2T_{k-1} + T_{k-1}^2$$

and
$$\Delta_{k+1}\Delta_k = (T_{k+1} - T_k)(T_k - T_{k-1})$$
$$= T_k - T_{k-1} - T_k^2 + T_{k-1}$$
$$= T_k - T_k^2.$$

Summing these together, we have $T_k - T_{k-1} = \Delta_k$.

At this stage, we have shown that
$$\sum_{k,\ell} \Delta_\ell \Delta_k(x) = \sum_{k=1}^{\infty} \Delta_k(x).$$

However,
$$\sum_{k=1}^{\infty} \Delta_k(x) = \sum_{k=1}^{\infty} [T_k(x) - T_{k-1}(x)] = \lim_{k \to \infty} T_k(x) = x,$$
which provides the stated equality.

It remains to show that $X \approx c_0((E_{k\ell}^*)_{k,\ell})$ is linearly isomorphic to c_0. By the Principle of Local Reflexivity (Example 8.29), we can identify each finite dimensional space $E_{k\ell}^*$ with a subspace of X. Since X is an $\mathcal{L}_\infty$-space, and since each $E_{k\ell}^*$ is finite dimensional, there exists a $\lambda \geq 1$ (which does not depend on the choice of k or ℓ) such that for each pair (k, ℓ), there is a finite dimensional subspace $F_{k\ell}$ of X such that $E_{k\ell}^* \subseteq F_{k\ell}$ and a linear isomorphism $T_{k\ell} : F_{k\ell} \to \ell_\infty^{N_{k\ell}}$ with $\|T_{k\ell}\| \|T_{k\ell}^{-1}\| \leq \lambda$.

We define a map $T : c_0((F_{k\ell})_{k,\ell}) \to c_0((\ell_\infty^{N_{k\ell}})_{k,\ell})$ by
$$T\Big((\xi_{k\ell})_{k,\ell}\Big) = \Big(T_{k\ell}(\xi_{k\ell})\Big)_{k\ell}.$$
Since the constant λ does not depend on the choice of pair (k, ℓ), the map T provides a linear isomorphism from $c_0((E_{k\ell}^*)_{k,\ell})$ into $c_0((\ell_\infty^{N_{k\ell}})_{k,\ell}) \approx c_0$.

For each (k, ℓ), let $P_{k\ell}$ be an orthogonal (and so norm 1) projection of $F_{k\ell}$ onto $E_{k\ell}^*$, which exists because $F_{k\ell}$ is finite dimensional and contains $E_{k\ell}^*$ as a subspace. Let P be the map from $c_0((F_{k\ell})_{k,\ell})$ onto $c_0((E_{k\ell}^*)_{k,\ell})$ given by
$$P\Big((\xi_{k\ell})_{k,\ell}\Big) = \Big(P_{k\ell}(\xi_{k\ell})\Big)_{k\ell}.$$
The map $T \circ P \circ T^{-1}$ is a projection of $c_0((\ell_\infty^{N_{k\ell}})_{k,\ell})$ onto $c_0((E_{k\ell}^*)_{k,\ell})$ having norm at most λ. It follows that the Banach space $X \approx c_0((E_{k\ell}^*)_{k,\ell})$ is a complemented subspace of $c_0 \approx c_0((\ell_\infty^{N_{k\ell}})_{k,\ell})$. Since c_0 is a prime Banach space, all complemented subspaces of c_0 are linearly isomorphic to c_0. Therefore, we conclude that X is linearly isomorphic to c_0, as required. $\square$

CHAPTER 11

Ultraproducts of Banach spaces

In this chapter, we are interested in a Banach space X and a non-principal ultrafilter $\mathcal{U}$ on $\mathbb{N}$. We will begin by defining a *filter*.

DEFINITION 11.1. A *filter* $\mathcal{F}$ on $\mathbb{N}$ is a nonempty subset of $2^{\mathbb{N}}$ such that:
   (i) $\emptyset \notin \mathcal{F}$.
   (ii) If $A \in \mathcal{F}$ and $A \subseteq B$, then $B \in \mathcal{F}$.
   (iii) If $A \in \mathcal{F}$ and $B \in \mathcal{F}$, then $A \cap B \in \mathcal{F}$.

A maximal filter is called an *ultrafilter*.

REMARK 11.2. Every filter is contained in an ultrafilter, by Zorn's lemma.

EXAMPLE 11.3. For $m \in \mathbb{N}$, the set $\mathcal{F}_m = \{A \subseteq \mathbb{N} : m \in A\}$ is a maximal filter, and so is an ultrafilter. A filter of the form $\mathcal{F}_m$ is called a *principal ultrafilter*.

EXAMPLE 11.4. Let $\mathcal{F}_\infty = \{A \subseteq \mathbb{N} : [m, \infty) \subseteq A \text{ for any } m \in \mathbb{N}\}$, where we are using $[m, \infty)$ to denote the set of integers $\{m, m+1, m+2, \ldots\}$. Then $\mathcal{F}_\infty$ is a filter, and must be contained in an ultrafilter. An ultrafilter that contains $\mathcal{F}_\infty$ is called a *non-principal ultrafilter*.

LEMMA 11.5. *A filter $\mathcal{F}$ is an ultrafilter if and only if for any $A \in 2^{\mathbb{N}}$ either $A \in \mathcal{F}$ or $A^c \in \mathcal{F}$.*

PROOF. Assume that $\mathcal{F}$ is a filter with the property that for any $A \in 2^{\mathbb{N}}$ either $A \in \mathcal{F}$ or $A^c \in \mathcal{F}$. Suppose $\mathcal{U}$ is an ultrafilter containing $\mathcal{F}$ which is strictly larger than $\mathcal{F}$. Then there exists some $A \in \mathcal{U}$ such that $A \notin \mathcal{F}$. But either A or A^c must be in $\mathcal{F}$, and so it follows that $A^c \in \mathcal{F} \subseteq \mathcal{U}$. Thus both A and A^c are in $\mathcal{U}$, and hence $A \cap A^c \in \mathcal{U}$. This is a contradiction, since an ultrafilter cannot contain the empty set. Consequently, $\mathcal{F}$ is an ultrafilter.

Now let $\mathcal{F}$ be an ultrafilter and suppose $A \subseteq \mathbb{N}$ is a set not in $\mathcal{F}$. Consider
$$\mathcal{G} = \{B \subseteq \mathbb{N} : \exists C \in \mathcal{F} \text{ with } B \supset A \cap C\}.$$
If for every $C \in \mathcal{F}$, the intersection $A \cap C$ is nonempty, then $\mathcal{G}$ is a filter that properly contains $\mathcal{F}$, contradicting the maximality of $\mathcal{F}$. Consequently, there exists a set $C \in \mathcal{F}$ for which $A \cap C = \emptyset$. It follows that $C \subseteq A^c$, and hence $A^c \in \mathcal{F}$. □

REMARK 11.6. The only principal ultrafilters are the ones of the form $\mathcal{F}_m$ for some positive integer m. If an ultrafilter is not a principal ultrafilter, then it must contain $\mathcal{F}_\infty$, and so an ultrafilter is either a principal ultrafilter or a non-principal ultrafilter.

DEFINITION 11.7. Suppose that $(\xi_n)_{n=1}^\infty$ is a sequence of real numbers and suppose that $\mathcal{F}$ is a filter on $\mathbb{N}$. The sequence $(\xi_n)_{n=1}^\infty$ is said to *converge through the filter $\mathcal{F}$ to the number L* if for each $\epsilon > 0$ we have
$$\{n \in \mathbb{N} : |\xi_n - L| < \epsilon\} \in \mathcal{F}.$$

In this case, we write
$$L = \lim_{\mathcal{F}} \xi_n$$
and say that L is the *limit of the sequence through the filter*.

EXAMPLE 11.8. If $\mathcal{F}_m$ is a principal ultrafilter (for some positive integer m), then the sequence $(\xi_n)_{n=1}^\infty$ of real numbers converges to a limit L through $\mathcal{F}_m$ if and only if $\xi_m = L$. In particular, every sequence of real numbers converges to a limit through a principal ultrafilter.

EXAMPLE 11.9. Let $\mathcal{F}_\infty = \{A \subseteq \mathbb{N} : [m, \infty) \subseteq A \text{ for any } m \in \mathbb{N}\}$. A sequence $(\xi_n)_{n=1}^\infty$ of real numbers converges to a limit L through $\mathcal{F}_\infty$ provided that for each $\epsilon > 0$, the set $\{n \in \mathbb{N} : |\xi_n - L| < \epsilon\}$ contains a set of the form $\{m, m+1, m+2, \ldots\}$. That is, for each $\epsilon > 0$, there is a positive integer m such that $|\xi_n - L| < \epsilon$ for all $n \geq m$. This is the usual definition of a convergent sequence of real numbers. Consequently,
$$\lim_{\mathcal{F}_\infty} \xi_n = L \quad \text{if and only if} \quad \lim_{n \to \infty} \xi_n = L.$$

LEMMA 11.10. *If $\xi = (\xi_n)_{n=1}^\infty \in \ell_\infty$ and $\mathcal{U}$ is an ultrafilter, then there exists a unique number $L = \lim_{\mathcal{U}} \xi_n$ such that*
$$\{n : |\xi_n - L| < \epsilon\} \in \mathcal{U}, \quad \epsilon > 0.$$
That is, every bounded sequence has a limit through an ultrafilter.

PROOF. Suppose that the limit does not exist. We will derive a contradiction. Let $M = \sup_{n \in \mathbb{N}} \|\xi_n\|$, which is assumed to exist. Since the sequence $(\xi_n)_{n=1}^\infty$ does not have a limit through $\mathcal{U}$, for each $x \in [-M, M]$, there is a set of the form
$$U_x = \{n : |\xi_n - x| < \epsilon_x\} \notin \mathcal{U}$$
for some $\epsilon_x > 0$ (which in general depends on x). The collection of intervals of the form $(x - \epsilon_x, x + \epsilon_x)$ for $x \in [-M, M]$ forms an open cover of the compact interval $[-M, M]$. Consequently, there is a finite subcover of $[-M, M]$, say
$$(x_1 - \epsilon_{x_1}, x_1 + \epsilon_{x_1}), \ldots, (x_k - \epsilon_{x_k}, x_k + \epsilon_{x_k}).$$
Consequently, each member of the sequence $(\xi_n)_{n=1}^\infty$ can be found in one of the sets in the finite subcover given above. That is, if ξ_n is a number in the sequence, then there is some j (which depends on n) such that $\xi_n \in (x_j - \epsilon_{x_j}, x_j + \epsilon_{x_j})$, where j comes from the finite set $\{1, \ldots, k\}$. That means that each positive integer n belongs to U_{x_j} for some j, and so
$$\mathbb{N} \subseteq \bigcup_{j=1}^k U_{x_j}.$$

For each $j \in \{1, \ldots, k\}$, the set U_{x_j} is not in $\mathcal{U}$. Therefore, by Lemma 11.5, the complement $U_{x_j}^c$ is in $\mathcal{U}$. By Definition 11.1(iii), the intersection of finitely many members of a filter is again in the filter, and so
$$U_{x_1}^c \cap \cdots \cap U_{x_k}^c \in \mathcal{U}.$$
However, the intersection $U_{x_1}^c \cap \cdots \cap U_{x_k}^c$ is empty (by De Morgan's law), because we have that $U_{x_1} \cup \cdots \cup U_{x_k} = \mathbb{N}$. Consequently, we have shown that $\emptyset \in \mathcal{U}$, which is a contradiction. (See Definition 11.1(i).) Therefore, the sequence $(\xi_n)_{n=1}^\infty$ has a limit through $\mathcal{U}$, as required. □

DEFINITION 11.11. Suppose X is a Banach space and $\mathcal{U}$ is an ultrafilter. Let $\ell_\infty(X)$ be the Banach space of bounded sequences in X, and let $c_{0,\mathcal{U}}(X)$ be the subspace of $\ell_\infty(X)$ consisting of bounded sequences $(x_n)_{n=1}^\infty$ in X such that $\lim_\mathcal{U} \|x_n\| = 0$. It can be shown that $c_{0,\mathcal{U}}(X)$ is a closed subspace of $\ell_\infty(X)$. The quotient
$$X_\mathcal{U} = \ell_\infty(X)/c_{0,\mathcal{U}}(X)$$
is called the *ultraproduct* of X (with respect to $\mathcal{U}$).

REMARK 11.12. If X is a Banach space and $\mathcal{U}$ is an ultrafilter on $\mathbb{N}$, then $X_\mathcal{U}$ is a Banach space, and it can be shown that the norm satisfies
$$\|(x_1, x_2, \ldots)\|_{X_\mathcal{U}} = \lim_\mathcal{U} \|x_n\|,$$
where $(x_1, x_2, \ldots)$ is a bounded sequence with terms in X.

PROPOSITION 11.13. *Suppose that X and Y are Banach spaces and suppose that Y is separable. If Y is crudely finitely representable in X, then Y is linearly isomorphic to a subspace of $X_\mathcal{U}$.*

PROOF. Suppose that $(E_n)_{n=1}^\infty$ is a sequence of finite dimensional subspaces of Y such that $E_1 \subset E_2 \subset E_3 \subset \cdots$ and such that
$$Y = \overline{\bigcup_{n=1}^\infty E_n}.$$
Because Y is crudely finitely representable in X, it is λ-finitely representable in X for some $\lambda > 1$. (See Definition 8.23 for the relevant definitions.) Thus, for each $k \in \mathbb{N}$ there exists a linear function $T_k : E_k \to X$ such that
$$\|y\| \leq \|T_k(y)\| \leq \lambda\|y\|, \quad y \in E_k.$$
For each $k \in \mathbb{N}$, extend T_k to $\bigcup_{n=1}^\infty E_n$ by setting $T_k(y) = 0$ for all $y \notin E_k$.

Define $T : Y \to X_\mathcal{U}$ by $T(y) = (T_k(y))_{k=1}^\infty$. The map T is linear and
$$\|y\| \leq \lim_\mathcal{U} \|T_k(y)\| \leq \lambda\|y\|, \quad y \in Y.$$
This completes the proof. □

THEOREM 11.14. *If X is a Banach space and $\mathcal{U}$ is an ultrafilter, then $X_\mathcal{U}$ is finitely representable in X.*

PROOF. Suppose E is a finite dimensional subspace of $X_\mathcal{U}$ and let $\epsilon > 0$.[1] Let δ be a positive number, the actual value of which we will choose later. We may choose a finite collection $\{e_1, \ldots, e_N\}$ of points in ∂B_E so that each $x \in \partial B_E$ is within δ from at least one of the elements in the set $\{e_1, \ldots, e_N\}$. That is,
$$\inf\{\|e_j - x\|_{X_\mathcal{U}} : 1 \leq j \leq N\} < \delta$$
for each $x \in \partial B_E$. (See Proposition 8.26.) For each of the indices $j \in \{1, \ldots, N\}$, let $e_j = (e_{jk})_{k=1}^\infty$. Then, since
$$\|e_j\|_{X_\mathcal{U}} = \lim_\mathcal{U} \|e_{jk}\| = 1$$
for each j, there is a set $A \in \mathcal{U}$ such that
$$1 - \delta < \|e_{jk}\| < 1 + \delta$$

---

[1] Kalton originally gave a proof more like that given by Heinrich in [46] (Proposition 6.1), but we have adopted a proof based on that given by Albiac and Kalton in [5] (Proposition 12.1.12(i)).

for all $j \in \{1, \ldots, N\}$ and all $k \in A$. Now let $k_0 \in A$ and define a map $T : E \to X$ by the formula $T(x) = x_{k_0}$ for all $x = (x_k)_{k=1}^\infty$ in E, which is linear. Then
$$1 - \delta < \|T(e_j)\| < 1 + \delta$$
for all $j \in \{1, \ldots, N\}$. Let $F = T(E)$. We claim that T is a linear isomorphism of E onto F with $\|T\|\|T^{-1}\| < 1 + \epsilon$.

Now let $x \in E$ with $\|x\|_{\mathcal{U}} = 1$. There is some $j \in \{1, \ldots, N\}$ for which $\|x - e_j\|_{\mathcal{U}} < \delta$. It follows that
$$\|T(x)\| \leq \|T(x) - T(e_j)\| + \|T(e_j)\| < \|T\|\delta + (1 + \delta),$$
and so
$$\|T\| < (1 + \delta)/(1 - \delta).$$
We also have
$$1 - \delta < \|T(e_j)\| \leq \|T(e_j) - T(x)\| + \|T(x)\| < \|T\|\delta + \|T(x)\|,$$
from which we conclude that
$$\|T(x)\| > (1 - \delta) - \|T\|\delta > (1 - \delta) - \frac{(1+\delta)\delta}{1 - \delta} = \frac{1 - 3\delta}{1 - \delta}.$$
Therefore,
$$\frac{1 - 3\delta}{1 - \delta} < \|T\| < \frac{1 + \delta}{1 - \delta},$$
and so
$$\|T\|\|T^{-1}\| \leq \frac{1 + \delta}{1 - \delta} \cdot \frac{1 - \delta}{1 - 3\delta} = \frac{1 + \delta}{1 - 3\delta} = 1 + \frac{4\delta}{1 - 3\delta}.$$
Consequently, if we choose δ so small that $\frac{4\delta}{1 - 3\delta} < \epsilon$, then $\|T\|\|T^{-1}\| < 1 + \epsilon$, which is the required result. $\square$

PROPOSITION 11.15. *Suppose X is a Banach space and $\mathcal{U}$ is an ultrafilter. View X as a subspace of $X_{\mathcal{U}}$ via the embedding $x \mapsto (x, x, x, \ldots)$.*

(i) *X is locally complemented in $X_{\mathcal{U}}$.*
(ii) *X is complemented in $X_{\mathcal{U}}$ if X is reflexive.*

PROOF. (i) Consider the short exact sequence
$$0 \longrightarrow X \xrightarrow{j} X_{\mathcal{U}} \longrightarrow X_{\mathcal{U}}/X \longrightarrow 0$$
where $j : X \to X_{\mathcal{U}}$ is the map $j(x) = (x, x, x, \ldots)$. In order to show that X is locally complemented in $X_{\mathcal{U}}$, we want to show that the dual sequence splits. Thus, we wish to find a linear lifting L from X^* to $X_{\mathcal{U}}^*$ so that the following diagram commutes.
$$0 \longrightarrow X^\perp \longrightarrow X_{\mathcal{U}}^* \xrightarrow{j^*} X^* \longrightarrow 0$$
$$L$$
That is, we want to find $L : X^* \to X_{\mathcal{U}}^*$ such that $j^* \circ L = \operatorname{Id}_{X^*}$.

For $x^* \in X^*$, define a linear functional $\widehat{x}^*$ on $(X_{\mathcal{U}})^*$ by
$$\widehat{x}^*\big((x_n)_{n=1}^\infty\big) = \lim_{\mathcal{U}} x^*(x_n).$$
Observe that
$$\|\widehat{x}^*\big((x_n)_{n=1}^\infty\big)\| \leq \|x^*\| \lim_{\mathcal{U}} \|x_n\| = \|x^*\| \|(x_n)_{n=1}^\infty\|_{X_{\mathcal{U}}}.$$

Now define a map $L : X^* \to X_{\mathcal{U}}^*$ by the rule $L(x^*) = \widehat{x}^*$ for each $x^* \in X^*$. The map L is a linear function with the property that $\|L(x^*)\| \le \|x^*\|$ for all $x^* \in X^*$, and so $\|L\| \le 1$. Finally, observe that for any $x \in X$, we have

$$(j^* \circ L)(x^*)(x) = L(x^*)(x,x,x,\ldots) = \lim_{\mathcal{U}} x^*\big((x,x,x,\ldots)\big) = x^*(x).$$

Therefore, we have that $(j^* \circ L)(x^*) = x^*$ for all $x^* \in X^*$, and so we conclude that the Banach space X is locally complemented in its ultraproduct $X_{\mathcal{U}}$.

(ii) We now assume that X is a reflexive Banach space. For each $(x_n)_{n=1}^\infty$ in $X_{\mathcal{U}}$, define a map $P\big((x_n)_{n=1}^\infty\big) : X^* \to \mathbb{R}$ by the rule

$$P\big((x_n)_{n=1}^\infty\big)(x^*) = \lim_{\mathcal{U}} x^*(x_n), \quad x^* \in X^*.$$

Then the map $P\big((x_n)_{n=1}^\infty\big)$ is a linear function and for each x^* in the dual space X^*, we have that

$$\left| P\big((x_n)_{n=1}^\infty\big)(x^*) \right| = \left| \lim_{\mathcal{U}} x^*(x_n) \right| \le \|x^*\| \lim_{\mathcal{U}} \|x_n\| = \|x^*\| \|(x_n)_{n=1}^\infty\|_{X_{\mathcal{U}}}.$$

Consequently, for each $(x_n)_{n=1}^\infty$ in $X_{\mathcal{U}}$,

$$\left\| P\big((x_n)_{n=1}^\infty\big) \right\| \le \|(x_n)_{n=1}^\infty\|_{X_{\mathcal{U}}},$$

and so we have that $P\big((x_n)_{n=1}^\infty\big)$ is an element of X^{**}. However, we assumed that X is reflexive, and so it follows that $P\big((x_n)_{n=1}^\infty\big)$ is an element of X.

We now define a map $P : X_{\mathcal{U}} \to X$ so that for each $(x_n)_{n=1}^\infty$ in $X_{\mathcal{U}}$, the image under P is the element $P\big((x_n)_{n=1}^\infty\big)$ we defined above. By what we showed in the previous paragraph, we know that P is a well defined bounded linear map into X. Furthermore, if $x \in X$, then for each $x^* \in X^*$, we have that

$$P(j(x))(x^*) = P\big((x,x,\ldots)\big)(x^*) = \lim_{\mathcal{U}} x^*(x) = x^*(x).$$

Therefore, we have shown that $P(j(x)) = x$ for each $x \in X$, and so P is a projection of $X_{\mathcal{U}}$ onto X in $X_{\mathcal{U}}$. (We are identifying X with $j(X)$.) $\square$

COMMENT 11.16. In the discussion below, we will use $\wedge$ and $\vee$ to denote infimum and supremum, respectively. That is,

$$x \wedge y = \inf\{x,y\} \quad \text{and} \quad x \vee y = \sup\{x,y\}.$$

We also recall that a vector space X with a partial ordering $\le$ is a *vector lattice* (also called a *Riesz space*) if for any pair of points x and y in X, (1) both $x \wedge y$ and $x \vee y$ exist (with respect to the partial ordering $\le$), (2) $x \le y$ implies that $x + z \le y + z$ for each $z \in X$, and (3) $x \le y$ and $a \ge 0$ imply $ax \le ay$.

DEFINITION 11.17. A Banach space X is a *Banach lattice* provided that X has a partial ordering $\le$ such that X is a vector lattice with respect to $\le$, and such that $|x| \le |y|$ implies $\|x\| \le \|y\|$ for any x and y in X, where $|x| = x \vee (-x)$.

We remark that if X is a Banach lattice, then $X_{\mathcal{U}}$ is a Banach lattice, where $(x_n)_{n=1}^\infty \le (y_n)_{n=1}^\infty$ if $\{n : x_n \le y_n\} \in \mathcal{U}$. It follows that the operations $\vee$ and $\wedge$ for $X_{\mathcal{U}}$ satisfy $(x_n)_{n=1}^\infty \vee (y_n)_{n=1}^\infty = (x_n \vee y_n)_{n=1}^\infty$ and $(x_n)_{n=1}^\infty \wedge (y_n)_{n=1}^\infty = (x_n \wedge y_n)_{n=1}^\infty$.

DEFINITION 11.18. A Banach space X is an AL_p-space (for $1 \leq p < \infty$) if X is a Banach lattice such that
$$\|x+y\|^p = \|x\|^p + \|y\|^p$$
whenever $x \geq 0$ and $y \geq 0$ and $x \wedge y = 0$.

EXAMPLE 11.19. If ξ and η are sequences in ℓ_p for $p \in [1, \infty)$ such that $\xi \geq 0$ and $\eta \geq 0$ and $\xi \wedge \eta = 0$, then
$$\|\xi + \eta\|^p = \|\xi\|^p + \|\eta\|^p.$$
Consequently, the classical sequence space ℓ_p is an AL_p-space whenever $1 \leq p < \infty$.

Theorem 11.20, given here without proof, characterizes the separable AL_p-spaces.[2]

THEOREM 11.20 (Kakutani–Bohnenblust [72]). *A separable AL_p-space is linearly isomorphic to one of ℓ_p, $L_p(0,1)$, $L_p(0,1) \oplus \ell_p$, ℓ_p^n, or $L_p(0,1) \oplus \ell_p^n$ where $n \in \mathbb{N}$ is such that $n > 1$, each of which is complemented in $L_p(0,1)$.*

One can use Theorem 11.20 to show that the class of L_p-spaces is stable under ultraproducts. That is, if $p \geq 1$ and $\mathcal{U}$ is a non-principal ultrafilter, then $(L_p(\mu))_{\mathcal{U}}$ is isometric to $L_p(\nu)$, where μ and ν are measures on a measure space.[3] The following related theorem (given here without proof) is due to Henson [47].

THEOREM 11.21 (Henson). *If $p \geq 1$ and $\mathcal{U}$ is a non-principal ultrafilter on $\mathbb{N}$, then $(L_p([0,1]))_{\mathcal{U}}$ is isometric to the ℓ_p-sum of $\mathfrak{c}$ copies of $L_p([0,1]^{\mathfrak{c}})$, where $\mathfrak{c} = 2^{\aleph_0}$ is the cardinality of the continuum.*

We now wish to address the following question.

QUESTION 11.22. *When is $X_{\mathcal{U}}$ reflexive?*

DEFINITION 11.23. A Banach space X is called *uniformly convex* if given $\epsilon > 0$, there exists a $\delta > 0$ such that if $\|x\| = \|y\| = 1$ and $\|x - y\| \geq \epsilon$, then
$$\left\| \frac{x+y}{2} \right\| \leq 1 - \delta.$$

The idea behind the definition is that if x and y are "far enough" apart on the surface of the unit ball, then the midpoint must be inside of the unit ball.

EXAMPLE 11.24. If $1 < p < \infty$, then the spaces ℓ_p and $L_p([0,1])$ are uniformly convex. (These examples were given by Clarkson in 1936 [24].)

COMMENT 11.25. The notion of a uniformly convex Banach space was introduced by Clarkson in a paper titled *Uniformly convex spaces* in 1936 [24]. It was in this paper that he showed that ℓ_p and $L_p([0,1])$ are uniformly convex for $1 < p < \infty$. This fact follows immediately from what are now known as *Clarkson's inequalities*, which appeared in the same 1936 paper.

---

[2]This theorem is often attributed to Kakutani and Bohnenblust, but there is more to be said about its history. The proof and a discussion of its origins can be found in Chapter 5 of [72].

[3]This result is usually attributed to Dacunha-Castelle and Krivine. See [46] for the proof.

THEOREM 11.26 (Clarkson's inequalities). *Let x and y be in ℓ_p for $1 < p < \infty$. If $p \geq 2$, then*
$$\left\| \frac{x+y}{2} \right\|_p^p + \left\| \frac{x-y}{2} \right\|_p^p \leq \frac{1}{2}\left(\|x\|_p^p + \|y\|_p^p \right)$$
and if $1 < p < 2$, then
$$\left\| \frac{x+y}{2} \right\|_p^q + \left\| \frac{x-y}{2} \right\|_p^q \leq \left(\frac{1}{2}\|x\|_p^p + \frac{1}{2}\|y\|_p^p \right)^{q/p},$$
where q is the exponent conjugate to p.

We have stated Clarkson's inequalities for sequence spaces, but they are also true for function spaces. It is a straightforward matter to use these inequalities to show that ℓ_p is uniformly convex for $1 < p < \infty$ (and so also $L_p([0,1])$). If we assume that $\|x\|_p = 1$ and $\|y\|_p = 1$, then the right side of both inequalities is equal to 1, and so when $\|x - y\| \geq \epsilon$, we have
$$\left\| \frac{x+y}{2} \right\|_p^p \leq 1 - \left\| \frac{x-y}{2} \right\|_p^p \leq 1 - \left(\frac{\epsilon}{2} \right)^p \quad \text{if } p \geq 2$$
and
$$\left\| \frac{x+y}{2} \right\|_p^q \leq 1 - \left\| \frac{x-y}{2} \right\|_p^q \leq 1 - \left(\frac{\epsilon}{2} \right)^q \quad \text{if } 1 < p < 2,$$
where q is the exponent conjugate to p. For the first case choose $\delta = (\epsilon/2)^p$ and for the second case choose $\delta = (\epsilon/2)^q$.

THEOREM 11.27 (Milman–Pettis). *A uniformly convex Banach space is necessarily reflexive.*

Theorem 11.27 (which we will not prove here) was proved independently by Milman in 1938 [**84**] and Pettis in 1939 [**95**]. The converse is not true, which was demonstrated by M. Day in 1941 [**26**]. In that work, Day gave examples of reflexive Banach spaces that are not isomorphic to any uniformly convex space.

Theorem 11.27 tells us that a uniformly convex Banach space is reflexive, but more is true. It turns out that any ultraproduct of a uniformly convex Banach space is also reflexive. To see why, we introduce to a related notion, called *super-reflexivity*.

DEFINITION 11.28. A Banach space X is called *super-reflexive* if every Banach space Y that is finitely representable in X is reflexive.

From Theorem 11.14, we know that any ultraproduct of a Banach space X is finitely representable in X. Consequently, if the Banach space X is super-reflexive, then $X_\mathcal{U}$ is reflexive for any ultrafilter $\mathcal{U}$. There is a close relationship between super-reflexivity and uniform convexity. In fact, Enflo [**34**] and Pisier [**97**] showed that a super-reflexive Banach space necessarily has an equivalent norm for which it is uniformly convex, showing that we can always think of super-reflexive spaces as uniformly convex spaces.

THEOREM 11.29 (Enflo–Pisier). *A Banach space is super-reflexive if and only if it has an equivalent norm in which the Banach space is uniformly convex.*

We will not prove the Enflo–Pisier theorem theorem here, but we will use it to draw the following conclusion.

COROLLARY 11.30. *If X is a uniformly convex Banach space, then every Banach space finitely representable in X is reflexive. In particular, every ultraproduct of a uniformly convex Banach space is reflexive.*

PROOF. In light of Theorem 11.29, we have that X is super-reflexive. Consequently, if Y is a Banach space that is finitely representable in X, then it is reflexive. Furthermore, since any ultraproduct of X is finitely representable in X, by Theorem 11.14, it follows that any ultraproduct of X is reflexive. □

REMARK 11.31. We can in fact show that if X is a uniformly convex Banach space, then every Banach space that is *crudely* finitely representable in X is reflexive, but this requires some tools that we have not developed. As before, we may assume that X is super-reflexive. Suppose that Y is crudely finitely representable in X. While we may not be able to assume that Y is finitely representable, any separable subspace Z of Y may be renormed so that Z (with the new norm) is finitely representable in X. (This can be found as Proposition 12.1.13 in [5], for example.) That means that any separable subspace of Y is reflexive, by Corollary 11.30. It is a consequence of James's criterion of reflexivity that a Banach space is reflexive if and only if every separable subspace of it is reflexive (see Theorem 1.13.8 in [83]), and so we conclude that Y is reflexive, as claimed.

Any super-reflexive Banach space is necessarily reflexive, but from the 1941 paper of M. Day we mentioned above [26], we know that not every reflexive Banach space is super-reflexive.

Ultraproducts of Banach spaces provided a notable connection between the uniform and Lipschitz structure of Banach spaces. Before presenting it, we recall the definition of the modulus of continuity from Chapter 1, but we recast it in terms of Banach spaces.

DEFINITION 11.32. Let X and Y be Banach spaces and suppose $f : X \to Y$ is a function. The *modulus of continuity* of f is the function $\omega_f : [0, \infty] \to [0, \infty]$ defined by the formula

$$\omega_f(t) = \sup \left\{ \|f(x) - f(y)\|_Y \ : \ \|x - y\|_X \leq t \right\},$$

where $t \geq 0$.

The modulus of continuity can be used to express different types of continuity. A function $f : X \to Y$ is *Lipschitz continuous* with *Lipschitz constant* K provided that $\omega_f(t) \leq Kt$ for all $t \geq 0$. More generally, a function $f : X \to Y$ is *uniformly continuous* if

$$\lim_{t \to 0^+} \omega_f(t) = 0.$$

DEFINITION 11.33. Two Banach spaces X and Y are said to be *uniformly homeomorphic* if there exists a bijection $f : X \to Y$ such that f and f^{-1} are uniformly continuous. Such a bijection f is called a *uniform homeomorphism* of Banach spaces.

LEMMA 11.34 (Corson–Klee[25]). *If X and Y are Banach spaces, and if $f : X \to Y$ is a uniformly continuous function, then there exists a constant $K > 0$ such that $\omega_f(t) \leq Kt$ whenever $t \geq 1$.*

COMMENT 11.35. We have defined uniform continuity in terms of the modulus of continuity, but it also has an ϵ-δ formulation: A function $f : X \to Y$ between Banach spaces is said to be uniformly continuous if for every $\epsilon > 0$ there exists a $\delta > 0$ such that
$$\|f(x) - f(y)\|_Y < \epsilon \quad \text{whenever} \quad \|x - y\|_X < \delta.$$
If a function $f : X \to Y$ satisfies the conclusion of Lemma 11.34, it is said to be a coarse Lipschitz function. That is, the function $f : X \to Y$ is *coarse Lipschitz* if there is a constant $K > 0$ so that $\omega_f(t) \leq Kt$ whenever $t \geq 1$.

A coarse Lipschitz function is also said to be *Lipschitz on large distances* (or *Lipschitz for large distances*) because a coarse Lipschitz function f satisfies the following condition: For every $\delta > 0$, there exists a $K_\delta > 0$ (that in general depends on δ) such that
$$\|f(x) - f(y)\|_Y \leq K_\delta \|x - y\|_X \quad \text{whenever} \quad \|x - y\| \geq \delta.$$
Consequently, Lemma 11.34 can be rephrased as

If $f : X \to Y$ is a uniformly continuous map between Banach spaces, then f is Lipschitz on large distances.

Indeed, this is how the result appears in the work of Corson and Klee [**25**], and in the classic paper of Heinrich and Mankiewicz [**45**].

Yet another way to formulate the coarse Lipschitz condition is the following: A function f is coarse Lipschitz if
$$\limsup_{t \to \infty} \frac{\omega_f(t)}{t} < \infty.$$
When $\omega_f(t) < \infty$ for each $t > 0$, we say that f is *coarsely continuous*. This terminology can cause some confusion, because a function need not be continuous in order to be coarsely continuous. A significant fact is that a function f between Banach spaces is coarsely continuous if and only if it is a coarse Lipschitz function. These two concepts can be defined in the more general context of metric spaces, and in general, these two concepts need not coincide, but they are the same when the spaces involved are metrically convex, and Banach spaces are metrically convex. (See [**62**] for more details.)

In view of all of the above, we can reformulate the Corson–Klee lemma (Lemma 11.34) as follows:

LEMMA 11.36 (Corson–Klee, Version 2). *If a function between Banach spaces is uniformly continuous, then it is coarsely continuous (or coarse Lipschitz).*

Usually (in this text), we will use the formulation in Lemma 11.36, and (since we are working with metrically convex spaces) we will use the terminology "coarse Lipschitz" to avoid the potential confusion that can arise with "coarsely continuous" (since a coarsely continuous function need not be continuous).

The defining condition for $f : X \to Y$ to be a coarse Lipschitz function between Banach spaces is that there is a constant $K > 0$ so that

(11.1) $\qquad \omega_f(t) \leq Kt \quad \text{whenever} \quad t \geq 1.$

This is equivalent to the condition that there are constants $K > 0$ and $C > 0$ such that $\omega_f(t) \leq Kt + C$ for all $t > 0$.

> In the context of metrically convex spaces (such as Banach spaces), the lower bound of 1 for t in (11.1) is in some sense arbitrary. We could equivalently say that f is a coarse Lipschitz map between Banach spaces X and Y if there is a constant K_{t_0} that depends on t_0 so that $\omega_f(t) \leq K_{t_0} t$ whenever $t \geq t_0$.

THEOREM 11.37. *Let X and Y be Banach spaces and suppose that $\mathcal{U}$ is a nonprincipal ultrafilter on $\mathbb{N}$. If X and Y are uniformly homeomorphic, then $X_\mathcal{U}$ and $Y_\mathcal{U}$ are Lipschitz isomorphic.*

PROOF. Let $f : X \to Y$ be a bijection such that f and f^{-1} are uniformly continuous, which exists by assumption. Define a function $\tilde{f} : X_\mathcal{U} \to Y_\mathcal{U}$ by

$$\tilde{f}\big((x_n)_{n=1}^\infty\big) = \left(\frac{1}{n} f(n x_n)\right)_{n=1}^\infty$$

where $(x_n)_{n=1}^\infty$ is a bounded sequence in with terms in X.

We need to show that $\tilde{f}$ is well-defined. Suppose that $(x_n)_{n=1}^\infty$ and $(y_n)_{n=1}^\infty$ are bounded sequences in X for which

$$\lim_\mathcal{U} \|x_n - y_n\| = 0.$$

For each $n \in \mathbb{N}$, we have

$$\left\| \frac{1}{n} f(nx_n) - \frac{1}{n} f(ny_n) \right\|_Y = \frac{1}{n} \big\| f(nx_n) - f(ny_n) \big\|_Y$$

$$\leq \frac{1}{n} \omega_f\big(\|nx_n - ny_n\|_X \big).$$

By assumption, the function f is uniformly continuous, and so by Lemma 11.34 there is a constant K such that $\omega_f(t) \leq Kt$ whenever $t \geq 1$. We can therefore find an additional constant C such that $\omega_f(t) \leq Kt + C$ for all t. Thus,

$$\frac{1}{n}\omega_f\big(\|nx_n - ny_n\|_X\big) \leq \frac{1}{n}\big(Kn\|x_n - y_n\|_X + C\big)$$

$$\leq K\|x_n - y_n\|_X + \frac{C}{n}.$$

Therefore,

$$\left\| \frac{1}{n} f(nx_n) - \frac{1}{n} f(ny_n) \right\|_Y \leq K\|x_n - y_n\|_X + \frac{C}{n}$$

for each $n \in \mathbb{N}$. Consequently, computing the limit through the ultrafilter $\mathcal{U}$, we have

(11.2) $$\lim_\mathcal{U} \left\| \frac{1}{n} f(nx_n) - \frac{1}{n} f(ny_n) \right\|_Y \leq \lim_\mathcal{U} \left(K\|x_n - y_n\|_X + \frac{C}{n} \right).$$

The term on the right side of (11.2) goes to zero, by assumption (and because $\mathcal{U}$ is a non-principal ultrafilter). Therefore, $\tilde{f}((x_n)_{n=1}^\infty) = \tilde{f}((y_n)_{n=1}^\infty)$, and so $\tilde{f}$ is well-defined. This also tells us that $\tilde{f}$ is one-to-one.

Notice also that (11.2) gives us the inequality

(11.3) $$\left\| \tilde{f}\big((x_n)_{n=1}^\infty\big) - \tilde{f}\big((y_n)_{n=1}^\infty\big) \right\|_{Y_\mathcal{U}} \leq K \left\| (x_n)_{n=1}^\infty - (y_n)_{n=1}^\infty \right\|_{X_\mathcal{U}},$$

which means that $\tilde{f}$ is a Lipschitz function. Since the inverse of $\tilde{f}$ is

$$\tilde{f}^{-1}\left((x_n)_{n=1}^\infty\right) = \left(\frac{1}{n}f^{-1}(ny_n)\right)_{n=1}^\infty,$$

or $\tilde{f}^{-1} = \widetilde{f^{-1}}$, we have shown that $X_{\mathcal{U}}$ and $Y_{\mathcal{U}}$ are Lipschitz isomorphic. $\square$

Theorem 11.37 tells us that the uniform structure of a Banach space determines the Lipschitz structure of the ultraproduct. To what extent does this determine the linear structure of the Banach space? We may try to apply the Heinrich–Mankiewicz theorem (Theorem 9.7), but we cannot apply the Heinrich–Mankiewicz theorem directly, because $X_{\mathcal{U}}$ and $Y_{\mathcal{U}}$ are not separable. However, we can use Theorem 9.7 to get the following local result.

COROLLARY 11.38. *If X and Y are uniformly homeomorphic Banach spaces, then X is crudely finitely representable in Y and Y is crudely finitely representable in X.*

PROOF. Suppose that X and Y are uniformly homeomorphic Banach spaces and let $\mathcal{U}$ be a non-principal ultrafilter on $\mathbb{N}$. By Theorem 11.37, the ultraproducts $X_{\mathcal{U}}$ and $Y_{\mathcal{U}}$ are Lipschitz isomorphic. Since X can be embedded into $X_{\mathcal{U}}$ via the map $x \to (x, x, x, \ldots)$, there exists a Lipschitz embedding $f : X \to Y_{\mathcal{U}}$.

We will show that X is crudely finitely representable in $(Y_{\mathcal{U}})^{**}$. Suppose that E is a finite dimensional subspace of X. Then $f|_E : E \to Y_{\mathcal{U}}$ is a Lipschitz embedding such that $\operatorname{Lip}(f|_E) \leq \operatorname{Lip}(f)$ and $\operatorname{Lip}(f|_E^{-1}) \leq \operatorname{Lip}(f^{-1})$. Since E is finite dimensional, it is certainly separable, and so we can apply Theorem 9.7 to find a linear embedding $T : E \to (Y_{\mathcal{U}})^{**}$. Let $F = T(E)$. Then F is a finite dimensional subspace of $(Y_{\mathcal{U}})^{**}$ and $T : E \to F$ is a linear isomorphism. Furthermore (see Comment 9.10), we also have that

$$\|T\|\|T^{-1}\| \leq \operatorname{Lip}(f|_E)\operatorname{Lip}(f|_E^{-1}) \leq \operatorname{Lip}(f)\operatorname{Lip}(f^{-1}).$$

The choice of E was arbitrary, and so it follows that X is crudely finitely representable in $(Y_{\mathcal{U}})^{**}$ with constant $\operatorname{Lip}(f)\operatorname{Lip}(f^{-1})$.

Finally, we note that $(Y_{\mathcal{U}})^{**}$ is finitely representable in $Y_{\mathcal{U}}$ (by the Principle of Local Reflexivity) and $Y_{\mathcal{U}}$ is finitely representable in Y (by Theorem 11.14). We conclude, therefore, that X is crudely finitely representable in Y, as required. The argument to show that Y is crudely finitely representable in X is similar. $\square$

REMARK 11.39. Let X and Y be two Banach spaces. We say that X and Y are uniformly homeomorphic if there exists a bijection $f : X \to Y$ such that f and f^{-1} are uniformly continuous. Furthermore, we say that the Banach spaces X and Y are *coarsely homeomorphic* if there exists a (not necessarily continuous) bijective function $f : X \to Y$ such that f and f^{-1} are coarsely continuous. It was for a long time an open question if coarsely homeomorphic Banach spaces are necessarily uniformly homeomorphic, until Kalton showed that they are not in [**64**].[4]

---

[4] At the time this lecture was given, Kalton said "We know of no examples that separate uniform homeomorphism from coarse homeomorphism." That is, at that time, there were no known examples of Banach spaces that were coarsely homeomorphic but not also uniformly homeomorphic, despite the fact that the topic had been studied for over 40 years. However, it was Kalton himself who found such an example. In [**64**] (published posthumously), Kalton showed the existence of two coarsely homeomorphic separable Banach spaces that are not uniformly homeomorphic (in Theorem 8.9).

CHAPTER 12

The uniform structure of Banach spaces

There exist many examples of separable Banach spaces which are uniformly homeomorphic, but not linearly isomorphic. (Such an example was first provided by Ribe in [**101**], and then Ribe's methods were modified by Aharoni and Lindenstrauss in [**4**] to create more examples.)

EXAMPLE 12.1. Suppose that X is a separable Banach space with norm $\|\cdot\|$. Define a metric d_θ on X by the rule

$$d_\theta(x,y) = \begin{cases} \|x-y\|^\theta & \text{if } \|x-y\| < 1 \\ \|x-y\| & \text{if } \|x-y\| \geq 1 \end{cases}$$

where θ is a fixed real number such that $0 < \theta < 1$. It can be readily verified that (X, d_θ) is uniformly homeomorphic to $(X, \|\cdot\|)$. If $Æ(X, d_\theta)$ is the Arens–Eells space for the metric space (X, d_θ), then the barycenter map $\beta : Æ(X, d_\theta) \to (X, d_\theta)$ is a linear quotient map such that $\beta \circ \delta = \mathrm{Id}_X$. (See Chapter 5, and especially Lemma 5.9, for the details.) If we define a map $\phi : (X, d_\theta) \oplus \ker(\beta) \to Æ(X, d_\theta)$ by

$$\phi(x, \mu) = \delta(x) + \mu$$

for each $x \in X$ and $\mu \in Æ(X, d_\theta)$, then ϕ is a Lipschitz isomorphism with

$$\phi^{-1}(\mu) = \big(\beta(\mu), \mu - (\delta \circ \beta)(\mu)\big)$$

for each $\mu \in Æ(X, d_\theta)$. It follows that $Æ(X, d_\theta)$ is uniformly homeomorphic to $(X, d_\theta) \oplus \ker(\beta)$, and so uniformly homeomorphic to $(X, \|\cdot\|) \oplus \ker(\beta)$.

It can be shown that $Æ(X, d_\theta)$ is a Schur space[1]. If we choose $X = c_0$, which is not a Schur space, then we have found a separable Schur space that is uniformly homeomorphic to a Banach space containing c_0. Consequently, the two spaces cannot be linearly isomorphic, even though they are uniformly homeomorphic.

COMMENT 12.2. We remind the reader that a Banach space X is called a *Schur space*, or alternately is said to have the *Schur property*, if the weak and norm topologies have the same convergent sequences. That is, the Banach space X is a Schur space if

$$\lim_{n \to \infty} \|x_n - x\| = 0$$

whenever the sequence $(x_n)_{n=1}^\infty$ converges to x in the weak topology on X.

The space c_0 of sequences converging to zero does not have the Schur property, and consequently cannot be (linearly isomorphic to) a subspace of a Banach space that does have the Schur property.

---

[1] See Theorem 4.6 in [**59**].

We will now show another method for constructing uniformly homeomorphic Banach spaces that are not linearly isomorphic. (These ideas can be found in [64] and [65], where they are more fully investigated.[2]) For the purpose of this discussion, if $H : X \to Y$ is a map between Banach spaces, we will let

$$\|H\| = \sup\{\|H(x)\|_Y : \|x\|_X \leq 1\}.$$

It can be shown that this defines a complete norm on the space $\mathcal{H}(X,Y)$ of all positively homogeneous bounded maps from X into Y. When we say that H is bounded in this context, we mean that $\|H\| < \infty$.

If H is a bounded linear map, then $\|H\|$ is the usual operator norm of H, but we do not require H to be linear for this definition.

DEFINITION 12.3. Let X and Y be real Banach spaces. A function $H : X \to Y$ is called *positively homogeneous* if $H(\alpha x) = \alpha H(x)$ whenever $\alpha \geq 0$. We say that the function H is *norm-preserving* if $\|H(x)\|_Y = \|x\|_X$ for all $x \in X$. (Note that if H is norm-preserving, then $\|H\| = 1$.) Finally, we say that H is of *CL-type* (L, ϵ) if

$$\omega_H(\delta) \leq L\delta + \epsilon,$$

where ω_H is the modulus of continuity for H, $L > 0$, and $\epsilon > 0$.

COMMENT 12.4. We recall that the *modulus of continuity* of $H : X \to Y$ is the function $\omega_H : [0, \infty] \to [0, \infty]$ defined by the formula

$$\omega_H(t) = \sup\{\|H(x) - H(y)\|_Y : \|x - y\|_X \leq t\},$$

where $t \geq 0$. We observe (as it will be needed in the proof of the next lemma) that

$$\|H(x) - H(y)\|_Y \leq \omega_H(\|x - y\|_X)$$

for each x and y in the Banach space X.

We also recall that a function $f : X \to Y$ is *coarse Lipschitz* if there is a constant $K > 0$ so that $\omega_f(t) \leq Kt$ whenever $t \geq 1$. This is equivalent to the condition that there are constants $K > 0$ and $C > 0$ such that $\omega_f(t) \leq Kt + C$ for all $t > 0$. This is where the "CL" comes from in "CL-type".

We can also define uniform continuity in terms of the modulus of continuity. The function $f : X \to Y$ is *uniformly continuous* on the set $E \subseteq X$ if $\lim_{t \to 0^+} \omega_{f|_E}(t) = 0$, where $f|_E$ represents the restriction of f to E, as usual.

In the following discussion, for simplicity of notation, we will denote the norms on both X and Y by $\|\cdot\|$, without any subscripts.

LEMMA 12.5. *If H is a positively homogeneous function of CL-type (L, ϵ), then*

$$\|H(x) - H(y)\| \leq 6 \max\left(L\|x - y\|, \epsilon\|x\|, \epsilon\|y\|\right).$$

In particular, if $L \geq 1$, then

$$\|H(x) - H(y)\| \leq 6L \max\left(\|x - y\|, \epsilon\|x\|, \epsilon\|y\|\right).$$

---

[2]Kalton's original lectures corresponding to this chapter were significantly different in parts. We have decided to follow more closely the treatment that appears in [64] and [65], both of which were written after the original lectures were given, and which we feel more accurately reflects how Kalton would prefer the topics to be presented.

PROOF. The second inequality is a trivial consequence of the first, so we will prove only the first. Without loss of generality, assume that $\|x\| \geq \|y\| > 0$. Using the positive homogeneity of the function H, we rewrite the difference $H(x) - H(y)$ in the following way:

$$H(x) - H(y) = \|x\| H\left(\frac{x}{\|x\|}\right) - \|y\| H\left(\frac{y}{\|y\|}\right)$$

$$= \|x\| \left[H\left(\frac{x}{\|x\|}\right) - H\left(\frac{y}{\|y\|}\right)\right] + (\|x\| - \|y\|) H\left(\frac{y}{\|y\|}\right).$$

Consequently, we have that

$$\|H(x) - H(y)\| \leq \|x\| \omega_H\left(\left\|\frac{x}{\|x\|} - \frac{y}{\|y\|}\right\|\right) + (\|x\| - \|y\|) \|H\|,$$

where we made use of the inequality from Comment 12.4 to get a bound in terms of the modulus of continuity. Note that we also used the fact that $y/\|y\|$ has norm equal to 1. Now we use the fact that H is of CL-type (L, ϵ) to achieve the bound

(12.1)
$$\|H(x) - H(y)\| \leq \|x\| \left(L\left\|\frac{x}{\|x\|} - \frac{y}{\|y\|}\right\| + \epsilon\right) + (\|x\| - \|y\|) \|H\|$$

$$\leq \|x\| \left(L \cdot \frac{2\|x - y\|}{\|x\|} + \epsilon\right) + \|x - y\| \|H\|.$$

(See Lemma 12.6, below, for an explanation of this last inequality.) Simplifying the last expression (the upper bound), and observing that $\|H\| \leq L$, we obtain the estimate

$$\|H(x) - H(y)\| \leq 3L\|x - y\| + \epsilon \|x\|.$$

To get the desired bound, we note that the sum of two terms is no more than twice the maximum of the two, which gives us

$$\|H(x) - H(y)\| \leq 6 \max\left(L\|x - y\|, \epsilon \|x\|\right).$$

Since we assumed that $\|x\| \geq \|y\|$, we have obtained the desired result. □

The last inequality in (12.1) follows from a useful inequality that may not be apparent. Since we will make use of it several times, we state it now as a lemma.

LEMMA 12.6. *Suppose that x and y are elements of a Banach space X. If $\|x\| \geq \|y\|$, then*

$$\left\|\frac{x}{\|x\|} - \frac{y}{\|y\|}\right\| \leq \frac{2\|x - y\|}{\|x\|} \leq \frac{2\|x - y\|}{\|y\|}.$$

PROOF. Note that the rightmost inequality follows from the assumption that $\|x\| \geq \|y\|$, and so we need only verify the leftmost inequality. Let $\alpha = \frac{\|x\|}{\|y\|} \geq 1$. Then

$$\|x\| \cdot \left\|\frac{x}{\|x\|} - \frac{y}{\|y\|}\right\| = \|x - \alpha y\| \leq \|x - y\| + \|y - \alpha y\|.$$

Since $\alpha \geq 1$,

$$\|y - \alpha y\| = (\alpha - 1)\|y\| = \left[\frac{\|x\|}{\|y\|} - 1\right] \|y\| = \|x\| - \|y\| \leq \|x - y\|.$$

It follows that
$$\|x\| \cdot \left\| \frac{x}{\|x\|} - \frac{y}{\|y\|} \right\| \leq 2\|x - y\|,$$
from which we obtain the desired inequality. $\square$

We now introduce a technical lemma in order to provide a tool that will enable us to show that two Banach spaces are uniformly homeomorphic

LEMMA 12.7. *Let X and Y be Banach spaces. Suppose there exist constants $L > 0$ and $c > 0$ and a family of positively homogeneous functions $(H_t : t \geq 1)$ from X to Y such that:*
  (1) *H_t is of CL-type $(L, c/t^2)$ for each $t \geq 1$.*
  (2) *Whenever $t \geq s \geq 1$,*

(12.2) $$\|H_t - H_s\| \leq c \left(\log(t/s) + \frac{1}{t^2} + \frac{1}{s^2} \right).$$

  (3) *H_t is uniformly continuous on B_X for each $t \geq 1$.*
  (4) *The map $t \to H_t$ is continuous from $[1, \infty)$ into $\mathcal{H}(X, Y)$.*

Then the function $f : X \to Y$ defined by
$$f(x) = \begin{cases} H_{\|x\|}(x) & \text{if } \|x\| \geq 1 \\ H_1(x) & \text{if } \|x\| < 1 \end{cases}$$
is uniformly continuous.

Furthermore, if H_t is a bijection for each $t \geq 1$ and the family $(H_t^{-1} : t \geq 1)$ satisfies conditions (1)–(4) (with the roles of X and Y reversed), then f is a uniform homeomorphism of X onto Y.

PROOF. For ease of notation, let $H_t = H_1$ for $0 \leq t < 1$. Then we can rewrite $f(x)$ as
$$f(x) = H_{\|x\|}(x)$$
for all $x \in X$. We claim that f is uniformly continuous. We will start by making some preliminary estimates. Let x and y be chosen in X so that $\|x\| \geq \|y\| > 0$. Then
$$\|f(x) - f(y)\| = \|H_{\|x\|}(x) - H_{\|y\|}(y)\|$$
$$\leq \underbrace{\|H_{\|x\|}(x) - H_{\|x\|}(y)\|}_{(A)} + \underbrace{\|H_{\|x\|}(y) - H_{\|y\|}(y)\|}_{(B)}.$$

We will bound the terms labeled (A) and (B) individually, starting with (A). Since H_t is of CL-type $(L, c/t^2)$ when $t \geq 1$, we have that

(12.3) $$\|H_{\|x\|}(x) - H_{\|x\|}(y)\| \leq 6 \max \left(L\|x - y\|, \frac{c}{\|x\|} \right),$$

by Lemma 12.5. That gives us the bound we will use for (A).

Next, we will determine a suitable bound for the term in (B). Using the inequality in (12.2), we have that

$$\|H_{\|x\|}(y) - H_{\|y\|}(y)\| \leq c \left[\log\left(\frac{\|x\|}{\|y\|}\right) + \min\left(\frac{1}{\|x\|^2}, 1\right) + \min\left(\frac{1}{\|y\|^2}, 1\right) \right] \|y\|.$$

The minima appearing in the above inequality come from the fact that $H_t = H_1$ whenever $t \leq 1$, and so for the cases where $\|x\| < 1$ or $\|y\| < 1$, the inequality in

(12.2) is applied with $t = 1$ or $s = 1$ (respectively). We assumed that $\|y\| \leq \|x\|$, and so we have that the term in (B) is bounded by

$$c\left[\|y\|\log\left(\frac{\|x\|}{\|y\|}\right) + \min\left(\frac{1}{\|x\|}, \|x\|\right) + \min\left(\frac{1}{\|y\|}, \|y\|\right)\right].$$

We can bound this further using an inequality from differential calculus. Recall that $\log(z) \leq z - 1$ for $z > 0$, and so

$$\log\left(\frac{\|x\|}{\|y\|}\right) \cdot \|y\| \leq \left(\frac{\|x\|}{\|y\|} - 1\right)\|y\| = \|x\| - \|y\| \leq \|x - y\|.$$

Therefore,

$$\|H_{\|x\|}(y) - H_{\|y\|}(y)\| \leq c\left[\|x - y\| + \min\left(\frac{1}{\|x\|}, \|x\|\right) + \min\left(\frac{1}{\|y\|}, \|y\|\right)\right],$$

and this gives us our desired bound for (B).

Putting together our bounds for (A) and (B), we have so far shown that the quantity $\|f(x) - f(y)\|$ is bounded by

(12.4) $\quad 6\max\left(L\|x - y\|, \frac{c}{\|x\|}\right) + c\left[\|x - y\| + \min\left(\frac{1}{\|x\|}, \|x\|\right) + \min\left(\frac{1}{\|y\|}, \|y\|\right)\right],$

which is further bounded by

$$6L\|x - y\| + \frac{6c}{\|x\|} + c\left(\|x - y\| + \frac{1}{\|x\|} + \frac{1}{\|y\|}\right).$$

Therefore,

(12.5) $\quad \|f(x) - f(y)\| \leq (6L + c)\|x - y\| + \frac{7c}{\|x\|} + \frac{c}{\|y\|}.$

Now let $\epsilon > 0$ be given. We claim that it is possible to find a $\delta > 0$ such that $\|f(x) - f(y)\| < \epsilon$ whenever $\|x - y\| < \delta$. Without loss of generality, we may assume that $\|x\| \geq \|y\| > 0$. We will consider two cases.

CASE 1. Suppose that $\|x\| \geq \|y\| > \max\left(\frac{7c}{\epsilon/3}, 1\right)$. If we choose $\delta_1 < \frac{\epsilon/3}{6L+c}$, then whenever $\|x - y\| < \delta_1$, we have that

$$\|x - y\| < \frac{\epsilon/3}{6L + c} \quad \text{and} \quad \frac{1}{\|x\|} \leq \frac{1}{\|y\|} < \frac{\epsilon/3}{7c},$$

and consequently we have $\|f(x) - f(y)\| < \epsilon$ by (12.5).

CASE 2. Suppose that $\|x\| \geq \|y\|$ and $\|y\| \leq \max\left(\frac{7c}{\epsilon/3}, 1\right)$. For ease of notation, let M_0 be the maximum value, so that $\|y\| \leq M_0$ and let $M = M_0 + 1$. The map $t \to H_t$ is continuous, and hence uniformly continuous on $[0, M+1]$, and so there exists some real number δ_2 with $0 < \delta_2 < 1$ so that

(12.6) $\quad\quad\quad\quad\quad\quad \|H_t - H_s\| < \frac{\epsilon}{3M}$

whenever $|t - s| < \delta_2$ and $0 \leq t, s \leq M + 1$. Now pick an integer $N > M$ so large that $\frac{M}{N} < \delta_2$.

Consider the collection of $N + 1$ functions

$$\left(H_{kM/N}\right)_{k=1}^{N} = \left(H_0, H_{M/N}, H_{2M/N}, \ldots, H_{(N-1)M/N}, H_M\right),$$

where $H_{kM/N} = H_1$ if $\frac{kM}{N} \leq 1$. Each of the functions $H_{kM/N}$ is uniformly continuous on B_X. Since there are finitely many of these function, we can find one real number δ_3 with $0 < \delta_3 < 1$ so that

(12.7) $$\|H_{kM/N}(u) - H_{kM/N}(v)\| < \frac{\epsilon}{3M}, \quad 0 \leq k \leq N,$$

for any u and v in B_X such that $\|u - v\| \leq \delta_3$.

Suppose that $\|x - y\| < \min(\frac{M}{N}, \delta_3)$. Since $\|y\| \leq M_0$, it follows that

$$\|x\| \leq \|x - y\| + \|y\| < \min(\tfrac{M}{N}, \delta_3) + M_0 < 1 + M_0 = M.$$

If $\|x\| \leq 1$, then it is also true that $\|y\| \leq 1$, and so

$$\|f(x) - f(y)\| = \|H_1(x) - H_1(y)\| = \|H_{M/N}(x) - H_{M/N}(y)\| \leq \frac{\epsilon}{3M} < \epsilon.$$

Thus, in this case, there is nothing else to show. Suppose instead that $\|x\| > 1$. Since $\|x - y\| < \frac{M}{N}$ and $\|x\| \geq \|y\|$, there is some $k_0 \in \{0, 1, \ldots, N\}$ such that

$$\left|\|x\| - \frac{k_0 M}{N}\right| \leq \frac{M}{N} \quad \text{and} \quad \left|\|y\| - \frac{k_0 M}{N}\right| \leq \frac{M}{N}.$$

Since $\frac{M}{N} < \delta_2$, the inequality in (12.6) that comes from the continuity assumption on the map $t \to H_t$ implies that

$$\|H_{\|x\|} - H_{k_0 M/N}\| < \frac{\epsilon}{3M} \quad \text{and} \quad \|H_{\|y\|} - H_{k_0 M/N}\| < \frac{\epsilon}{3M}.$$

Then,

$$\|f(x) - f(y)\|$$
$$\leq \underbrace{\|H_{\|x\|}(x) - H_{\frac{k_0 M}{N}}(x)\|}_{< \frac{\epsilon}{3M} \cdot \|x\|} + \|H_{\frac{k_0 M}{N}}(x) - H_{\frac{k_0 M}{N}}(y)\| + \underbrace{\|H_{\frac{k_0 M}{N}}(y) - H_{\|y\|}(y)\|}_{< \frac{\epsilon}{3M} \cdot \|y\|}$$

and so

$$\|f(x) - f(y)\| \leq \frac{\epsilon \|x\|}{3M} + \|H_{\frac{k_0 M}{N}}(x) - H_{\frac{k_0 M}{N}}(y)\| + \frac{\epsilon \|y\|}{3M}.$$

Since $\|y\| \leq \|x\| < M$, we conclude that

$$\|f(x) - f(y)\| \leq \frac{\epsilon}{3} + \|H_{\frac{k_0 M}{N}}(x) - H_{\frac{k_0 M}{N}}(y)\| + \frac{\epsilon}{3}.$$

We claim that the middle term in the previous inequality is also less than $\epsilon/3$. To see why, we start by using the positive homogeneity of $H_{k_0 M/N}$ to get

$$\|H_{\frac{k_0 M}{N}}(x) - H_{\frac{k_0 M}{N}}(y)\| = \|x\| \left\|H_{\frac{k_0 M}{N}}\left(\tfrac{x}{\|x\|}\right) - H_{\frac{k_0 M}{N}}\left(\tfrac{y}{\|x\|}\right)\right\|.$$

If we let $u = \frac{x}{\|x\|}$ and $v = \frac{y}{\|x\|}$, then $\|u\| = 1$ and $\|v\| \leq 1$. In particular, both u and v are in the unit ball B_X. Furthermore, since we have assumed that $\|x\| > 1$, we have that

$$\|u - v\| = \left\|\frac{x}{\|x\|} - \frac{y}{\|x\|}\right\| = \frac{\|x - y\|}{\|x\|} < \|x - y\| < \delta_3.$$

This means we can use the inequality in (12.7) that comes from the assumption that $H_{k_0 M/N}$ is uniformly continuous on B_X to conclude that

$$\|H_{\frac{k_0 M}{N}}(x) - H_{\frac{k_0 M}{N}}(y)\| < \|x\| \cdot \frac{\epsilon}{3M}.$$

Once again we use the fact that $\|x\| < M$, and we see that
$$\|H_{\frac{k_0 M}{N}}(x) - H_{\frac{k_0 M}{N}}(y)\| < M \cdot \frac{\epsilon}{3M} = \frac{\epsilon}{3}.$$
It follows that $\|f(x) - f(y)\| < \epsilon$.

Conclusion. Let $\delta < \min(\delta_1, \frac{M}{N}, \delta_3)$. Then $\|f(x) - f(y)\| < \epsilon$ whenever x and y are points in X such that $\|x - y\| < \delta$, and so $f : X \to Y$ is a uniformly continuous function, as claimed.

If we assume additionally that $H_t : X \to Y$ is a bijection for each $t \geq 1$, then the function f has an inverse $f^{-1} : Y \to X$ given by the formula
$$f^{-1}(y) = H_{\|y\|}^{-1}(y), \quad y \in Y.$$

We can show that f^{-1} is uniformly continuous using an argument similar to the one we used for f, and so under this additional assumption, we conclude that f is a uniform homeomorphism of the Banach spaces X and Y. □

REMARK 12.8. Consider the function $f : X \to Y$ defined in the statement of Lemma 12.7, as well as the assumptions given in (1)–(4). We used the assumptions in (3) and (4) to show that f is a uniformly continuous function. If we do not make the assumptions in (3) and (4), however, we can still conclude that f is a *coarse Lipschitz* function, as we now demonstrate.

Suppose that x and y are two points in X for which $\|y\| \leq \|x\|$. If $\|x\| \leq 1$, then we can use the fact that H_1 is of CL-type (L, c) to conclude that
$$\|f(x) - f(y)\| = \|H_1(x) - H_1(y)\| \leq L\|x - y\| + c.$$

On the other hand, if $\|x\| \geq 1$, then the CL-type assumption in (1) and the inequality in (2) can be used to give us the bound in (12.4), which is

$$\|f(x) - f(y)\|$$
$$\leq 6 \max\left(L\|x-y\|, \frac{c}{\|x\|}\right) + c\left[\|x-y\| + \min\left(\frac{1}{\|x\|}, \|x\|\right) + \min\left(\frac{1}{\|y\|}, \|y\|\right)\right].$$

Consequently,
$$\|f(x) - f(y)\| \leq \left(6L\|x-y\| + 6c\right) + c\left(\|x-y\| + 1 + 1\right)$$
$$\leq (6L + c)\|x - y\| + 6c + 2.$$

Therefore, there are positive constants K and C such that
$$\|f(x) - f(y)\| \leq K\|x - y\| + C, \quad x \in X, y \in X,$$
for all x and y in X, and so f is a coarse Lipschitz function.

Shortly, we will come to our main result in this chapter. Before stating it, however, we introduce some new notation and terminology.

DEFINITION 12.9. For any $H \in \mathcal{H}(X, Y)$, if $\epsilon > 0$, then we define $\|H\|_\epsilon$ to be the least constant λ such that $\lambda \geq \|H\|$ and
$$\|H(x) - H(y)\| \leq \lambda \max\left(\|x - y\|, \epsilon \|x\|, \epsilon \|y\|\right).$$

A straightforward argument shows that $\|\cdot\|_\epsilon$ is a norm on $\mathcal{H}(X,Y)$ for each $\epsilon > 0$, and a norm that is equivalent to the original norm. In particular, we have

$$\|H\| \leq \|H\|_\epsilon \leq 2\epsilon^{-1}\|H\|,$$

for all $H \in \mathcal{H}(X,Y)$.

REMARK 12.10. If H is a positively homogeneous function of CL-type (L, ϵ), then $\|H\|_\epsilon \leq 6(L+1)$, by Lemma 12.5. (The constant $6(L+1)$ is not optimal.)

REMARK 12.11. In Lemma 12.7, we assumed the function H_t is of CL-type $(L, c/t^2)$ for each $t \geq 1$. In the proof of the lemma, we only used this assumption once, when finding the bound in (12.3) via Lemma 12.5. Instead of the CL-type assumption, it would be sufficient to assume that there exists some constants $c > 0$ and $L > 0$ so that

$$\|H_t\|_{c/t^2} \leq L$$

for all $t \geq 1$.

In light of Remark 12.11, we now reformulate Lemma 12.7 in terms of the $\|\cdot\|_\epsilon$ norm. (We will state it for the bijection case.)

LEMMA 12.12. *Let X and Y be Banach spaces. Suppose there exist constants $L > 0$ and $c > 0$ and a family of positively homogeneous bijections $(H_t : t \geq 1)$ from X to Y such that:*
  (1) $\|H_t\|_{c/t^2} \leq L$ and $\|H_t^{-1}\|_{c/t^2} \leq L$ for each $t \geq 1$.
  (2) $\max(\|H_t - H_s\|, \|H_t^{-1} - H_s^{-1}\|) \leq c\left(\log(t/s) + \frac{1}{t^2} + \frac{1}{s^2}\right)$ if $t > s$.
  (3) H_t and H_t^{-1} are uniformly continuous on B_X and B_Y, respectively, for each $t \geq 1$.
  (4) The maps $t \to H_t$ and $t \to H_t^{-1}$ are continuous from $[1,\infty)$ into $\mathcal{H}(X,Y)$ and $\mathcal{H}(Y,X)$, respectively.

Then X and Y are uniformly homeomorphic.

PROOF. The proof is essentially the same as that of Lemma 12.7, except we replace the bound in (12.3) with

$$\|H_{\|x\|}(x) - H_{\|x\|}(y)\| \leq L\max\left(\|x-y\|, \frac{c}{\|x\|}\right).$$

Since the remainder of the proof is essentially the same (with some constants changed), we omit the details. □

We saw in Lemma 12.12 that the requirements in Lemma 12.7 can be relaxed in some ways. Specifically, we replaced the CL-type assumption on the maps H_t and H_t^{-1} with an assumption on the size of $\|H_t\|_\epsilon$ and $\|H_t^{-1}\|_\epsilon$. Another way to relax the assumptions involves changing the domains for H_t and H_t^{-1}.

Instead of requiring each H_t to be a bijection from the Banach space X to the Banach space Y, it is sufficient to find bijections from the unit sphere ∂B_X to the unit sphere ∂B_Y, provided we keep the other assumptions. To see why this is the case, we provide a construction to extend a map $H : \partial B_X \to \partial B_Y$ to a positively homogeneous norm-preserving function $\widetilde{H} : X \to Y$.

12. THE UNIFORM STRUCTURE OF BANACH SPACES

PROPOSITION 12.13. *Let X and Y be Banach spaces and let $H : \partial B_X \to \partial B_Y$ be a function between unit spheres. Define a function $\widetilde{H} : X \to Y$ by the rule*

$$\widetilde{H}(x) = \begin{cases} \|x\| H\left(\frac{x}{\|x\|}\right), & x \neq 0, \\ 0, & x = 0. \end{cases}$$

Then the map $\widetilde{H}$ is a positively homogeneous norm-preserving function. If H is of CL-type (L, ϵ), then $\|\widetilde{H}\|_\epsilon < 4L + 2$. Furthermore, if H is uniformly continuous on ∂B_X, then $\widetilde{H}$ is uniformly continuous on B_X.

PROOF. It is trivial to check that $\widetilde{H}$ is a positively homogeneous norm-preserving function from X to Y. Suppose now that $H : \partial B_X \to \partial B_Y$ is of CL-type (L, ϵ), where $L > 0$. Then for any x and y in X such that $\|x\| \geq \|y\| > 0$, we have

$$\|\widetilde{H}(x) - \widetilde{H}(y)\| = \left\| \|x\| H\left(\frac{x}{\|x\|}\right) - \|y\| H\left(\frac{y}{\|y\|}\right) \right\|.$$

Using the triangle inequality, we can bound this by

$$\left\| \|x\| H\left(\frac{x}{\|x\|}\right) - \|y\| H\left(\frac{x}{\|x\|}\right) \right\| + \left\| \|y\| H\left(\frac{x}{\|x\|}\right) - \|y\| H\left(\frac{y}{\|y\|}\right) \right\|$$

$$\leq \|x - y\| \cdot \|H\| + \|y\| \cdot \omega_H\left(\left\|\frac{x}{\|x\|} - \frac{y}{\|y\|}\right\|\right).$$

Since we assumed that H is of CL-type (L, ϵ), and since $\|H\| = 1$, this can be bounded further by

$$\|x - y\| + \|y\| \cdot \left(L \left\|\frac{x}{\|x\|} - \frac{y}{\|y\|}\right\| + \epsilon\right).$$

We now use the inequality in Lemma 12.6, and the fact that $\|x\| \geq \|y\|$ to bound this by

$$\|x - y\| + \|y\| \cdot \left(\frac{2L\|x - y\|}{\|x\|} + \epsilon\right) \leq (1 + 2L)\|x - y\| + \epsilon \|x\|.$$

Since the sum of two terms is less than twice the maximum of the two terms, we conclude then that

$$\|\widetilde{H}(x) - \widetilde{H}(y)\| \leq 2\max\left((1 + 2L)\|x - y\|, \epsilon \|x\|\right).$$

It follows that $\|\widetilde{H}\|_\epsilon \leq 2 + 4L$.

Now suppose that $H : \partial B_X \to \partial B_Y$ is a uniformly continuous function on its domain ∂B_X. We claim that the extension $\widetilde{H} : X \to Y$ is uniformly continuous on B_X. To that end, let x and y be points in B_X. Without loss of generality, we assume that $\|x\| \geq \|y\| > 0$. By the triangle inequality, we can bound $\|\widetilde{H}(x) - \widetilde{H}(y)\|$ by

$$\left\| \|x\| H\left(\frac{x}{\|x\|}\right) - \|y\| H\left(\frac{x}{\|x\|}\right) \right\| + \left\| \|y\| H\left(\frac{x}{\|x\|}\right) - \|y\| H\left(\frac{y}{\|y\|}\right) \right\|,$$

which in turn is bounded by

$$\|x - y\| \|H\| + \|y\| \left\| H\left(\frac{x}{\|x\|}\right) - H\left(\frac{y}{\|y\|}\right) \right\|.$$

Observing that $\|H\| = 1$ and using the inequality from Comment 12.4, we conclude that

$$\|\widetilde{H}(x) - \widetilde{H}(y)\| \leq \|x - y\| + \|y\| \omega_H\left(\left\|\frac{x}{\|x\|} - \frac{y}{\|y\|}\right\|\right).$$

The modulus of continuity is monotone, and so by Lemma 12.6, we have that
$$\omega_H\left(\left\|\frac{x}{\|x\|} - \frac{y}{\|y\|}\right\|\right) \leq \omega_H\left(\frac{2\|x-y\|}{\|x\|}\right).$$
In addition to being monotone, the modulus of continuity is also subadditive, and so

(12.8) $$\omega_H\left(\frac{2\|x-y\|}{\|x\|}\right) \leq \left(1 + \frac{2}{\|x\|}\right)\omega_H(\|x-y\|).$$

(See Comment 12.14, for more on the inequality in (12.8).) Thus,
$$\|\widetilde{H}(x) - \widetilde{H}(y)\| \leq \|x-y\| + \|y\|\left(1 + \frac{2}{\|x\|}\right)\omega_H(\|x-y\|)$$
$$= \|x-y\| + \left(\|y\| + \frac{2\|y\|}{\|x\|}\right)\omega_H(\|x-y\|).$$

Since $\|y\| \leq \|x\| \leq 1$, it follows that
$$\|\widetilde{H}(x) - \widetilde{H}(y)\| \leq \|x-y\| + 3\omega_H(\|x-y\|).$$

Now suppose that $\epsilon > 0$ is given. Because H is uniformly continuous, we can choose $\delta_0 > 0$ so small that $\omega_H(\|x-y\|) < \frac{\epsilon}{6}$ whenever $\|x-y\| < \delta_0$. Then, if we let $\delta > 0$ be chosen so that $\delta < \min(\frac{\epsilon}{2}, \delta_0)$, then
$$\|\widetilde{H}(x) - \widetilde{H}(y)\| < \frac{\epsilon}{2} + 3\cdot\frac{\epsilon}{6} = \epsilon,$$
whenever $\|x-y\| < \delta$, and so $\widetilde{H}$ is uniformly continuous on B_X. □

COMMENT 12.14. The last inequality in (12.8) follows from the monotonicity and subadditivity of the modulus of continuity. Suppose that $f : X \to Y$ is a uniformly continuous function between Banach spaces. If $c > 0$, then for $t \geq 0$,
$$\omega_f(ct) \leq (1+c)\,\omega_f(t).$$
To see why this is true, let $c > 0$ be given. There exists some $n \in \mathbb{N}$ such that $n \leq c < n+1$. Then, because ω_f is monotone and subadditive,
$$\omega_f(ct) \leq \omega_f\big((n+1)t\big) \leq (n+1)\,\omega_f(t) \leq (c+1)\,\omega_f(t).$$

REMARK 12.15. If the map $H : \partial B_X \to \partial B_Y$ in Proposition 12.13 is a bijection, then we can similarly define an extension $\widetilde{H^{-1}} : Y \to X$ of $H^{-1} : \partial B_Y \to \partial B_X$ by
$$\widetilde{H^{-1}}(y) = \begin{cases} \|y\|H^{-1}\left(\frac{y}{\|y\|}\right), & y \neq 0, \\ 0, & y = 0. \end{cases}$$
A straightforward computation shows that $\widetilde{H^{-1}} = \widetilde{H}^{-1}$.

Consequently, if the map $H : \partial B_X \to \partial B_Y$ is a uniform homeomorphism of the unit spheres with H and H^{-1} of CL-type (L, ϵ), then $\widetilde{H}$ and $\widetilde{H}^{-1}$ are positively homogeneous norm-preserving extensions of H and H^{-1}, respectively, that are uniformly continuous on their unit balls, and
$$\max\left(\|\widetilde{H}\|_\epsilon, \|\widetilde{H}^{-1}\|_\epsilon\right) \leq 2 + 4L.$$

In light of Remark 12.15, if in Lemma 12.7 we do not assume that H_t and H_t^{-1} ($t \geq 1$) are bijections defined on all of X and Y (respectively), but instead from ∂B_X and ∂B_Y (respectively), then the fact that H_t and H_t^{-1} are of CL-type $(L, c/t^2)$ implies that the extensions $\widetilde{H}_t$ and $\widetilde{H}_t^{-1}$ ($t \geq 1$) have $\|\cdot\|_{c/t^2}$-norm bounded by $2 + 4L$. We can then still conclude that X and Y are uniformly homeomorphic, by Lemma 12.12. We state this now as a corollary.

COROLLARY 12.16. *Let X and Y be Banach spaces and suppose that for each $t \geq 1$, there is a bijection $H_t : \partial B_X \to \partial B_Y$. Suppose also that there exist constants $L > 0$ and $c > 0$ such that:*
  (1) *H_t and H_t^{-1} are of CL-type $(L, c/t^2)$ for each $t \geq 1$.*
  (2) *$\max(\|H_t - H_s\|, \|H_t^{-1} - H_s^{-1}\|) \leq c \left(\log(t/s) + \frac{1}{t^2} + \frac{1}{s^2}\right)$ if $t > s$.*
  (3) *H_t and H_t^{-1} are uniformly continuous for each $t \geq 1$.*
  (4) *The functions $t \to H_t$ and $t \to H_t^{-1}$ are continuous from $[1, \infty)$ into $\mathcal{H}(\partial B_X, \partial B_Y)$ and $\mathcal{H}(\partial B_Y, \partial B_X)$, respectively.*
Then X and Y are uniformly homeomorphic.

PROOF. We claim the maps $\widetilde{H}_t : X \to Y$ and $\widetilde{H}_t^{-1} : Y \to X$ ($t \geq 1$) constructed in Proposition 12.13 satisfy the conditions of Lemma 12.12. We already know from Proposition 12.13 (and Remark 12.15) that the maps $\widetilde{H}_t$ and $\widetilde{H}_t^{-1}$ are positively homogeneous norm-preserving bijections that satisfy conditions (1) and (3) of Lemma 12.12. ((1) is satisfied with the constant $4L + 2$ in place of L.) Conditions (2) and (4) from Lemma 12.12 are satisfied because $\widetilde{H}_t|_{\partial B_X} = H_t$ and $\widetilde{H}_t^{-1}|_{\partial B_Y} = H_t^{-1}$ for each $t \geq 1$, and those conditions are about the properties of the maps on their respective unit spheres. Therefore, the Banach spaces X and Y are uniformly homeomorphic, by Lemma 12.12. □

In order to state our main result, we need to introduce further terminology.

DEFINITION 12.17. We say that two Banach spaces X and Y are *uniformly close* if there exists a number $\lambda \geq 1$ so that whenever $0 < \epsilon < 1$, we can find a positively homogeneous bijection $H : X \to Y$ such that H and H^{-1} are uniformly continuous on B_X and B_Y, respectively, and such that $\|H\|_\epsilon \leq \lambda$ and $\|H^{-1}\|_\epsilon \leq \lambda$. When such a λ exists, we say that X and Y are λ-*uniformly close* or *uniformly close with constant λ*.

We also have a notion of "closeness" for coarse Lipschitz maps. In order to define it, we need to extend our definition of CL-type to maps between metric spaces. (In Definition 12.3, we defined CL-type for Banach spaces.)

DEFINITION 12.18. A function f between metric spaces is of *CL-type (L, ϵ)* for $L > 0$ and $\epsilon > 0$ if $\omega_f(\delta) \leq L\delta + \epsilon$ for $t \geq 0$.

DEFINITION 12.19. Suppose that X and Y are metric spaces. A bijection $f : X \to Y$ is called a *coarse Lipschitz homeomorphism* if both f and f^{-1} are coarse Lipschitz functions. We say that f is a *CL-homeomorphism of type (L, ϵ)* if both f and f^{-1} are of CL-type (L, ϵ). If there exists some $L > 0$ such that for each $\epsilon > 0$ there exists a CL-homeomorphism $f : X \to Y$ of type (L, ϵ), then we say that X and Y are *almost Lipschitz isomorphic*. If in addition we can find a CL-homeomorphism of type (L, ϵ) that is uniformly continuous, then we say X and Y are *uniformly almost Lipschitz isomorphic*.

Due to Proposition 12.13, it is perhaps not surprising that there is a natural relationship between uniform closeness of Banach spaces and uniformly almost Lipschitz isomorphism of their unit spheres. We make the relationship explicit in Proposition 12.20.

PROPOSITION 12.20. *Two Banach spaces X and Y are uniformly close if and only if ∂B_X and ∂B_Y are uniformly almost Lipschitz isomorphic.*

PROOF. Suppose that X and Y are λ-uniformly close. We wish to find a constant $L > 0$ so that for every $\epsilon > 0$, there exists a uniformly continuous CL-homeomorphism from ∂B_X to ∂B_Y of type (L, ϵ). To that end, let $\epsilon > 0$ be given.

By assumption, if $\epsilon_0 > 0$, then there exists a positively homogeneous bijection $H : X \to Y$ such that H and H^{-1} are uniformly continuous on B_X and B_Y, respectively, and such that $\|H\|_{\epsilon_0} \leq \lambda$ and $\|H^{-1}\|_{\epsilon_0} \leq \lambda$. Define a normalized version of H to be the map $\widehat{H} : \partial B_X \to \partial B_Y$ given by the formula

$$\widehat{H}(x) = \frac{H(x)}{\|H(x)\|}, \quad x \in \partial B_X.$$

Then $\widehat{H}$ is a uniformly continuous function with uniformly continuous inverse

$$\widehat{H}^{-1}(y) = \frac{H^{-1}(y)}{\|H^{-1}(y)\|}, \quad y \in \partial B_Y.$$

We claim that for a suitably chosen value of ϵ_0, the map $\widehat{H}$ is a CL-homeomorphism of type $(2\lambda^2, \epsilon)$. To see this, suppose that x_1 and x_2 are in ∂B_X. Then, using the inequality in Lemma 12.6,

$$\|\widehat{H}(x_1) - \widehat{H}(x_2)\| = \left\|\frac{H(x_1)}{\|H(x_1)\|} - \frac{H(x_2)}{\|H(x_2)\|}\right\| \leq \frac{2\|H(x_1) - H(x_2)\|}{\|H(x_1)\|}.$$

We did not specify which of $\|H(x_1)\|$ and $\|H(x_2)\|$ is larger, but we do not actually need to know that in order to apply Lemma 12.6 to draw this conclusion. (Note that neither of $\|H(x_1)\|$ and $\|H(x_2)\|$ can be zero.) Since

$$1 = \|x_1\| = \|H^{-1}(H(x_1))\| \leq \|H^{-1}\| \|H(x_1)\|,$$

we have that

$$\|H(x_1)\| \geq \frac{1}{\|H^{-1}\|} \geq \frac{1}{\lambda},$$

and so

$$\frac{2\|H(x_1) - H(x_2)\|}{\|H(x_1)\|} \leq 2\lambda \|H(x_1) - H(x_2)\| \leq 2\lambda^2 \max(\|x_1 - x_2\|, \epsilon_0)$$

$$\leq 2\lambda^2 \Big(\|x_1 - x_2\| + \epsilon_0\Big).$$

Consequently, the normalized map $\widehat{H}$ is of CL-type $(2\lambda^2, 2\lambda^2 \epsilon_0)$. The same argument gives us the same conclusion for the normalized inverse map $\widehat{H}^{-1}$, and so we conclude that $\widehat{H}$ is a CL-homeomorphism of type $(2\lambda^2, 2\lambda^2 \epsilon_0)$. If we go back to the start of the argument and choose $\epsilon_0 < \frac{\epsilon}{2\lambda^2}$, then the resulting map $\widehat{H}$ is a CL-homeomorphism of type $(2\lambda^2, \epsilon)$, and that is what we wanted to find.

Now suppose that ∂B_X and ∂B_Y are uniformly almost Lipschitz isomorphic with CL-homeomorphism type (L, ϵ). We want to show that there exists a $\lambda > 0$ such that for each $\epsilon > 0$, there is a positively homogeneous bijection from X to Y

such that it and its inverse are uniformly continuous on their respective unit balls and have $\|\cdot\|_\epsilon$ bounded by λ.

Let $\epsilon > 0$ be given. By assumption, there exists a uniformly continuous CL-homeomorphism $h : \partial B_X \to \partial B_Y$ of type (L, ϵ). We can extend h to a positively homogeneous norm-preserving function $\widetilde{h} : X \to Y$ defined by

$$\widetilde{h}(x) = \begin{cases} \|x\| h\left(\frac{x}{\|x\|}\right), & x \neq 0, \\ 0, & x = 0. \end{cases}$$

We saw in Proposition 12.13 and Remark 12.15 that $\widetilde{h}$ is a positively homogeneous norm-preserving bijection such that $\widetilde{h}$ and $\widetilde{h}^{-1}$ are uniformly continuous on B_X and B_Y, respectively, and that

$$\max\left(\|\widetilde{h}\|_\epsilon, \|\widetilde{h}^{-1}\|_\epsilon\right) \leq 4L + 2.$$

Therefore, the map $\widetilde{h}$ satisfies the necessary conditions with $\lambda = 4L + 2$, and so X and Y are uniformly close, as claimed. $\square$

We are now ready to state the main theorem of this chapter, although we are not quite yet ready to prove it.

THEOREM 12.21. *If X and Y are uniformly close Banach spaces such that $X \approx X^2$ and $Y \approx Y^2$, then X and Y are uniformly homeomorphic.*

We remind the reader that for Banach spaces V and W, we write $V \approx W$ to mean that V and W are linearly isomorphic. We are using the notation W^2 to mean $W \oplus W$. For convenience, we will treat all direct sums as ℓ_∞ sums.

Before proving Theorem 12.21, we will provide some preliminary results.

LEMMA 12.22. *Suppose that X, Y, and Z are Banach spaces. If $f \in \mathcal{H}(X, Y)$ and $g \in \mathcal{H}(Y, Z)$, then for each $0 < \epsilon < 1$, we have $\|g \circ f\|_\epsilon \leq \|g\|_\epsilon \|f\|_\epsilon$.*

The proof of Lemma 12.22 is immediate, and so we omit the details.

LEMMA 12.23. *Suppose that W, X, Y, and Z are Banach spaces and suppose that $f \in \mathcal{H}(W, X)$ and $h \in \mathcal{H}(Y, Z)$. If g_1 and g_2 are any functions in $\mathcal{H}(X, Y)$, and if $0 < \epsilon < 1$, then*

$$\|hg_1 f - hg_2 f\| \leq \|h\|_\epsilon \|f\|_\epsilon \left(\|g_1 - g_2\| + \epsilon K\right),$$

where $K = \max(\|g_1\|, \|g_2\|)$.

PROOF. Suppose that $x \in B_W$. We observe that for functions g_1 and g_2 in $\mathcal{H}(X, Y)$, we have that

$$\|(g_1 f)(x)\| \leq K \|f(x)\| \quad \text{and} \quad \|(g_2 f)(x)\| \leq K \|f(x)\|,$$

where $K = \max(\|g_1\|, \|g_2\|)$. Let ϵ be a real number in the interval $(0, 1)$. Then, by the definition of $\|h\|_\epsilon$, we have that

$\|(hg_1 f)(x) - (hg_2 f)(x)\|$

$\leq \|h\|_\epsilon \max\left(\|(g_1 f)(x) - (g_2 f)(x)\|, \epsilon \|(g_1 f)(x)\|, \epsilon \|(g_2 f)(x)\|\right)$

$\leq \|h\|_\epsilon \left(\|g_1 - g_2\| \|f(x)\| + \epsilon K \|f(x)\|\right).$

The result then follows from the fact that $\|f(x)\| \leq \|f\| \leq \|f\|_\epsilon$. $\square$

We need one more technical, but significant result before we are prepared to prove Theorem 12.21.

THEOREM 12.24. *Let X and Y be Banach spaces. If X and $X \oplus Y$ are uniformly close and Y and Y^2 are linearly isomorphic, then X^2 and $X^2 \oplus Y$ are uniformly homeomorphic.*

PROOF. We assumed that X and $X \oplus Y$ are uniformly close, and so there exists some $\lambda \geq 1$ such that whenever $0 < \epsilon < 1$, we can find a positively homogeneous bijection $H : X \to X \oplus Y$ (that depends on ϵ) such that H and H^{-1} are uniformly continuous on B_X and $B_{X \oplus Y}$, respectively, and such that both $\|H\|_\epsilon \leq \lambda$ and $\|H^{-1}\|_\epsilon \leq \lambda$. Furthermore, since Y and Y^2 are assumed to be linearly isomorphic Banach spaces, there exists a linear isomorphism $T : Y \to Y^2$. Let $L \geq \max(\lambda, \|T\|, \|T^{-1}\|)$.

STEP 1. Suppose that $G : X \to X \oplus Y$ is any positively homogeneous bijection such that G and G^{-1} are uniformly continuous on B_X and $B_{X \oplus Y}$, respectively, and such that $\|G\|_\epsilon \leq L$ and $\|G^{-1}\|_\epsilon \leq L$ for some $\epsilon \in (0,1)$. Let $G_X : X \to X$ and $G_Y : X \to Y$ be functions such that
$$G(x) = (G_X(x), G_Y(x)), \quad x \in X.$$
Similarly, let T_1 and T_2 be the bounded linear maps from Y to Y so that
$$T(y) = (T_1(y), T_2(y)), \quad y \in Y.$$
Now define an "intertwining" function $\phi_G : X \to X \oplus Y$ by
$$\phi_G(x) = \Big(G^{-1}\big(G_X(x), (T_1 \circ G_Y)(x)\big), (T_2 \circ G_Y)(x)\Big), \quad x \in X.$$
It is readily verified that ϕ_G is a positively homogeneous bijection with
$$\phi_G^{-1}(x,y) = G^{-1}\Big(G_X(x), T^{-1}\big(G_Y(x), y\big)\Big), \quad (x,y) \in X \oplus Y.$$
By Lemma 12.22, we see that $\|\phi_G\|_\epsilon \leq L^3$ and $\|\phi_G^{-1}\|_\epsilon \leq L^3$.

STEP 2. Suppose that $G : X \to X \oplus Y$ and $H : X \to X \oplus Y$ are any positively homogeneous bijections such that both G and H are uniformly continuous on B_X, both G^{-1} and H^{-1} are uniformly continuous on $B_{X \oplus Y}$, and
$$\max\big(\|G\|_\epsilon, \|G^{-1}\|_\epsilon, \|H\|_\epsilon, \|H^{-1}\|_\epsilon\big) \leq L$$
for some $\epsilon \in (0,1)$.

CLAIM (Linking Lemma). *There exists a family $(F_t : 0 \leq t \leq 1)$ of positively homogeneous bijections, where $F_t : X^2 \to X^2 \oplus Y$ for $0 \leq t \leq 1$, such that:*
  (1) $\|F_t\|_\epsilon \leq \sqrt{2}L^3$ *and* $\|F_t^{-1}\|_\epsilon \leq \sqrt{2}L^3$ *for* $0 \leq t \leq 1$.
  (2) $\max(\|F_t - F_s\|, \|F_t^{-1} - F_s^{-1}\|) \leq 2\pi L^3 \sqrt{2}|t-s| + 4L^3\sqrt{2}\epsilon$ *if* $0 \leq s < t \leq 1$.
  (3) F_t *and* F_t^{-1} *are uniformly continuous on* B_{X^2} *and* $B_{X^2 \oplus Y}$, *respectively, for each* $0 \leq t \leq 1$.
  (4) $t \to F_t$ *and* $t \to F_t^{-1}$ *are continuous from* $[0,1]$ *into* $\mathcal{H}(X^2, X^2 \oplus Y)$ *and* $\mathcal{H}(X^2 \oplus Y, X^2)$, *respectively.*

Furthermore,
$$F_0(x_1, x_2) = (x_1, \phi_G(x_2)), \quad (x_1, x_2) \in X^2,$$
and
$$F_1(x_1, x_2) = (x_1, \phi_H(x_2)), \quad (x_1, x_2) \in X^2,$$

where ϕ_G and ϕ_H are the functions we construct in Step 1.

Proof of claim. We wish to find a family of functions that "link" F_0 to F_1. To that end, we introduce the *Linking Principle*[3].

Linking Principle

Let W be any Banach space and let $J_W : W \oplus W \to W \oplus W$ be the interchange operator given by the formula $J_W(w_1, w_2) = (w_2, -w_1)$. Now, for each $\theta \in [0, \frac{\pi}{2}]$, let $\Psi(\theta)$ be the linear operator on $W \oplus W$ represented by the matrix

$$\Psi(\theta) = \begin{pmatrix} \cos(\theta) & \sin(\theta) \\ -\sin(\theta) & \cos(\theta) \end{pmatrix},$$

so that

$$\Psi(\theta)(w_1, w_2) = \big(w_1 \cos(\theta) + w_2 \sin(\theta), -w_1 \sin(\theta) + w_2 \cos(\theta)\big)$$

for all $(w_1, w_2) \in W \oplus W$. Then $\|\Psi(\theta)\| \leq \sqrt{2}$ and $\|\Psi(\theta)^{-1}\| \leq \sqrt{2}$ for each $\theta \in [0, \frac{\pi}{2}]$, the maps $\theta \to \Psi(\theta)$ and $\theta \to \Psi(\theta)^{-1}$ are both Lipschitz continuous functions with constant $\sqrt{2}$, and we have $\Psi(0) = \mathrm{Id}_{W \oplus W}$ and $\Psi(\frac{\pi}{2}) = J_W$. Thus, Φ allows us to "smoothly link" the identity to J_W.

We comment that $\theta \to \Psi(-\theta)$ links the identity to the inverse of J_W in the sense that $\Psi(0) = \mathrm{Id}_{W \oplus W}$ and $\Psi(-\frac{\pi}{2}) = -J_W = (J_W)^{-1}$. We will refer to this as a *backwards link*.

We will define three maps. For the following definitions, x_1, x_2 are two points in X and y_1, y_2, y_3 are three points in Y. Let

$$M_1 : X^2 \to (X \oplus Y)^2,$$
$$M_2 : (X \oplus Y)^2 \to (X \oplus Y)^2 \oplus Y, \text{ and}$$
$$M_3 : (X \oplus Y)^2 \oplus Y \to X^2 \oplus Y$$

be given by the rules

$$M_1(x_1, x_2) = \big(H(x_1), G(x_2)\big) = \big(H_X(x_1), H_Y(x_1), G_X(x_2), G_Y(x_2)\big),$$
$$M_2(x_1, y_1, x_2, y_2) = \big(x_1, y_1, x_2, T_1(y_2), T_2(y_2)\big), \text{ and}$$
$$M_3(x_1, y_1, x_2, y_2, y_3) = \big(H^{-1}(x_1, y_1), G^{-1}(x_2, y_2), y_3\big).$$

Because of the assumptions on G, H, and T, and because we are using the ℓ_∞ sum for the direct sums, we have that M_i is invertible and both $\|M_i\|_\epsilon \leq L$ and $\|M_i^{-1}\|_\epsilon \leq L$ for each $i \in \{1, 2, 3\}$.

We claim that $F_0(x_1, x_2) = (M_3 M_2 M_1)(x_1, x_2)$. We can see this by direct computation:

$$(M_3 M_2 M_1)(x_1, x_2) = (M_3 M_2)\big(H_X(x_1), H_Y(x_1), G_X(x_2), G_Y(x_2)\big)$$
$$= M_3\Big(H_X(x_1), H_Y(x_1), G_X(x_2), T_1(G_Y(x_2)), T_2(G_Y(x_2))\Big)$$
$$= \Big(x_1, G^{-1}\big(G_X(x_2), T_1(G_Y(x_2))\big), T_2(G_Y(x_2))\Big)$$
$$= (x_1, \phi_G(x_2)) = F_0(x_1, x_2).$$

---

[3]The Linking Principle can be found in this formulation in [**65**].

We can achieve $F_1(x_1, x_2)$ as a similar composition, but for that we must introduce several interchange operators:
$$J_1(x_1, x_2) = (x_2, -x_1),$$
$$J_2(x_1, y_1, x_2, y_2) = (x_1, -y_2, x_2, y_1),$$
$$J_3(x_1, y_1, x_2, y_2, y_3) = (x_1, y_2, x_2, -y_1, y_3),$$
$$J_4(x_1, x_2, y_3) = (-x_2, x_1, y_3).$$

These are all interchange operators: J_1 is an operator on X^2, J_2 is an operator on $(X \oplus Y)^2$, J_3 is an operator on $(X \oplus Y)^2 \oplus Y$, and J_4 is an operator on $X^2 \oplus Y$. Using these operators, we can show that $F_1(x_1, x_2) = J_4 M_3 J_3 M_2 J_2 M_1 J_1$. The verification of this is like the one we did for F_0, albeit it a little more technically involved, and so we will omit the details.

We wish to link F_0 to F_1, and we will do this through repeated application of the Linking Principle. Start by making the following definitions:
$$F_{1/4} = M_3 M_2 M_1 J_1,$$
$$F_{1/2} = M_3 M_2 J_2 M_1 J_1,$$
$$F_{3/4} = M_3 J_3 M_2 J_2 M_1 J_1.$$

First, we link F_0 to $F_{1/4}$. Note that $J_1 = J_X$ (using the notation from when we introduced the Linking Principle). Consequently, we may link F_0 to $F_{1/4}$ using the linking map Ψ. For each $t \in (0, \frac{1}{4})$, let
$$F_t = M_3 M_2 M_1 \Psi(2\pi t).$$

The map J_2 is an operator on $(X \oplus Y)^2$ that acts as the identity on the two copies of X, but acts like $-J_Y$ on the two copies of Y. Thus, in order to link $F_{1/4}$ to $F_{1/2}$, we need a modification of the backwards link. Let
$$\Psi_{2,4}(\theta)(x_1, y_1, x_2, y_2) = \big(x_1, y_1 \cos(\theta) + y_2 \sin(\theta), x_2, -y_1 \sin(\theta) + y_2 \cos(\theta)\big).$$
This is the "forward link" map, so we will need to use $\theta \to \Psi_{2,4}(-\theta)$ to get the backwards link. With that in mind, for each $t \in (\frac{1}{4}, \frac{1}{2})$, let
$$F_t = M_3 M_2 \Psi_{2,4}\left(-2\pi t + \frac{\pi}{2}\right) M_1 J_1.$$

Next, we need to link $F_{1/2}$ to $F_{3/4}$. For this, we need to link $\mathrm{Id}_{(X \oplus Y)^2 \oplus Y}$ to J_3. The operator J_3 acts as J_Y on the second and fourth entries, and acts as the identity on the other three. Consequently, we can link the identity to J_3 using $\theta \to \Psi_{2,4}(\theta) \oplus \mathrm{Id}_Y$. Thus, for each $t \in (\frac{1}{2}, \frac{3}{4})$, let
$$F_t = M_3 \Big[\Psi_{2,4}\big(2\pi t - \pi\big) \oplus \mathrm{Id}_Y\Big] M_2 J_2 M_1 J_1.$$

The final step is to link $F_{3/4}$ to F_1. Notice in this case that $J_4 = -J_X \oplus \mathrm{Id}_Y$. Consequently, for each $t \in (\frac{3}{4}, 1)$, let
$$F_t = \left[\Psi\left(-2\pi t + \frac{3\pi}{2}\right) \oplus \mathrm{Id}_Y\right] M_3 J_3 M_2 J_2 M_1 J_1.$$

We need to verify that the family of functions $(F_t : 0 \le t \le 1)$ satisfies the conditions of the claim. By construction, it is evident that each of the functions is a positively homogeneous bijection such that each F_t is uniformly continuous on B_{X^2} and each F_t^{-1} is uniformly continuous on $B_{X^2 \oplus Y}$, and the maps $t \to F_t$ and $t \to F_t^{-1}$ are continuous. The bounds in (1) are also easily verified, since each F_t

is a composition of three maps having $\|\cdot\|_\epsilon$ bound L, a number of isometries, and a linear operator that comes from the linking process that has norm bounded by $\sqrt{2}$. Putting all of this together, we get that $\|F_t\|_\epsilon \leq \sqrt{2}L^3$ and $\|F_t^{-1}\|_\epsilon \leq \sqrt{2}L^3$ for each $t \in [0,1]$.

It remains to verify the inequalities in (2). For this, we will use Lemma 12.23. In hopes of maintaining clarity, we will introduce some extra notation. For each $k \in \{1,2,3,4\}$, let Φ_k be the linking operator on the interval $[\frac{k-1}{4}, \frac{k}{4}]$, so that

$$\Phi_1(t) = \Psi(2\pi t), \quad 0 \leq t \leq \frac{1}{4},$$

$$\Phi_2(t) = \Psi_{2,4}\left(-2\pi t + \frac{\pi}{2}\right), \quad \frac{1}{4} \leq t \leq \frac{1}{2},$$

$$\Phi_3(t) = \Psi_{2,4}\left(2\pi t - \pi\right) \oplus \mathrm{Id}_Y, \quad \frac{1}{2} \leq t \leq \frac{3}{4},$$

$$\Phi_4(t) = \Psi\left(-2\pi t + \frac{3\pi}{2}\right) \oplus \mathrm{Id}_Y, \quad \frac{3}{4} \leq t \leq 1.$$

With that in mind, we suppose that t and s belong to an interval $[\frac{k-1}{4}, \frac{k}{4}]$ for some $k \in \{1,2,3,4\}$. Then

$$\|F_t - F_s\| \leq \|M_3\|_\epsilon \|M_2\|_\epsilon \|M_1\|_\epsilon \left(\|\Phi_j(t) - \Phi_j(s)\| + \epsilon K\right),$$

where $K = \max(\|\Phi_j(t)\|, \|\Phi_j(s)\|) = \sqrt{2}$. Since the map $\theta \to \Psi(\theta)$ is Lipschitz continuous with constant $\sqrt{2}$, we have that $\|\Phi_j(t) - \Phi_j(s)\| \leq 2\pi\sqrt{2}|t-s|$, and so

$$\|F_t - F_s\| \leq L^3\left(2\pi\sqrt{2}|t-s| + \epsilon\sqrt{2}\right) = 2\pi L^3\sqrt{2}|t-s| + \epsilon L^3\sqrt{2}.$$

The above estimate applies when s and t are both in the same interval $[\frac{k-1}{4}, \frac{k}{4}]$. If we allow s and t to be any numbers in $[0,1]$, then we have to increase the bound. The most extreme case is when one of s or t is in $[0, \frac{1}{4})$ and the other is in $(\frac{3}{4}, 1]$. In the original statement of property (2), we supposed that $s < t$, and so assume that $0 \leq s < \frac{1}{4} < \frac{3}{4} < t \leq 1$. For ease of notation, let $c = 2\pi L^3\sqrt{2}$ and $k = L^3\sqrt{2}$. Then

$$\|F_t - F_s\| \leq \|F_t - F_{3/4}\| + \|F_{3/4} - F_{1/2}\| + \|F_{1/2} - F_{1/4}\| + \|F_{1/4} - F_s\|$$
$$\leq [c(t - \tfrac{3}{4}) + k\epsilon] + [c(\tfrac{1}{4}) + k\epsilon] + [c(\tfrac{1}{4}) + k\epsilon] + [c(\tfrac{1}{4} - s) + k\epsilon]$$
$$= c(t-s) + 4k\epsilon,$$

and so $\|F_t - F_s\| \leq 2\pi L^3\sqrt{2}(t-s) + 4L^3\sqrt{2}\epsilon$. Since this is the most extreme case, we conclude that the bound must hold for all $1 \geq t > s \geq 0$.

A similar argument shows the same bound for $\|F_t^{-1} - F_s^{-1}\|$, and so the proof of the claim is complete. $\diamond$

STEP 3. Recall that X and $X \oplus Y$ are assumed to be λ-uniformly close with $\lambda \leq L$. Thus, for each $n \in \{0,1,2,\ldots\}$, we can find a positively homogeneous bijection $G_n : X \to X \oplus Y$ such that G_n and G_n^{-1} are uniformly continuous on B_X and $B_{X \oplus Y}$, respectively, and so that

$$\|G_n\|_{e^{-2n-2}} \leq L \quad \text{and} \quad \|G_n^{-1}\|_{e^{-2n-2}} \leq L.$$

Next, for each $n \in \{0,1,2,\ldots\}$, let $F_n = \mathrm{Id}_X \oplus \phi_{G_n}$, where ϕ_{G_n} is the "intertwining" function that we constructed in Step 1.

Now let $n \in \{0,1,2,\ldots\}$. Applying the Linking Lemma (from Step 2), but translating the domain from $[0,1]$ to $[n, n+1]$, we can construct a family of functions

$(F_t : n \leq t \leq n+1)$ such that $F_n = \mathrm{Id}_X \oplus \phi_{G_n}$ and $F_{n+1} = \mathrm{Id}_X \oplus \phi_{G_{n+1}}$, and such that properties (1)–(4) from the Linking Lemma are satisfied; in particular, we have that
$$\max\left(\|F_t\|_{e^{-2n-2}},\ \|F_t^{-1}\|_{e^{-2n-2}}\right) \leq \sqrt{2}L^3$$
and
$$\max\left(\|F_t - F_s\|,\ \|F_t^{-1} - F_s^{-1}\|\right) \leq 2\pi L^3 \sqrt{2}|t-s| + 4L^3\sqrt{2}e^{-2n-2}$$
for $n \leq s < t \leq n+1$.

In order to apply the Linking Lemma in the way we did above, it is important to observe that
$$\|G_n\|_{e^{-2n-2}} \leq L \quad \text{and} \quad \|G_{n+1}\|_{e^{-2n-2}} \leq \|G_{n+1}\|_{e^{-2n-4}} \leq L$$
(and the same for the inverses), and so on the interval $[n, n+1]$, we can apply the Linking Lemma with $\epsilon_n = e^{-2n-2}$.

In general, the $\|\cdot\|_\epsilon$-norms decrease as ϵ increases. That is, if $\epsilon > \epsilon'$, then $\|\cdot\|_\epsilon \leq \|\cdot\|_{\epsilon'}$. (The infimum is taken over a larger set.) In particular if $n \leq t \leq n+1$, then $e^{-2t} \geq e^{-2n-2}$, and so
$$\|F_t\|_{e^{-2t}} \leq \|F_t\|_{e^{-2n-2}} \leq \sqrt{2}L^3.$$

The same estimate is true for the inverse function, and so for each $t \geq 1$, we have that
$$\max\left(\|F_t\|_{e^{-2t}},\ \|F_t^{-1}\|_{e^{-2t}}\right) \leq \sqrt{2}L^3.$$

Now suppose that $t > s \geq 1$. If there is a positive integer n such that s and t are in the same interval $[n, n+1]$, then
$$\|F_t - F_s\| \leq 2\pi L^3 \sqrt{2}|t-s| + 4L^3\sqrt{2}e^{-2n-2} \leq 2\pi L^3\sqrt{2}|t-s| + 4L^3\sqrt{2}e^{-2t}.$$

If instead s and t are not in the same interval, then we can find positive integers n and m for which $m \leq s \leq m+1 \leq n \leq t \leq n+1$. In order to estimate $\|F_t - F_s\|$ in this case, we use the triangle inequality repeatedly to get

$\|F_t - F_s\|$
$$\leq \|F_t - F_n\| + \|F_n - F_{n-1}\| + \cdots + \|F_{m+2} - F_{m+1}\| + \|F_{m+1} - F_s\|$$
$$\leq 2\pi L^3 \sqrt{2}(t-n) + 4L^3\sqrt{2}e^{-2n-2}$$
$$\quad + 2\pi L^3\sqrt{2} + 4L^3\sqrt{2}e^{-2(n-1)-2}$$
$$\quad \vdots$$
$$\quad + 2\pi L^3\sqrt{2} + 4L^3\sqrt{2}e^{-2(m+1)-2}$$
$$\quad + 2\pi L^3\sqrt{2}(m+1-s) + 4L^3\sqrt{2}e^{-2m-2}$$
$$= 2\pi L^3\sqrt{2}(t-s) + 4L^3\sqrt{2}\left(\sum_{k=m}^{n} e^{-2k-2}\right).$$

The sum in the last line is a finite geometric series, which we can bound using an infinite series:
$$\sum_{k=m}^{n} e^{-2k-2} \leq \sum_{k=m}^{\infty} e^{-2k-2} = e^{-2m-2}\sum_{k=0}^{\infty}(e^{-2})^k = \frac{e^{-2m-2}}{1-e^{-2}} \leq \frac{e^{-2s}}{1-e^{-2}}.$$

Therefore, if we assume that $t > s$, then
$$\|F_t - F_s\| \leq 2\pi L^3 \sqrt{2}(t-s) + \frac{4L^3\sqrt{2}}{1-e^{-2}} e^{-2s}.$$
In general, if we do not assume that $t > s$, then we have that
$$\|F_t - F_s\| \leq 2\pi L^3 \sqrt{2}|t-s| + \frac{4L^3\sqrt{2}}{1-e^{-2}} \max\left(e^{-2t}, e^{-2s}\right)$$
$$\leq 2\pi L^3 \sqrt{2}|t-s| + \frac{4L^3\sqrt{2}}{1-e^{-2}} \left(e^{-2t} + e^{-2s}\right).$$
We can derive the same bound for $\|F_t^{-1} - F_s^{-1}\|$ using the same argument.

STEP 4. We will construct a family of bijections $(H_t : t \geq 1)$ from X^2 to $X \oplus Y$ that satisfies the conditions of Lemma 12.12. For each $t \geq 1$, let
$$H_t(x_1, x_2) = F_{\ln(t)}(x_1, x_2), \quad (x_1, x_2) \in X^2.$$
We immediately have that H_t and H_t^{-1} ($t \geq 1$) are uniformly continuous on B_{X^2} and $B_{X \oplus Y}$, respectively, and the maps $t \to H_t$ and $t \to H_t^{-1}$ are continuous, since this is true for the collection of functions $(F_t : t \geq 0)$ and $(F_t^{-1} : t \geq 0)$. Thus, conditions (3) and (4) of Lemma 12.12 are satisfied.

We still need to verify that conditions (1) and (2) are satisfied. To that end, let $t \geq 1$. Then
$$\|H_t\|_{1/t^2} = \|F_{\ln(t)}\|_{t^{-2}} = \|F_{\ln(t)}\|_{e^{-2\ln(t)}} \leq \sqrt{2}L^3.$$
So if we choose $c > 1$, then $\|H_t\|_{c/t^2} \leq \sqrt{2}L^3$, and that will verify condition (1) with $\lambda = \sqrt{2}L^3$. For (2), we first suppose that $t > s \geq 1$. Then
$$\|H_t - H_s\| = \|F_{\ln(t)} - F_{\ln(s)}\|$$
$$\leq 2\pi L^3 \sqrt{2}|\ln(t) - \ln(s)| + \frac{4L^3\sqrt{2}}{1-e^{-2}} \left(e^{-2\ln(t)} + e^{-2\ln(s)}\right)$$
$$\leq 2\pi L^3 \sqrt{2}\ln(t/s) + \frac{4L^3\sqrt{2}}{1-e^{-2}} \left(\frac{1}{t^2} + \frac{1}{s^2}\right).$$
Let $\lambda = \sqrt{2}L^3$ and let $c = 2\pi L^3 \sqrt{2} > \frac{4L^3\sqrt{2}}{1-e^{-2}}$. Then we have
$$\|H_t\|_{c/t^2} \leq \lambda$$
and
$$\|H_t - H_s\| \leq c\left(\ln(t/s) + \frac{1}{t^2} + \frac{1}{s^2}\right).$$
The same estimates are true for the inverse functions. Therefore, we may apply Lemma 12.12 and conclude that X^2 is uniformly homeomorphic to $X \oplus Y$, as required. $\square$

We can now easily prove Theorem 12.21.

PROOF OF THEOREM 12.21. Let X and Y be uniformly close Banach spaces such that $X \approx X^2$ and $Y \approx Y^2$. Since X is uniformly close to Y, we conclude that $X \approx X^2 = X \oplus X$ is uniformly close to $X \oplus Y$. We also assumed that Y and Y^2 are linearly isomorphic, and so Theorem 12.24 tells us that X^2 and $X^2 \oplus Y$ are uniformly homeomorphic. Since X is linearly isomorphic to X^2, by assumption, we conclude that X is uniformly homeomorphic $X \oplus Y$.

Using the same argument, but changing the roles of X and Y, we can also show that Y is uniformly homeomorphic to $X \oplus Y$. It follows that X is uniformly homeomorphic Y, as claimed. □

Shortly, we will use Theorem 12.21 to prove a theorem of Ribe from 1984 [**101**]. To that end, we introduce what are known as the *Mazur maps*.

DEFINITION 12.25. Suppose that (Ω, μ) is a measure space, and let p and q be numbers in the interval $[1, \infty)$. We define the *Mazur map* from $\partial B_{L_p(\Omega,\mu)}$ to $\partial B_{L_q(\Omega,\mu)}$ to be the function $M_{p,q} : \partial B_{L_p(\Omega,\mu)} \to \partial B_{L_q(\Omega,\mu)}$ given by the formula

$$M_{p,q}(f) = |f|^{p/q} \operatorname{sign}(f), \quad f \in L_p(\Omega, \mu).$$

We note that

$$\|M_{p,q}(f)\|_q^q = \int_\Omega \left(|f(x)|^{p/q} |\operatorname{sign}(f(x))| \right)^q \mu(dx) = \|f\|_p^p,$$

and so the map $M_{p,q}$ is well-defined. We also observe that μ can be counting measure on $\mathbb{N}$, so that in that case, the Mazur map $M_{p,q}$ is a map from ∂B_{ℓ_p} to ∂B_{ℓ_q}.

COMMENT 12.26. We intend to show that the Mazur maps are uniformly continuous on their respective domains. In order to do so, we will make use of several inequalities.

LEMMA 12.27. *If u and v are positive numbers, then*

$$|u^\theta - v^\theta| \le |u - v|^\theta, \quad 0 < \theta < 1,$$

and

$$|u^\theta - v^\theta| \le \theta |u - v|(u + v)^{\theta - 1}, \quad \theta > 1.$$

PROOF. To see why the first inequality is true, consider the function

$$Q(x) = \frac{1 - x^\theta}{(1 - x)^\theta}, \quad 0 \le x < 1.$$

A simple computation shows that

$$Q'(x) = \frac{\theta(1 - x^{\theta-1})}{(1 - x)^{\theta+1}}, \quad 0 < x < 1,$$

which is negative on the interval $(0, 1)$. Therefore, the function Q is decreasing on that interval, and since $Q(0) = 1$, we have the desired inequality.

To prove the second inequality, let $f(x) = x^\theta$, where $\theta > 1$, and suppose that u and v are two distinct positive real numbers. Then f is differentiable on any open interval containing u and v, so by the mean value theorem, there exists some c between u and v so that

$$\frac{u^\theta - v^\theta}{u - v} = f'(c) = \theta c^{\theta - 1} \le \theta \Big(\max(u, v)\Big)^{\theta - 1} \le \theta(u + v)^{\theta - 1},$$

from which we get the desired inequality. □

The inequalities in the previous lemma allow us to estimate $|u^\theta - v^\theta|$ for positive numbers u and v, where $\theta > 0$. It will also be helpful to estimate $|u^\theta + v^\theta|$.

LEMMA 12.28. *If u and v are positive numbers, then*
$$|u^\theta + v^\theta| \le 2^{1-\theta}|u+v|^\theta, \quad 0 < \theta < 1,$$
and
$$|u^\theta + v^\theta| \ge 2^{1-\theta}|u+v|^\theta, \quad \theta > 1.$$

PROOF. Both of these inequalities can be demonstrated by looking at the function
$$Q(x) = \frac{1+x^\theta}{(1+x)^\theta}.$$
By once again looking at the derivative, we can see that this function has a maximum value at $x = 1$ when $0 < \theta < 1$, but it has a minimum value at $x = 1$ when $\theta > 1$. (We will omit the details in this case.) Notice that $Q(1) = 2^{1-\theta}$, which is why that constant appears in the inequality. □

The preceding two lemmas will allow us to prove the following inequalities of Mazur [80].

PROPOSITION 12.29. *If a and b are two real numbers, then*
$$\left||a|^\theta \operatorname{sign}(a) - |b|^\theta \operatorname{sign}(b)\right| \le 2|a-b|^\theta, \quad 0 < \theta < 1,$$
and
$$\left||a|^\theta \operatorname{sign}(a) - |b|^\theta \operatorname{sign}(b)\right| \le \theta|a-b|(|a|+|b|)^{\theta-1}, \quad \theta > 1.$$

The proof of this proposition follows immediately from the inequalities in Lemma 12.27 and Lemma 12.28 (although we do not need the second inequality from Lemma 12.28), and so we leave out the details.

THEOREM 12.30. *Suppose that (Ω, μ) is a measure space, and let p and q be numbers in the interval $[1, \infty)$. The Mazur map $M_{p,q}: \partial B_{L_p(\Omega,\mu)} \to \partial B_{L_q(\Omega,\mu)}$ is a uniformly continuous bijection and $M_{p,q}^{-1} = M_{q,p}$.*

PROOF. It is not difficult to see that $M_{p,q}^{-1} = M_{q,p}$, and so we will show that $M_{p,q}$ is uniformly continuous. If $p = q$, then $M_{p,q}$ is the identity map, and so nothing needs to be shown in that case. We will check the cases $p > q$ and $p < q$ separately.[4]

CASE 1. Suppose that $p > q$, so that $p/q > 1$, and let f and g be in $\partial B_{L_p(\Omega,\mu)}$. Then, using the second inequality in Proposition 12.29 (with $\theta = p/q$), we have

$$\|M_{p,q}(f) - M_{p,q}(g)\|_q^q = \int_\Omega \left||f(x)|^{\frac{p}{q}} \operatorname{sign}(f(x)) - |g(x)|^{\frac{p}{q}} \operatorname{sign}(g(x))\right|^q \mu(dx)$$
$$\le \left(\frac{p}{q}\right)^q \int_\Omega \left|f(x) - g(x)\right|^q \left(|f(x)| + |g(x)|\right)^{p-q} \mu(dx).$$

---

[4]We follow the approach of Mazur in [80]. For an alternate approach that uses the Gâteaux derivative of the Mazur map $M_{p,q}$ when $p > q$, see Theorem 9.1 in [13].

By Hölder's inequality (using the conjugate exponents $\frac{p}{q}$ and $\frac{p}{p-q}$), this is bounded by

$$\left(\frac{p}{q}\right)^q \|f-g\|_p^q \left[\int_\Omega \left(|f(x)|+|g(x)|\right)^p \mu(dx)\right]^{\frac{p-q}{p}}$$

$$\leq \left(\frac{p}{q}\right)^q \|f-g\|_p^q \left(\|f\|_p + \|g\|_p\right)^{p-q}.$$

Since $\|f\|_p = 1$ and $\|g\|_p = 1$ (by assumption), we have that

$$\|M_{p,q}(f) - M_{p,q}(g)\|_q^q \leq 2^{p-q} \left(\frac{p}{q}\right)^q \|f-g\|_p^q,$$

and consequently

$$\|M_{p,q}(f) - M_{p,q}(g)\|_q \leq 2^{(p-q)/q} \left(\frac{p}{q}\right) \|f-g\|_p.$$

Therefore, the map $M_{p,q}$ is Lipschitz continuous, and hence (in particular) it is uniformly continuous.

CASE 2. If $p < q$, then for f and g in $L_p(\Omega, \mu)$, we use the first inequality in Proposition 12.29 (with $\theta = p/q < 1$) to see that

$$\|M_{p,q}(f) - M_{p,q}(g)\|_q = \left(\int_\Omega \left||f(x)|^{\frac{p}{q}} \text{sign}(f(x)) - |g(x)|^{\frac{p}{q}} \text{sign}(g(x))\right|^q \mu(dx)\right)^{\frac{1}{q}}$$

$$\leq \left(\int_\Omega 2^q |f(x) - g(x)|^p \mu(dx)\right)^{\frac{1}{q}}$$

$$= 2\|f-g\|_p^{p/q}.$$

Therefore, $M_{p,q}$ is uniformly continuous in this case. (Indeed, in this case, it is Hölder continuous with exponent $\theta = \frac{p}{q}$.) □

We wish to investigate the CL-type for the Mazur maps. We saw in the proof of Theorem 12.30 that if $p > q$, then $M_{p,q}$ is a Lipschitz function with Lipschitz constant $2^{(p/q)-1}(p/q)$, and so in this case, $M_{p,q}$ is of CL-type $(2^{(p/q)-1}(p/q), \epsilon)$ for any $\epsilon > 0$. In the case where $p < q$, the Mazur map $M_{p,q}$ is Hölder continuous with exponent $\theta = \frac{p}{q}$. In order to determine the CL-type for this case, we use Lemma 12.31.

LEMMA 12.31. *If $0 \leq \theta \leq 1$, then $x^\theta \leq x + \epsilon(\theta)$ for $0 \leq x \leq 1$, where*

$$\epsilon(\theta) = \left(\frac{1}{\theta}\right)^{\frac{\theta}{\theta-1}} - \left(\frac{1}{\theta}\right)^{\frac{1}{\theta-1}}$$

and $\epsilon(\theta) \to 0$ as $\theta \to 1$.

PROOF. We wish to find the value of $\epsilon(\theta)$ for which $x^\theta - x \leq \epsilon(\theta)$ for $0 \leq x \leq 1$. If $f(x) = x^\theta - x$, then $f'(x) = 0$ when $x = (1/\theta)^{1/(\theta-1)}$. A simple check shows that this provides the maximum value for $f(x)$. Let $\epsilon(\theta) = f((1/\theta)^{1/(\theta-1)})$. Then $\epsilon(\theta)$ has the value indicated in the statement of the lemma. □

We will now collect what we have learned about Mazur maps in the following Corollary.

COROLLARY 12.32. *Suppose (Ω, μ) is a measure space, and let p and q be numbers in $[1, \infty)$ such that $p \neq q$. The Mazur map $M_{p,q} : \partial B_{L_p(\Omega,\mu)} \to \partial B_{L_q(\Omega,\mu)}$ is a uniformly continuous bijection with inverse $M_{p,q}^{-1} = M_{q,p}$ and having CL-type $(L_{p,q}, \epsilon_{p,q})$, where:*

- *If $p > q$, then $L_{p,q} = 2^{(p/q)-1}\left(\frac{p}{q}\right)$ and $\epsilon_{p,q} > 0$.*
- *If $p < q$, then $L_{p,q} = 2$ and*

$$\epsilon_{p,q} = 2\left(\frac{1}{p/q}\right)^{\frac{p/q}{(p/q)-1}} - 2\left(\frac{1}{p/q}\right)^{\frac{1}{(p/q)-1}},$$

and $\epsilon_{p,q} \to 0$ as $p/q \to 1$.

PROOF. We already know from Theorem 12.30 that $M_{p,q}$ is a uniformly continuous bijection and $M_{p,q}^{-1} = M_{q,p}$. It remains to show that $M_{p,q}$ is of CL-type (L, ϵ) for some L and ϵ that depend on p and q (or more precisely on the ratio p/q). We will consider the two cases $p > q$ and $p < q$ separately.

If $p > q$, then the map $M_{p,q}$ is Lipschitz continuous and

$$\|M_{p,q}(f) - M_{p,q}(g)\|_q \leq 2^{(p-q)/q}\left(\frac{p}{q}\right)\|f - g\|_p,$$

as we saw in the proof of Theorem 12.30. Thus, $M_{p,q}$ is of CL-type (L, ϵ) with $L = 2^{(p/q)-1}\left(\frac{p}{q}\right)$ and $\epsilon > 0$. (So ϵ can be any positive number in this case.)

Suppose instead that $p < q$. In this case, we saw in the proof of Theorem 12.30 that the map $M_{p,q}$ is Hölder continuous with constant 2 and exponent $\theta = \frac{p}{q}$. If we suppose that $\|f - g\| \geq 1$, then

$$\|M_{p,q}(f) - M_{p,q}(g)\|_q \leq 2\|f - g\|_p^{p/q} \leq 2\|f - g\|_p,$$

because $p/q < 1$. (This is not surprising, since $M_{p,q}$ is uniformly continuous, and consequently it is Lipschitz at large distances.) On the other hand, if $\|f - g\|_p \leq 1$, then by Lemma 12.31, we have that

$$\|M_{p,q}(f) - M_{p,q}(g)\|_q \leq 2\|f - g\|_p^{p/q} \leq 2\|f - g\|_p + 2\epsilon(p/q),$$

where $\epsilon(p/q)$ has the value given in the statement of Lemma 12.31. Then $M_{p,q}$ is of CL-type $(2, 2\epsilon(p/q))$. □

We are going to use the Mazur maps to find two Banach spaces that are uniformly homeomorphic but not linearly isomorphic. We recall that the notation $A \stackrel{\text{unif}}{\approx} B$ means that A and B are uniformly homeomorphic.

THEOREM 12.33 (Ribe). *There exists a sequence of positive numbers $(p_n)_{n=1}^{\infty}$ with $p_n > 1$ for each $n \in \mathbb{N}$ such that*

$$\oplus_2 \ell_{p_n} \stackrel{\text{unif}}{\approx} \left(\oplus_2 \ell_{p_n}\right) \oplus \ell_1,$$

where

$$\oplus_2 \ell_{p_n} = \left\{(x_1, x_2, x_3, \ldots) : x_n \in \ell_{p_n} \text{ for each } n \in \mathbb{N} \text{ and } \sum_{n=1}^{\infty} \|x_n\|_{p_n}^2 < \infty\right\}.$$

Consequently, there are Banach spaces that are uniformly homeomorphic, but not linearly isomorphic.

PROOF. Let $(p_n : n \in \mathbb{N})$ be a strictly decreasing sequence of positive numbers such that $2 > p_n > 1$ for all $n \in \mathbb{N}$ and such that $p_n \to 1$ as $n \to \infty$. Suppose also that $p_n^2 < p_{n-1}$ for each $n \in \mathbb{N}$. For each $n \in \mathbb{N}$, let $\epsilon_n = 2\epsilon(p_{n+1}/p_n)$, where $\epsilon(\theta)$ is given by the formula in Lemma 12.31. By choosing p_n so that $p_n^2 < p_{n-1}$, we can be sure that $p_n/p_{n-1} < p_{n+1}/p_n$, and so $\epsilon_n < \epsilon_{n-1}$ for each n. Since $\epsilon_n \to 0$ as $n \to \infty$, we may (and do) choose p_n so that $\epsilon_n \le 2^{-n}$ for each $n \in \mathbb{N}$. For notational simplicity, we let $X = \oplus_2 \ell_{p_n}$ and we let $\boldsymbol{x} = (x_1, x_2, \ldots)$ denote an element in X. Our objective is to show that X is uniformly close to $X \oplus_2 \ell_1$, but first we set the stage.

For each p and q, let $\widetilde{M}_{p,q} : \ell_p \to \ell_q$ be the norm-preserving positively homogeneous extension of $M_{p,q}$ given by the formula

$$\widetilde{M}_{p,q}(x) = \begin{cases} \|x\| M_{p,q}\left(\frac{x}{\|x\|}\right), & x \ne 0, \\ 0, & x = 0. \end{cases}$$

By Proposition 12.13, we know that $\widetilde{M}_{p,q}$ is uniformly continuous on the unit ball B_{ℓ_p}, and hence uniformly continuous on bounded subsets of ℓ_p, and for any $\epsilon > 0$, we have that $\|\widetilde{M}_{p,q}\|_\epsilon < 4\|M_{p,q}\|_\epsilon + 2$.

For each $n \in \mathbb{N}$, we chose p_n and p_{n+1} so that $2 > p_n > p_{n+1} > 1$. Consequently, the map M_{p_{n+1}, p_n} is of CL-type $(2, \epsilon_n)$. The inverse map $M_{p_n, p_{n+1}}$ is Lipschitz with constant

$$L_{p_n, p_{n+1}} = 2^{(p_n/p_{n+1})-1}\left(\frac{p_n}{p_{n+1}}\right) \le 4.$$

Thus $M_{p_n, p_{n+1}}$ is of CL-type $(4, \epsilon)$ for any $\epsilon > 0$. In particular, $M_{p_n, p_{n+1}}$ is of CL-type $(4, \epsilon_n)$. Then, for each $n \in \mathbb{N}$, the norm-preserving positively homogeneous extensions $\widetilde{M}_{p_n, p_{n+1}}$ and $\widetilde{M}_{p_{n+1}, p_n}$ both satisfy the inequalities

$$\|\widetilde{M}_{p_n, p_{n+1}}\|_{\epsilon_n} \le 18 \quad \text{and} \quad \|\widetilde{M}_{p_{n+1}, p_n}\|_{\epsilon_n} \le 18.$$

We are now ready to construct a family of bijections between X and $X \oplus_2 \ell_1$. For notational convenience, we will write $\boldsymbol{x} \oplus w$ for an element of $X \oplus_2 \ell_1$, where $\boldsymbol{x} \in X$ and $w \in \ell_1$. Now, for each $n \in \mathbb{N}$, define a map $g_n : X \to X \oplus_2 \ell_1$ by

$$g_n(\boldsymbol{x}) = \left(x_1, \ldots, x_{n-1}, \widetilde{M}_{p_{n+1}, p_n}(x_{n+1}), \widetilde{M}_{p_{n+2}, p_{n+1}}(x_{n+2}), \ldots\right) \oplus \widetilde{M}_{p_n, 1}(x_n).$$

(We are using the $\oplus$ symbol to set apart the term in ℓ_1 instead of writing it inside the parentheses.) Then g_n is a bijection with inverse

$$g_n^{-1}(\boldsymbol{x} \oplus w) = \left(x_1, \ldots, x_{n-1}, \widetilde{M}_{1, p_n}(w), \widetilde{M}_{p_n, p_{n+1}}(x_n), \widetilde{M}_{p_{n+1}, p_{n+2}}(x_{n+1}), \ldots\right),$$

where $\boldsymbol{x} \in X$ and $w \in \ell_1$. Note that g_n and g_n^{-1} are norm-preserving, because the extended Mazur maps are norm-preserving.

When constructing g_n and g_n^{-1}, in addition to the Mazur maps of the form $M_{p_k, p_{k+1}}$ and M_{p_{k+1}, p_k}, we also needed $M_{p_n, 1}$ and M_{1, p_n}—or rather their extensions. Let us determine the CL-type of $M_{p_n, 1}$ and M_{1, p_n}. Recall that $(p_n)_{n=1}^\infty$ is a strictly decreasing sequence with limit 1. In particular, we have that $p_n > 1$ for each $n \in \mathbb{N}$. Thus, the map M_{1, p_n} is of CL-type $(2, 2\epsilon(1/p_n))$, where

$$2\epsilon(1/p_n) < 2\epsilon(p_n/p_{n-1}) = \epsilon_{n-1} \le 2^{-n+1}.$$

The first of these inequalities follows from the assumption that $p_n^2 < p_{n-1}$. The inverse map $M_{p_n,1}$ is Lipschitz with constant
$$L_{p_n,1} = 2^{p_n-1} p_n \le 2.$$
Thus $M_{p_n,1}$ is of CL-type $(2,\epsilon)$ for any $\epsilon > 0$, but in particular it is of CL-type $(2, \epsilon_{n-1})$. Thus, the extension maps satisfy
$$\|\widetilde{M}_{1,p_n}\|_{\epsilon_{n-1}} \le 10 \quad \text{and} \quad \|\widetilde{M}_{p_n,1}\|_{\epsilon_{n-1}} \le 10.$$
Now suppose that $\boldsymbol{x}$ and $\boldsymbol{y}$ are two elements of X. Then
$$g_n(\boldsymbol{x}) - g_n(\boldsymbol{y})$$
$$= \left(x_1 \ldots, x_{n-1}, \widetilde{M}_{p_{n+1},p_n}(x_{n+1}), \widetilde{M}_{p_{n+2},p_{n+1}}(x_{n+2}), \ldots \right) \oplus \widetilde{M}_{p_n,1}(x_n)$$
$$- \left(y_1 \ldots, y_{n-1}, \widetilde{M}_{p_{n+1},p_n}(y_{n+1}), \widetilde{M}_{p_{n+2},p_{n+1}}(y_{n+2}), \ldots \right) \oplus \widetilde{M}_{p_n,1}(y_n),$$
and so
$$\|g_n(\boldsymbol{x}) - g_n(\boldsymbol{y})\|_{X \oplus_2 \ell_1}^2$$
$$= \sum_{k=1}^{n-1} \|x_k - y_k\|_{\ell_{p_k}}^2 + \sum_{k=n+1}^{\infty} \|\widetilde{M}_{p_k,p_{k-1}}(x_k) - \widetilde{M}_{p_k,p_{k-1}}(y_k)\|_{\ell_{p_{k-1}}}^2$$
$$+ \|\widetilde{M}_{p_n,1}(x_n) - \widetilde{M}_{p_n,1}(y_n)\|_{\ell_1}^2.$$
Since $\|\widetilde{M}_{p_n,p_{n-1}}\|_{\epsilon_{n-1}} \le 18$ and $\|\widetilde{M}_{p_n,1}\|_{\epsilon_{n-1}} \le 10 < 18$, this is bounded by
$$\sum_{k=1}^{n-1} \|x_k - y_k\|_{\ell_{p_k}}^2 + \sum_{k=n+1}^{\infty} 18^2 \max\left(\|x_k - y_k\|_{\ell_{p_k}}, \epsilon_{k-1}\|x_k\|_{\ell_{p_k}}, \epsilon_{k-1}\|y_k\|_{\ell_{p_k}} \right)^2$$
$$+ 18^2 \max\left(\|x_n - y_n\|_{\ell_{p_n}}, \epsilon_{n-1}\|x_n\|_{\ell_{p_n}}, \epsilon_{n-1}\|y_n\|_{\ell_{p_n}} \right)^2$$
$$= \sum_{k=1}^{n-1} \|x_k - y_k\|_{\ell_{p_k}}^2 + 18^2 \sum_{k=n}^{\infty} \max\left(\|x_k - y_k\|_{\ell_{p_k}}, \epsilon_{k-1}\|x_k\|_{\ell_{p_k}}, \epsilon_{k-1}\|y_k\|_{\ell_{p_k}} \right)^2.$$
The square of a maximum of three numbers is the maximum of the respective three squares, and also the maximum of three numbers is less than the sum of those three numbers, and so we have that
$$\|g_n(\boldsymbol{x}) - g_n(\boldsymbol{y})\|_{X \oplus_2 \ell_1}^2$$
$$\le \sum_{k=1}^{n-1} \|x_k - y_k\|_{\ell_{p_k}}^2 + 18^2 \sum_{k=n}^{\infty} \left(\|x_k - y_k\|_{\ell_{p_k}}^2 + \epsilon_{k-1}^2 \|x_k\|_{\ell_{p_k}}^2 + \epsilon_{k-1}^2 \|y_k\|_{\ell_{p_k}}^2 \right)$$
$$\le 18^2 \left(\sum_{k=1}^{\infty} \|x_k - y_k\|_{\ell_{p_k}}^2 + \sum_{k=n}^{\infty} \epsilon_{k-1}^2 \|x_k\|_{\ell_{p_k}}^2 + \sum_{k=n}^{\infty} \epsilon_{k-1}^2 \|y_k\|_{\ell_{p_k}}^2 \right).$$
Since $(\epsilon_k)_{k=1}^{\infty}$ is a decreasing sequence, we have that $\epsilon_{k-1} \le \epsilon_{n-1}$ for all $k \ge n$, and thus
$$\|g_n(\boldsymbol{x}) - g_n(\boldsymbol{y})\|_{X \oplus_2 \ell_1}^2 \le 18^2 \left(\|\boldsymbol{x} - \boldsymbol{y}\|_X^2 + \epsilon_{n-1}^2 \sum_{k=n}^{\infty} \|x_k\|_{\ell_{p_k}}^2 + \sum_{k=n}^{\infty} \|y_k\|_{\ell_{p_k}}^2 \right)$$
$$\le 18^2 \left(\|\boldsymbol{x} - \boldsymbol{y}\|_X^2 + \epsilon_{n-1}^2 \|\boldsymbol{x}\|_X^2 + \epsilon_{n-1}^2 \|\boldsymbol{y}\|_X^2 \right).$$

Finally, we observe that the sum of three numbers is less than three times the largest of the three numbers, and so

$$\|g_n(\boldsymbol{x}) - g_n(\boldsymbol{y})\|^2_{X \oplus_2 \ell_1} \le 18^2 \cdot 3 \max\left(\|\boldsymbol{x} - \boldsymbol{y}\|^2_X,\ \epsilon^2_{n-1}\|\boldsymbol{x}\|^2_X,\ \epsilon^2_{n-1}\|\boldsymbol{y}\|^2_X\right).$$

Therefore,

$$\|g_n(\boldsymbol{x}) - g_n(\boldsymbol{y})\|_{X \oplus_2 \ell_1} \le 18\sqrt{3} \max\left(\|\boldsymbol{x} - \boldsymbol{y}\|_X,\ \epsilon_{n-1}\|\boldsymbol{x}\|_X,\ \epsilon_{n-1}\|\boldsymbol{y}\|_X\right),$$

and so $\|g_n\|_{\epsilon_{n-1}} \le 18\sqrt{3}$ for each $n \in \mathbb{N}$.

We can use a similar argument to show that $\|g_n^{-1}\|_{\epsilon_{n-1}} \le 18\sqrt{3}$ (although since the Mazur maps in that case are Lipschitz we can do better). However, since the argument is similar, we omit the details.

It remains to show that each g_n is uniformly continuous on the unit ball. This follows from the uniform continuity of the extended Mazur maps. Suppose that $\epsilon > 0$ is given. Suppose that $\boldsymbol{x}$ and $\boldsymbol{y}$ in B_X. Earlier, we saw that

$$\|g_n(\boldsymbol{x}) - g_n(\boldsymbol{y})\|^2_{X \oplus_2 \ell_1}$$
$$= \sum_{k=1}^{n-1} \|x_k - y_k\|^2_{\ell_{p_k}} + \sum_{k=n+1}^{\infty} \|\widetilde{M}_{p_k, p_{k-1}}(x_k) - \widetilde{M}_{p_k, p_{k-1}}(y_k)\|^2_{\ell_{p_{k-1}}}$$
$$+ \|\widetilde{M}_{p_n, 1}(x_n) - \widetilde{M}_{p_n, 1}(y_n)\|^2_{\ell_1}.$$

For ease of notation, let

$$\Delta_k(x_k, y_k) = \begin{cases} \|\widetilde{M}_{p_k, p_{k-1}}(x_k) - \widetilde{M}_{p_k, p_{k-1}}(y_k)\|_{\ell_{p_{k-1}}} & \text{if } k \ne n, \\ \|\widetilde{M}_{p_n, 1}(x_n) - \widetilde{M}_{p_n, 1}(y_n)\|_{\ell_1} & \text{if } k = n. \end{cases}$$

Then for each $k \ge n$,

$$\Delta_k(x_k, y_k) \le 18 \max\left(\|x_k - y_k\|_{\ell_{p_k}},\ 2^{-k+1}\|x_k\|_{\ell_{p_k}},\ 2^{-k+1}\|y_k\|_{\ell_{p_k}}\right).$$

There exists an $N \in \mathbb{N}$ such that $2^{-N+1} < \frac{\epsilon}{45}$. We may assume that $N > n$. Then by uniform continuity of the extended Mazur maps, we can find a $\delta_0 > 0$ such that $\Delta_k(x_k, y_k) \le \frac{\epsilon}{\sqrt{3N}}$ whenever $\|x_k - y_k\| < \delta_0$, for each $k \in \{n+1, \cdots N\}$.

Let J be the collection of indices $k > N$ for which

$$\|x_k - y_k\|_{\ell_{p_k}} \ge \max\left(2^{-k+1}\|x_k\|_{\ell_{p_k}},\ 2^{-k+1}\|y_k\|_{\ell_{p_k}}\right),$$

and let J' be the collection of indices $k > N$ for which that inequality is not true. Then

$$\sum_{k=N+1}^{\infty} \Delta(x_k, y_k)^2 = \sum_{k \in J} \Delta(x_k, y_k)^2 + \sum_{k \in J'} \Delta(x_k, y_k)^2$$
$$\le 18^2 \sum_{k \in J} \|x_k - y_k\|^2_{\ell_{p_k}} + 18^2 \cdot 2^{-2N+2}\left(\sum_{k \in J'} \|x_k\|^2_{\ell_{p_k}} + \sum_{k \in J'} \|y_k\|^2_{\ell_{p_k}}\right)$$
$$\le 18^2 \|\boldsymbol{x} - \boldsymbol{y}\|^2_X + 18^2 \cdot 2^{-2N+2}\left(\|\boldsymbol{x}\|^2_X + \|\boldsymbol{y}\|^2_X\right).$$

Consequently, since $\|\boldsymbol{x}\|_X = 1$ and $\|\boldsymbol{y}\|_X = 1$,

$$\|g_n(\boldsymbol{x}) - g_n(\boldsymbol{y})\|_{X\oplus_2\ell_1}^2 \leq \sum_{k=1}^{n} \|x_k - y_k\|_{\ell_{p_k}}^2 + \sum_{k=n+1}^{N} \Delta(x_k, y_k)^2 + \sum_{k=N+1}^{\infty} \Delta(x_k, y_k)^2$$

$$\leq 325\|\boldsymbol{x} - \boldsymbol{y}\|_X^2 + \sum_{k=n+1}^{N} \Delta(x_k, y_k)^2 + 648 \cdot (2^{-N+1})^2.$$

Now, choose $\delta > 0$ so that $\delta < \min(\delta_0, \frac{\epsilon}{32})$. If $\|\boldsymbol{x} - \boldsymbol{y}\|_X < \delta$, then

$$\|g_n(\boldsymbol{x}) - g_n(\boldsymbol{y})\|_{X\oplus_2\ell_1}^2 < 325\left(\frac{\epsilon}{32}\right)^2 + (N-n-1)\left(\frac{\epsilon}{\sqrt{3N}}\right)^2 + 648\left(\frac{\epsilon}{45}\right)^2 < \epsilon^2,$$

and so

$$\|g_n(\boldsymbol{x}) - g_n(\boldsymbol{y})\|_{X\oplus_2\ell_1} < \epsilon.$$

We conclude that g_n is uniformly continuous on the unit ball of X. A similar argument shows that g_n^{-1} is uniformly continuous on the unit ball (although again we can do better, since the Mazur maps are Lipschitz in this case).

Now for any $\epsilon > 0$, we can find an $n \in \mathbb{N}$ so that both $\|g_n\|_\epsilon < 18\sqrt{3}$ and $\|g_n^{-1}\|_\epsilon < 18\sqrt{3}$, and so the family of function $(g_n)_{n=1}^\infty$ is what we need to show that X and $X \oplus_2 \ell_1$ are uniformly close. Then, by Theorem 12.21, we conclude that

$$\oplus_2 \ell_{p_n} \stackrel{\text{unif}}{\approx} \left(\oplus_2 \ell_{p_n}\right) \oplus \ell_1.$$

Finally, we note that X is reflexive, but $X\oplus_2\ell_1$ is not, and so the two Banach spaces are not linearly isomorphic, even though they are uniformly homeomorphic. □

COMMENT 12.34. One of the motivating questions of nonlinear functional analysis is "In what ways does the nonlinear geometry of a Banach space determine the linear structure?" We have found examples of Lipschitz isomorphic spaces that are not linearly isomorphic, but those examples were not separable. In Theorem 12.33, we gave two separable Banach spaces that are uniformly homeomorphic, but not linearly isomorphic.

Theorem 12.33 was provided by Ribe in 1984 [**101**], although the proof was different. One interesting fact we we observed in Theorem 12.33 is that the property of reflexivity is not preserved by uniform homeomorphism, since $\oplus_2\ell_{p_n}$ is reflexive, but $\left(\oplus_2 \ell_{p_n}\right) \oplus_2 \ell_1$ is not.

In 1985, Aharoni and Lindenstrauss extended the method of Ribe to show a similar result if ℓ_1 is replaced with ℓ_p for $p > 1$, and the sequence $(p_n : n \in \mathbb{N})$ is chosen so that it converges to p, but none of the terms are p [**4**]. This provides an example of two separable reflexive Banach spaces that are uniformly homeomorphic but not linearly isomorphic.

An important ingredient that we used in Theorem 12.33 is the fact that the Mazur maps $M_{p,q} : \partial B_{\ell_p} \to \partial B_{\ell_q}$ $(1 < p < \infty, 1 < q < \infty)$ are uniform homeomorphisms. It might be tempting to guess from this fact that ℓ_p and ℓ_q are uniformly homeomorphic, but this is not the case (if $p \neq q$). We will address this question in Chapter 13.

CHAPTER 13

The unique uniform structure of sequence spaces

In this chapter, we will investigate the following theorem of Johnson, Lindenstrauss, and Schechtman [50].

THEOREM 13.1 (Johnson, Lindenstrauss, and Schechtman). *If X is uniformly homeomorphic to ℓ_p, where $1 < p < \infty$, then X is linearly isomorphic to ℓ_p.*

REMARK 13.2. One consequence of Theorem 13.1 is that ℓ_p and ℓ_q are uniformly homeomorphic only if $p = q$. For function spaces, this question was settled in the 1960s by Lindenstrauss [75] and Enflo [32], when they showed that $L_p(0,1)$ and $L_q(0,1)$ are uniformly homeomorphic only if $p = q$, where p and q are in the interval $[1, \infty)$. (Lindenstrauss showed it when $\max(p,q) > 2$ and Enflo showed it for the case where $\max(p,q) \leq 2$.)

The next task was to determine if ℓ_p and $L_p(0,1)$ are uniformly homeomorphic. When $p = 2$, because ℓ_2 and $L_2(0,1)$ are both separable Hilbert spaces, they are not only uniformly homeomorphic but linearly isomorphic.[1] This is the only case where this is true, however. In 1970, Enflo showed that ℓ_1 is not uniformly homeomorphic to $L_1(0,1)$ (in an unpublished work[2]). Later, in 1987, Bourgain showed that ℓ_p is not uniformly homeomorphic to $L_p(0,1)$ for $1 < p < 2$ [19]. This circle of ideas was completed in 1994, when Gorelik showed that ℓ_p is not uniformly homeomorphic to $L_p(0,1)$ for $2 < p < \infty$ [41].

In [50], Johnson, Lindenstrauss, and Schechtman gave a proof of Theorem 13.1 using the Gorelik principle (Theorem 10.6), which was developed from the techniques introduced by Gorelik in [41]. We will not prove Theorem 13.1 in this way. Instead, we will follow the development of Kalton and Randrianarivony in [67].

Before proving Theorem 13.1, we will start with a reduction.

THEOREM 13.3. *If X is uniformly homeomorphic to ℓ_p, where $1 < p < \infty$, then X is linearly isomorphic to a complemented subspace of $L_p(\Omega, \mu)$ for some measure space (Ω, μ).*

PROOF. Assume X is a separable Banach space that is uniformly homeomorphic to ℓ_p for some p such that $1 < p < \infty$. We claim that X is reflexive. To see this, observe that since X is uniformly homeomorphic to ℓ_p, it is crudely finitely representable in ℓ_p, by Corollary 11.38. But we also know that ℓ_p is uniformly convex (see Example 11.24), and so X is reflexive (see Remark 11.31).

---

[1] In fact, in [33], Enflo proved that every Banach space uniformly homeomorphic to a Hilbert space is actually isomorphic to it.
[2] While Enflo did not publish this result, it can be found in a collection of notes by Benyamini [14] (the example following Theorem 5.4) and in [13] (as part of Theorem 10.13). We also include the proof at the end of this chapter. (See Theorem 13.40 in Comment 13.39.)

Now choose $\mathcal{U}$ to be a non-principal ultrafilter. Another consequence of X being uniformly homeomorphic to ℓ_p is that $X_\mathcal{U}$ is Lipschitz isomorphic to $(\ell_p)_\mathcal{U}$. (See Theorem 11.37.)

Let $F : X_\mathcal{U} \to (\ell_p)_\mathcal{U}$ be a Lipschitz isomorphism, which exists by assumption. We can view X as a subspace of $X_\mathcal{U}$, via the map $x \to (x,x,x,\ldots)$, and so we can view $F(X)$ is a subset of $(\ell_p)_\mathcal{U}$. The space $(\ell_p)_\mathcal{U}$ is a large (i.e., nonseparable) $\mathcal{L}_p$-space[3], but $F(X)$ is separable. Let Y_1 be a closed, separable sublattice of $(\ell_p)_\mathcal{U}$ that contains $F(X)$. Now let X_1 be a closed separable subspace of $X_\mathcal{U}$ that contains $F^{-1}(Y_1)$. Next, let Y_2 be a closed, separable sublattice of $(\ell_p)_\mathcal{U}$ that contains $F(X_1)$, and let X_2 be a closed separable subspace of $X_\mathcal{U}$ that contains $F^{-1}(Y_2)$.

Continuing inductively, construct an increasing sequence of sets $(X_n : n \in \mathbb{N})$ and an increasing sequence of sets $(Y_n : n \in \mathbb{N})$ such that the set X_n is a separable closed subspace of $X_\mathcal{U}$ for each n, the set Y_n is a separable closed sublattice of $(\ell_p)_\mathcal{U}$ for each n, and such that

$$F(X_{n-1}) \subseteq Y_n, \quad F^{-1}(Y_n) \subseteq X_n$$

for each n (where $X_0 = X$). Let

$$X_\infty = \overline{\bigcup_{n=1}^\infty X_n} \quad \text{and} \quad Y_\infty = \overline{\bigcup_{n=1}^\infty Y_n}.$$

Then $F|_{X_\infty} : X_\infty \to Y_\infty$ is a Lipschitz isomorphism, where Y_∞ is a separable $\mathcal{L}_p$-space, and thus a complemented subspace of $L_p(\Omega, \mu)$ for some measure space (Ω, μ). Consequently, by Theorem 9.28, the space X_∞ is linearly isomorphic to a complemented subspace of Y_∞, and so X_∞ is linearly isomorphic to a complemented subspace of $L_p(\Omega, \mu)$.

It remains only to show that X is complemented in X_∞. We will find a projection $P : X_\infty \to X$. For any $(x_n)_{n=1}^\infty \in X_\infty$, the limit

$$\lim_\mathcal{U} x^*(x_n)$$

through the ultrafilter $\mathcal{U}$ exists and

$$\left|\lim_\mathcal{U} x^*(x_n)\right| \leq \|x^*\| \lim_\mathcal{U} \|x_n\| = \|x^*\| \|(x_n)_{n=1}^\infty\|_{X_\mathcal{U}}.$$

Consequently, the sequence $(x_n)_{n=1}^\infty$ determines an element of X^{**}, which we will call $\lim_\mathcal{U} x_n$, where

$$\left(\lim_\mathcal{U} x_n\right)(x^*) = \lim_\mathcal{U} x^*(x_n)$$

and

$$\|\lim_\mathcal{U} x_n\|_{X^{**}} \leq \|(x_n)_{n=1}^\infty\|_{X_\mathcal{U}}.$$

Since X is reflexive, we have in fact that $\lim_\mathcal{U} x_n$ is an element of X. Now we define $P : X_\infty \to X$ by

$$P\big((x_n)_{n=1}^\infty\big) = \lim_\mathcal{U} x_n.$$

Since $P\big((x_n)_{n=1}^\infty\big) = x$ for any $x \in X$, it follows that P is a projection, which completes the proof.[4] $\square$

---

[3]In [**47**], Henson showed (see Theorem 2.4 in that document) that $(\ell_p)_\mathcal{U}$ is isometric to $\ell_p((\mathbb{N})_\mathcal{U}) \oplus_p (L_p([0,1]))_\mathcal{U}$, and also that $(L_p([0,1]))_\mathcal{U}$ is isometric to the ℓ_p-sum of $\mathfrak{c}$ copies of $L_p([0,1]^\mathfrak{c})$. We gave the latter half of that as Theorem 11.21 in the current text.

[4]The argument that X is complemented in X_∞ is similar to the proof of Proposition 11.15(ii).

COMMENT 13.4. In the proof of Theorem 13.3, we are making free use of some classic results about $\mathcal{L}_p$-spaces that were provided by Lindenstrauss and Pełczyński in their highly influential paper of 1968 [**73**].

We recall (see also Comment 9.26) that for $p \geq 1$ (including $p = \infty$), a Banach space X is called an $\mathcal{L}_p$-*space* if there exists some $\lambda \geq 1$ such that for every finite dimensional subspace F of X, there is a finite dimensional subspace G containing F such that $d_{\mathrm{BM}}(G, \ell_p^n) \leq \lambda$, where $n = \dim(G)$ and $d_{\mathrm{BM}}(G, \ell_p^n)$ is the Banach–Mazur distance between G and ℓ_p^n:

$$d_{\mathrm{BM}}(G, \ell_p^n) = \inf \left\{ \|T\| \|T^{-1}\| \ \Big| \ T: G \to \ell_p^n \text{ is an isomorphism} \right\}.$$

The specific result from [**73**] that we are using in the proof of Theorem 13.3 is the following.

THEOREM 13.5 (Theorem 7.1 in [**73**]). *If $1 < p < \infty$, then an $\mathcal{L}_p$-space is isomorphic to a complemented subspace of $L_p(\Omega, \mu)$ for some measure space (Ω, μ).*

More generally, in their 1968 paper, Lindenstrauss and Pełczyński showed that an $\mathcal{L}_p$-space X (for $1 \leq p \leq \infty$) is isomorphic to a complemented subspace of $L_p(\Omega, \mu)$ for some measure space (Ω, μ) if and only if X is complemented in X^{**}. When $1 < p < \infty$, this is immediate since an $\mathcal{L}_p$-space is reflexive when $1 < p < \infty$.

Theorem 13.3 tells us that if X is uniformly homeomorphic to ℓ_p $(1 < p < \infty)$, then X is an $\mathcal{L}_p$-space. For the next step in the proof of Theorem 13.1, we need the following theorem of Johnson and Odell from 1974 [**52**], given here without proof.[5]

THEOREM 13.6 (Johnson–Odell). *Suppose that $1 < p < \infty$, $p \neq 2$. If X is a separable $\mathcal{L}_p$-space, then either*

(a) *X has a subspace linearly isomorphic to ℓ_2 or*
(b) *X is linearly isomorphic to ℓ_p.*

So in order to show that X is linearly isomorphic to ℓ_p (for $p \neq 2$), it suffices to show that X does not contain a subspace isomorphic to ℓ_2. For the case $p = 2$, we already know that a separable $\mathcal{L}_2$-space is a separable Hilbert space, and hence a separable $\mathcal{L}_2$-space is linearly isomorphic to ℓ_2. That means that if X is uniformly homeomorphic to ℓ_2, it is isomorphic to ℓ_2.

We should emphasize that Theorem 13.6 is for $p \in (1, \infty)$. There exists a separable $\mathcal{L}_1$-space that is not isomorphic to a complemented subspace of $L_1(\Omega, \mu)$ for any measure space (Ω, μ). It can be shown that a space X is a separable $\mathcal{L}_1$-space if and only if X^* is isomorphic to ℓ_∞. Similarly, a space X is an $\mathcal{L}_\infty$-space if and only if X^{**} is isomorphic to a space of the type $\mathcal{C}(K)$, where K is a compact Hausdorff space. (These results can be found in Lindenstrauss and Pełczyński's landmark 1968 paper [**73**].)

Returning to the proof of Theorem 13.1, we have determined that if X is uniformly homeomorphic to ℓ_p, where $1 < p < \infty$ and $p \neq 2$, then either (a) X has a subspace isomorphic to ℓ_2 or (b) X is linearly isomorphic to ℓ_p. Consequently, our objective is to show that ℓ_2 is not isomorphic to a subspace of X.

Suppose to the contrary that ℓ_2 is isomorphic to a subspace of X, where X is uniformly homeomorphic to ℓ_p, where $p \neq 2$ (and $1 < p < \infty$). It follows that

---
[5]See Corollary 1 in [**52**].

there exists an invertible uniformly continuous function from ℓ_2 into ℓ_p. By the Corson–Klee lemma[6] (Lemma 11.36), there exists a function $f : \ell_2 \to \ell_p$ that is a *coarse Lipschitz embedding*. That is, there exist constants $C > 0$ and $c > 0$ such that

$$c\|x - y\| \leq \|f(x) - f(y)\| \leq C\|x - y\| \quad \text{whenever} \quad \|x - y\| \geq 1.$$

We will show that this is a contradiction by proving the following theorem of Kalton and Randrianarivony [**67**].

THEOREM 13.7 (Kalton–Randrianarivony). *There is no coarse Lipschitz embedding of ℓ_2 into ℓ_p if $p \neq 2$.*

In order to prove Theorem 13.7, we will employ the techniques developed in [**67**]. We will use what is known as the *midpoint technique* for metric spaces for the case $1 < p < 2$ and *asymptotic smoothness techniques* for the case where $p > 2$. We will start with the case where $1 < p < 2$.

DEFINITION 13.8. Let M be a metric space with metric d, and let x and y be points in M. If $\delta > 0$, then the *approximate metric midpoints* between x and y with error δ are the points in the set

$$\text{Mid}(x, y, \delta) = \left\{ z \in M : \max\left(d(x, z), d(y, z)\right) \leq \frac{1 + \delta}{2} d(x, y) \right\}.$$

For the case $\delta = 0$, the points in the set

$$\text{Mid}(x, y, 0) = \left\{ z \in M : d(x, z) = d(y, z) = \frac{d(x, y)}{2} \right\}.$$

are called the *metric midpoints* of x and y.[7]

EXAMPLE 13.9. In the metric space $L_1(0, 1)$, the set of metric midpoints for $\chi_{[0,1]}$ and 0 includes all indicator functions χ_A for which the measure of A is $\frac{1}{2}$.

LEMMA 13.10. *Let M be a metric space with metric d, and let x and y be points in M. If $z \in \text{Mid}(x, y, \delta)$ for $0 < \delta < 1$, then*

$$\min\left(d(x, z), d(y, z)\right) \geq \frac{1 - \delta}{2} d(x, y).$$

PROOF. This is an application of the triangle inequality. We will show the lower bound for $d(x, z)$, since the other is similar. By assumption, the point z is an approximate metric midpoint between x and y with error δ, and so (by definition) we know that

$$d(y, z) \leq \frac{1 + \delta}{2} d(x, y).$$

Now, using the triangle inequality, we have

$$d(x, z) \geq d(x, y) - d(y, z) \geq d(x, y) - \frac{1 + \delta}{2} d(x, y) = \frac{1 - \delta}{2} d(x, y),$$

which is the desired lower bound. To obtain the lower bound for $d(y, z)$, simply interchange the roles of x and y. □

---

[6]We recall that the Corson–Klee lemma states that a uniformly continuous function between Banach spaces is a coarse Lipschitz function; and we also recall that a coarse Lipschitz function is "Lipschitz at large distances."

[7]Metric midpoints are discussed in the context of Banach spaces in Remark 4.5.

13. THE UNIQUE UNIFORM STRUCTURE OF SEQUENCE SPACES

DEFINITION 13.11. Let X be a Banach space with norm $\|\cdot\|$ and let M be a metric space with metric d. If f is a coarse Lipschitz function, let

$$\mathrm{Lip}_t(f) = \sup\left\{\frac{d(f(x), f(y))}{\|x - y\|} : \|x - y\| \geq t\right\}.$$

Furthermore, we let

$$\mathrm{Lip}_\infty(f) = \inf_{t>0} \mathrm{Lip}_t(f).$$

We remark that $\mathrm{Lip}_\infty(f) < \infty$ since f is a coarse Lipschitz function, and

$$\mathrm{Lip}_\infty(f) = \lim_{t\to\infty} \mathrm{Lip}_t(f)$$

(which is where we get the notation), because $\mathrm{Lip}_t(f)$ decreases as t increases. It is possible to define $\mathrm{Lip}_t(f)$ for X a metric space (instead of a Banach space), but for our purposes, it will suffice to have X a Banach space.

PROPOSITION 13.12. *Let X be a Banach space with norm $\|\cdot\|$ and let M be a metric space with metric d. Suppose that $f : X \to M$ is a coarse Lipschitz function with $\mathrm{Lip}_\infty(f) > 0$. For all $t > 0$ and $\epsilon > 0$, if $0 < \delta < 1$, then there are points x and y in X with $\|x - y\| \geq t$ such that*

$$f\big(\mathrm{Mid}(x, y, \delta)\big) \subseteq \mathrm{Mid}\big(f(x), f(y), (1 + \epsilon)\delta\big).$$

PROOF. Let $t > 0$ and $\epsilon > 0$ be given and suppose $0 < \delta < 1$. For each $\nu > 0$, there exists an $s > t$ such that

(13.1) $$\mathrm{Lip}_s(f) < (1 + \nu)\mathrm{Lip}_\infty(f).$$

For this choice of s, choose x and y so that

$$\|x - y\| \geq \frac{2s}{1 - \delta}$$

and

(13.2) $$d(f(x), f(y)) > (1 - \nu)\mathrm{Lip}_\infty(f)\|x - y\|.$$

Pick u in the set $M(x, y, \delta)$ of approximate metric midpoints between x and y with error δ. Then, by definition,

(13.3) $$\max\big(\|x - u\|, \|y - u\|\big) \leq \frac{1 + \delta}{2}\|x - y\|.$$

By Lemma 13.10,

$$\min\big(\|x - u\|, \|y - u\|\big) \geq \frac{1 - \delta}{2}\|x - y\| \geq \left(\frac{1 - \delta}{2}\right) \cdot \frac{2s}{1 - \delta} = s.$$

In particular, we have that $\|x - u\| \geq s$ and $\|y - u\| \geq s$, and so combining (13.1) and (13.3) we conclude that

$$d(f(x), f(u)) \leq (1 + \nu)\mathrm{Lip}_\infty(f)\|x - u\|$$
$$\leq \frac{1}{2}(1 + \nu)(1 + \delta)\mathrm{Lip}_\infty(f)\|x - y\|$$

and

$$d(f(y), f(u)) \leq (1 + \nu)\mathrm{Lip}_\infty(f)\|y - u\|$$
$$\leq \frac{1}{2}(1 + \nu)(1 + \delta)\mathrm{Lip}_\infty(f)\|x - y\|.$$

Finally, we use (13.2) to bound the quantity $\operatorname{Lip}_\infty(f)\|x-y\|$ in each of the two inequalities above, and we have that
$$\max\Big(d\big(f(x),f(u)\big),\, d\big(f(y),f(u)\big)\Big) \le \frac{1}{2}\left(\frac{1+\nu}{1-\nu}\right)(1+\delta)\, d\big(f(x),f(y)\big).$$

If we choose $\nu = \frac{\delta\epsilon}{2(1+\delta)+\delta\epsilon}$, then we have obtained the desired bound to show that $f(u)$ is an approximate metric midpoint between $f(x)$ and $f(y)$ with error $(1+\epsilon)\delta$. That is, $f(u) \in \operatorname{Mid}\big(f(x),f(y),(1+\epsilon)\delta\big)$, which completes the proof. $\square$

We will also make use of the following Lemmas 13.13 and 13.15.

LEMMA 13.13. *Let x and y be in ℓ_p, where $1 \le p < \infty$. If δ is a real number such that $0 < \delta < 1$, then there exists a compact set K such that*
$$\operatorname{Mid}(x,y,\delta) \subseteq K + 2\delta^{1/p}\|v\|B_{\ell_p},$$
where $v = \tfrac{1}{2}(x-y)$.

PROOF. First, suppose that $y = -x$, so that $v = x$. Since $x \in \ell_p$, we may write $x = (\xi_j)_{j=1}^\infty$ where $\sum_{j=1}^\infty |\xi_j|^p < \infty$. Let $\nu > 0$ be given. (The value of ν will be determined later.) Then there is a positive integer N for which
$$\sum_{j=N+1}^\infty |\xi_j|^p < \nu^p.$$

Let $(e_j)_{j=1}^\infty$ be the standard basis for ℓ_p. That is, e_j is the sequence having 1 in the jth position and zeros everywhere else. Let $[e_j]_{j=1}^N$ denote the closed linear span of the vectors $e_1, \ldots, e_N$ in ℓ_p, and let $[e_j]_{j=N+1}^\infty$ denote the closed linear span of the vectors $e_{N+1}, e_{N+2}, \ldots$ in ℓ_p. Let $F = [e_j]_{j=1}^N$ and $E = [e_j]_{j=N+1}^\infty$.

Let $z \in \operatorname{Mid}(x,-x,\delta)$. Then

(13.4) $$\max\Big(\|z-x\|,\, \|z+x\|\Big) \le \frac{1+\delta}{2}\|x-(-x)\| = (1+\delta)\|x\|.$$

Consequently, since
$$z = \frac{1}{2}(z-x) + \frac{1}{2}(z+x),$$
it follows that
$$\|z\| \le \frac{1}{2}\|z-x\| + \frac{1}{2}\|z+x\| \le (1+\delta)\|x\|.$$
Write $z = z' + z''$, where $z' \in F = [e_j]_{j=1}^N$ and $z'' \in E = [e_j]_{j=N+1}^\infty$. Then
$$\|z'\| \le \|z\| \le (1+\delta)\|x\|.$$
Let
$$K = (1+\delta)\|x\|B_F.$$
Then $z' \in K$. It remains to show that $\|z''\| \le 2\delta^{1/p}\|x\|$. Once again using (13.4), we have
$$\frac{1}{2}\Big(\|x-z\|^p + \|x+z\|^p\Big) \le (1+\delta)^p\|x\|^p.$$
But

(13.5) $$\|x\|^p - \nu^p + \|z''\|^p \le \frac{1}{2}\Big(\|x-z\|^p + \|x+z\|^p\Big),$$

and so it follows that
$$\|x\|^p - \nu^p + \|z''\|^p \le (1+\delta)^p\|x\|^p.$$

Therefore,
$$\|z''\|^p \le (1+\delta)^p \|x\|^p - \|x\|^p + \nu^p = \left[(1+\delta)^p - 1\right] \|x\|^p + \nu^p.$$

If $0 < \delta < 1$, then $(1+\delta)^p - 1 < 2^p \delta$. Consequently, we may choose ν so small that
$$\left[(1+\delta)^p - 1\right] \|x\|^p + \nu^p < 2^p \delta \|x\|^p.$$

We conclude that for the specified ν, we have $\|z''\|^p < 2^p \delta \|x\|^p$, which completes the proof in this case.

Now let x and y be any elements in ℓ_p. If $u = \frac{x+y}{2}$ and $v = \frac{x-y}{2}$, then
$$\mathrm{Mid}(x,y,\delta) = \frac{x+y}{2} + \mathrm{Mid}\left(\frac{x-y}{2}, \frac{y-x}{2}, \delta\right) = u + \mathrm{Mid}(v,-v,\delta).$$

By the earlier part of this proof, we know that there is a compact set K' such that
$$\mathrm{Mid}(v,-v,\delta) \subseteq K' + 2\delta^{1/p}\|v\| B_{\ell_p}.$$

Therefore, if we let $K = u + K'$, then
$$\mathrm{Mid}(x,y,\delta) \subseteq K + 2\delta^{1/p}\|v\| B_{\ell_p},$$

as required. $\square$

COMMENT 13.14. In case the inequality in (13.5) is not immediately obvious, let's take a closer look at why it is true. Note first that
$$\|x\|^p - \nu^p + \|z''\|^p \le \sum_{j=1}^{N} |\xi_j|^p + \sum_{j=N+1}^{\infty} |z_j|^p,$$
where $z = (z_j)_{j=1}^{\infty}$. Write $x = x' + x''$ so that $x' \in F$ and $x'' \in E$ (to follow the example of what we did with z). Then the quantity on the right of the above inequality is
$$\|x' + z''\|^p = \|(\xi_1, \ldots, \xi_N, z_{N+1}, \ldots)\|^p.$$
If we let η_j denote the jth component of the vector $x' + z''$, then by convexity we know that
$$|\eta_j|^p \le \frac{1}{2}\Big(|\xi_j - z_j|^p + |\xi_j + z_j|^p\Big),$$
and so if we sum over all $j \in \mathbb{N}$, we get that
$$\|x' + z''\|^p \le \sum_{j=1}^{\infty} \frac{1}{2}\Big(|\xi_j - z_j|^p + |\xi_j + z_j|^p\Big) = \frac{1}{2}\Big(\|x - z\|^p + \|x + z\|^p\Big).$$
From there, we get the inequality in (13.5).

LEMMA 13.15. *Let x and y be in ℓ_p, where $1 \le p < \infty$. Let $(e_j)_{j=1}^{\infty}$ be the standard basis for ℓ_p. If δ is a real number such that $0 < \delta < 1$, then there exists a positive integer N such that*
$$u + \delta^{1/p}\|v\| B_E \subseteq \mathrm{Mid}(x,y,\delta),$$
where $u = \frac{1}{2}(x+y)$ and $v = \frac{1}{2}(x-y)$, and $E = [e_j]_{j=N+1}^{\infty}$ is the closed linear span of the vectors $e_{N+1}, e_{N+2}, \ldots$ in ℓ_p.

PROOF. First, suppose that $y = -x$ (so that $u = 0$ and $v = x$). The case $p = 1$ is immediate, and so we may assume that $p > 1$. Following the notation used in the proof of Lemma 13.13, let $x = (\xi_j)_{j=1}^\infty$ where $\sum_{j=1}^\infty |\xi_j|^p < \infty$. Let $\nu > 0$ be given. Then there is a positive integer N for which

$$\sum_{j=N+1}^\infty |\xi_j|^p < \nu^p.$$

Let $F = [e_j]_{j=1}^N$ and $E = [e_j]_{j=N+1}^\infty$, where $(e_j)_{j=1}^\infty$ is the standard basis for ℓ_p.

We wish to show that

$$\delta^{1/p}\|x\|B_E \subseteq \mathrm{Mid}(x, -x, \delta).$$

To that end, let z be an element of E for which $\|z\| \leq \delta^{1/p}\|x\|$. Because $z \in E$, we can write $z = (z_j)_{j=1}^\infty$, where $z_j = 0$ for $j \in \{1, \ldots, N\}$ and $\sum_{j=N+1}^\infty |z_j|^p < \infty$. Then,

$$\|x \pm z\|^p = \sum_{j=1}^\infty |\xi_j \pm z_j|^p = \sum_{j=1}^N |\xi_j|^p + \sum_{j=N+1}^\infty |\xi_j \pm z_j|^p.$$

Note that, by Minkowski's inequality,

$$\sum_{j=N+1}^\infty |\xi_j \pm z_j|^p = \Big(\sum_{j=N+1}^\infty |\xi_j \pm z_j|^p\Big)^{\frac{1}{p} \cdot p} < (\nu + \|z\|)^p.$$

Consequently,

$$\|x \pm z\|^p < \|x\|^p + (\nu + \|z\|)^p.$$

Now choose $\nu > 0$ so that

$$\nu < \big((1+\delta)^p - 1\big)^{1/p}\|x\| - \delta^{1/p}\|x\|.$$

Then

$$(\nu + \|z\|)^p < \Big(\big((1+\delta)^p - 1\big)^{1/p}\|x\|\Big)^p = \big((1+\delta)^p - 1\big)\|x\|^p,$$

and from that it follows that

$$\|x \pm z\|^p \leq \|x\|^p + \big((1+\delta)^p - 1\big)\|x\|^p = (1+\delta)^p\|x\|^p.$$

Therefore, $\|x \pm z\| \leq (1+\delta)\|x\|$, and so $z \in \mathrm{Mid}(x, -x, \delta)$, which completes the proof in this case.

Now let x and y be any elements in ℓ_p. If $u = \frac{x+y}{2}$ and $v = \frac{x-y}{2}$, then

$$\mathrm{Mid}(x, y, \delta) = \frac{x+y}{2} + \mathrm{Mid}\Big(\frac{x-y}{2}, \frac{y-x}{2}, \delta\Big) = u + \mathrm{Mid}(v, -v, \delta).$$

By the first part of this proof, we know that

$$\delta^{1/p}\|v\|B_E \subseteq \mathrm{Mid}(v, -v, \delta).$$

Consequently,

$$u + \delta^{1/p}\|v\|B_E \subseteq u + \mathrm{Mid}(v, -v, \delta) = \mathrm{Mid}(x, y, \delta),$$

which completes the proof. □

We now put these preliminary results together to prove the following significant theorem, which is the first step in proving Theorem 13.7.

THEOREM 13.16. *There is no coarse Lipschitz embedding of ℓ_r into ℓ_p if $r > p$.*

PROOF. Suppose that $f : \ell_r \to \ell_p$ is a coarse Lipschitz embedding, where $r > p$. Then there exist constants $C > 0$ and $c > 0$ such that

(13.6) $\quad c\|x - y\| \leq \|f(x) - f(y)\| \leq C\|x - y\| \quad \text{whenever} \quad \|x - y\| \geq 1.$

Since f is a Lipschitz embedding, it follows that $\operatorname{Lip}_\infty(f) > 0$. (See Comment 13.17 for a proof of this assertion.) Consequently, by Proposition 13.12, we can find points x and y in X with $\|x - y\|$ as large as we like such that

$$f\big(\operatorname{Mid}(x, y, \delta)\big) \subseteq \operatorname{Mid}\big(f(x), f(y), 2\delta\big),$$

where $0 < \delta < 1$. (We pick $\epsilon = 1$ when applying Proposition 13.12, and leave t as-yet undetermined.)

By Lemma 13.15, there exists a closed subspace E having finite co-dimension in ℓ_r such that

$$u + \delta^{1/r} \left\| \frac{x - y}{2} \right\| B_E \subseteq \operatorname{Mid}(x, y, \delta),$$

where $u = \frac{1}{2}(x + y)$. By Lemma 13.13, there exists a compact set K in ℓ_p such that

$$\operatorname{Mid}\big(f(x), f(y), 2\delta\big) \subseteq K + 2(2\delta)^{1/p} \left\| \frac{f(x) - f(y)}{2} \right\| B_{\ell_p}.$$

Putting all of these results together, we have that

(13.7) $\quad f\left(u + \frac{1}{2}\delta^{1/r} \|x - y\| B_E\right) \subseteq K + (2\delta)^{1/p} \|f(x) - f(y)\| B_{\ell_p}.$

Choose a sequence $(z_j)_{j=1}^\infty$ of points in B_E with $\|z_j - z_k\| \geq 1$ whenever $j \neq k$. Then, because f is a coarse Lipschitz embedding, we have that

(13.8) $\quad \left\| f\left(u + \frac{1}{2}\delta^{1/r} \|x - y\| z_j\right) - f\left(u + \frac{1}{2}\delta^{1/r} \|x - y\| z_k\right) \right\| \geq c \cdot \frac{1}{2}\delta^{1/r} \|x - y\|,$

by (13.6), provided that the arguments of f are at least 1 unit apart. In order to ensure that, we choose x and y so that

$$\frac{1}{2}\delta^{1/r}\|x - y\| > 1,$$

which we can do by choosing a large enough value for t when applying Proposition 13.12. (Pick t so that $t > 2\delta^{-1/r}$.)

We now use (13.7). For each $j \in \mathbb{N}$, we have that $z_j \in B_E$, and so by (13.7),

$$f\left(u + \frac{1}{2}\delta^{1/r}\|x - y\|z_j\right) \in K + (2\delta)^{1/p}\|f(x) - f(y)\| B_{\ell_p}.$$

That means that for each $j \in \mathbb{N}$, we can find a $w_j \in K$ and a $v_j \in B_{\ell_2}$ so that

$$f\left(u + \frac{1}{2}\delta^{1/r}\|x - y\|z_j\right) = w_j + (2\delta)^{1/p}\|f(x) - f(y)\|v_j.$$

Therefore, by (13.8),

$$\left\| \big(w_j + (2\delta)^{1/p}\|f(x) - f(y)\|v_j\big) - \big(w_k + (2\delta)^{1/p}\|f(x) - f(y)\|v_k\big) \right\|$$
$$\geq \frac{c}{2}\delta^{1/r}\|x - y\|.$$

For any $\gamma > 0$, we can find an infinite subset of $\mathbb{N}$ for which

$$(2\delta)^{1/p}\|f(x) - f(y)\|\|v_j - v_k\| \geq \frac{c}{2}\delta^{1/r}\|x - y\| - \gamma,$$

whenever $j \neq k$. Consequently, we may pass to an infinite subset $\mathbb{M}$ of $\mathbb{N}$ such that

$$(2\delta)^{1/p}\|f(x) - f(y)\|\|v_j - v_k\| \geq \frac{c}{4}\delta^{1/r}\|x - y\|,$$

whenever j and k are in $\mathbb{M}$ and $j \neq k$. Since $v_j \in B_{\ell_p}$ for each j, it follows that $\|v_j - v_k\| \leq 2$, and so

$$(2\delta)^{1/p}\|f(x) - f(y)\| \geq \frac{c}{8}\delta^{1/r}\|x - y\|.$$

Now we use the upper bound in (13.6) to conclude that

$$C(2\delta)^{1/p}\|x - y\| \geq \frac{c}{8}\delta^{1/r}\|x - y\|.$$

Since $2^{1/p} < 2$, and since $x \neq y$, we have that

$$2C\delta^{1/p} \geq \frac{c}{8}\delta^{1/r},$$

which can be rewritten as

$$\delta^{\frac{1}{p} - \frac{1}{r}} \geq \frac{c}{16C}.$$

But δ can be chosen arbitrarily close to 0, and if δ is close enough to 0, then this leads to a contradiction. We conclude, therefore, that there is no coarse Lipschitz embedding of ℓ_r into ℓ_p if $r > p$. □

COMMENT 13.17. In the proof of Theorem 13.16, we commented that since f is a coarse Lipschitz embedding, it follows that $\operatorname{Lip}_\infty(f) > 0$. This may not be completely obvious. Suppose that $f : X \to Y$ is a coarse Lipschitz embedding between Banach spaces X and Y. Then there exist constants $C > 0$ and $c > 0$ such that
$$c\|x - y\| \leq \|f(x) - f(y)\| \leq C\|x - y\| \quad \text{whenever} \quad \|x - y\| \geq 1.$$
In particular, we have that
$$\|x - y\| \leq \frac{1}{c}\|f(x) - f(y)\|$$
whenever $\|x - y\| \geq 1$. Consequently, if s is any number such that $\|x - y\| \geq s > 1$, then
$$\|f(x) - f(y)\| \leq \operatorname{Lip}_s(f)\|x - y\| \leq \operatorname{Lip}_s(f) \cdot \frac{1}{c}\|f(x) - f(y)\|.$$
It follows that $\operatorname{Lip}_s(f) \geq c$ for any $s > 1$, and so $\operatorname{Lip}_\infty(f) \geq c$, as well, since it is the infimum of $\operatorname{Lip}_s(f)$ for $s > 1$. In particular, if f is a coarse Lipschitz embedding, then $\operatorname{Lip}_\infty(f) > 0$, as claimed.

An alternate approach to showing that $\operatorname{Lip}_\infty(f) > 0$ when f is a coarse Lipschitz embedding is to prove the following more general result, which is interesting in its own right. (This is Lemma 2.2 in [67].)

PROPOSITION 13.18. *Let X, Y, and Z be Banach spaces. If $f : X \to Y$ and $g : Y \to Z$ are coarse Lipschitz, then $g \circ f : X \to Z$ is coarse Lipschitz and $\operatorname{Lip}_\infty(g \circ f) \leq \operatorname{Lip}_\infty(f)\operatorname{Lip}_\infty(g)$.*

PROOF. Let $\alpha > \operatorname{Lip}_\infty(f)$ and $\beta > \operatorname{Lip}_\infty(g)$. Then there exist constants A and B (that depend on α and β, respectively) so that
$$\|f(x_1) - f(x_2)\|_Y \leq \alpha \|x_1 - x_2\|_X + A,$$
for all x_1 and x_2 in X, and
$$\|g(y_1) - g(y_2)\|_Z \leq \beta \|y_1 - y_2\|_Y + B,$$
for all y_1 and y_2 in Y. Consequently,
$$\|g(f(x_1)) - g(f(x_2))\|_Z \leq \beta \|f(x_1) - f(x_2)\|_Y + B \leq \beta\Big(\alpha\|x_1 - x_2\|_X + A\Big) + B$$
$$= \beta\alpha \|x_1 - x_2\|_X + \Big(\beta A + B\Big),$$
for all x_1 and x_2 in X. It follows that $g \circ f$ is coarse Lipschitz and $\operatorname{Lip}_\infty(g \circ f) \leq \alpha\beta$. Since α and β were any constants such that $\alpha > \operatorname{Lip}_\infty(f)$ and $\beta > \operatorname{Lip}_\infty(g)$, it follows that $\operatorname{Lip}_\infty(g \circ f) \leq \operatorname{Lip}_\infty(f) \operatorname{Lip}_\infty(g)$. $\square$

If f is a coarse Lipschitz embedding from a Banach space X into a Banach space Y, then f is a coarse Lipschitz homeomorphism onto its image $f(X)$. Specifically, we have that $f : X \to f(X)$ and $f^{-1} : f(X) \to X$ are coarse Lipschitz and $f^{-1} \circ f = \operatorname{Id}|_X$. Since $\operatorname{Lip}_\infty(\operatorname{Id}|_X) = 1$, it follows that $\operatorname{Lip}_\infty(f) \operatorname{Lip}_\infty(f^{-1}) \geq 1$, and in particular neither $\operatorname{Lip}_\infty(f)$ nor $\operatorname{Lip}_\infty(f^{-1})$ is zero.

Within the proof of Theorem 13.16, there is a useful result that we wish to isolate for future use. The proof is quite similar to the first part of the proof of Theorem 13.16, but we will include it for completeness.[8]

PROPOSITION 13.19. *Suppose $f : \ell_r \to \ell_p$ is a coarse Lipschitz function, where $r > p$. Given $t > 0$ and $\delta > 0$, there exists a positive number $\tau > t$, a point $x \in \ell_r$, a subspace E of ℓ_r having the form*
$$E = \{(\xi_j)_{j=1}^\infty \in \ell_r : \xi_1 = 0, \ldots, \xi_N = 0\},$$
and a compact subset K of ℓ_p such that
$$f(x + \tau B_E) \subseteq K + \delta\tau B_{\ell_p}.$$

PROOF. Let $t > 0$ and $\delta > 0$ be given. We will first consider the case where $\operatorname{Lip}_\infty(f) = 0$. Since $\delta > 0$, there exists some constant $C > 0$ (that depends on δ) such that
$$\|f(x) - f(y)\| \leq \left(\frac{\delta}{2}\right) \|x - y\| + C,$$
for all x and y in ℓ_r. Choose $\tau > t$ so large that $\tau > \frac{2C}{\delta}$. Consequently, if x and y are any elements of ℓ_r for which $\|x - y\| \leq \tau$, then
$$\|f(x) - f(y)\| \leq \left(\frac{\delta}{2}\right)\tau + \frac{\delta\tau}{2} = \delta\tau.$$
Pick $x = 0$. Then if $\|y\| \leq \tau$, it follows that $\|f(0) - f(y)\| \leq \delta\tau$, and so if we let $K = \{f(0)\}$, then
$$f(y) \in K + \delta\tau B_{\ell_p} \quad \text{whenever} \quad y \in \tau B_{\ell_r},$$
which gives the desired result.

---
[8]Proposition 13.19 is a special case of Proposition 3.5 in [**67**].

Now assume that $\text{Lip}_\infty(f) > 0$. Then, by Proposition 13.12, we can find points x and y in ℓ_r with $\|x - y\| \geq T$, with T as large as we like, such that
$$f\big(\text{Mid}(x, y, \eta)\big) \subseteq \text{Mid}\big(f(x), f(y), 2\eta\big),$$
where $0 < \eta < 1$ is chosen sufficiently small. (We pick $\epsilon = 1$ when applying Proposition 13.12, and use η instead of δ, since δ already has a role in this proof. We leave the lower bound on T and the upper bound on η undetermined for now, and we will determine appropriate bounds later in the proof.)

By Lemma 13.15, there exists a closed subspace E having finite co-dimension in ℓ_r of the form $\{(\xi_j)_{j=1}^\infty \in \ell_r : \xi_1 = 0, \ldots, \xi_N = 0\}$ for some $N \in \mathbb{N}$ such that
$$u + \eta^{1/r}\|v\|B_E \subseteq \text{Mid}(x, y, \eta),$$
where $u = \frac{1}{2}(x + y)$ and $v = \frac{1}{2}(x - y)$. By Lemma 13.13, there exists a compact set K in ℓ_p such that
$$\text{Mid}\big(f(x), f(y), 2\eta\big) \subseteq K + 2(2\eta)^{1/p}\left\|\frac{f(x) - f(y)}{2}\right\|B_{\ell_p}.$$
Putting all of these results together, we have that
$$(13.9) \qquad f\Big(u + \eta^{1/r}\|v\|B_E\Big) \subseteq K + (2\eta)^{1/p}\|f(x) - f(y)\|B_{\ell_p}.$$

We assumed that $\text{Lip}_\infty(f) > 0$, and so there exists some real number $s > 0$ for which $0 < \text{Lip}_s(f) < \infty$. In particular, that means that if $\|x - y\| \geq s$, then we can be sure that $\|f(x) - f(y)\| \leq \text{Lip}_s(f)\|x - y\|$. Now we choose the desired values for η and T. Let η be chosen in the interval $(0, 1)$ so that
$$4\eta^{1/p - 1/r}\text{Lip}_s(f) < \delta,$$
and pick T so that $T > \max(s, 2t\eta^{-1/r})$. Let $\tau = \eta^{1/r}\|v\|$, where we recall that $v = \frac{1}{2}(x - y)$. Then
$$\tau = \eta^{1/r}\|v\| = \frac{1}{2}\eta^{1/r}\|x - y\| \geq \frac{1}{2}\eta^{1/r}T > \frac{1}{2}\eta^{1/r} \cdot 2t\eta^{-1/r} = t,$$
as required. Furthermore, since $\|x - y\| \geq T$ and $T > s$, we have that
$$(2\eta)^{1/p}\|f(x) - f(y)\| \leq (2\eta)^{1/p}\text{Lip}_s(f)\|x - y\| = (2\eta)^{1/p}\text{Lip}_s(f) \cdot 2\tau\eta^{-1/r}.$$
The last equality follows from the fact that $\|x - y\| = 2\tau\eta^{-1/r}$. Consequently, since $2^{1/p} < 2$, we have that
$$(2\eta)^{1/p}\|f(x) - f(y)\| \leq 4\eta^{1/p - 1/r}\text{Lip}_s(f)\tau < \delta\tau.$$
The above inequality, combined with (13.9) gives us
$$f\Big(u + \tau B_E\Big) \subseteq K + \delta\tau B_{\ell_p},$$
as required. (We just rename u as x to get the desired form of the inclusion, as it appears in the statement of the proposition.) $\square$

We can now prove the first part of the theorem of Johnson, Lindenstrauss, and Schechtman (Theorem 13.1), which we state as a separate corollary.

COROLLARY 13.20 (J–L–S, $1 < p < 2$). *If p is a number such that $1 < p < 2$, then ℓ_p has unique uniform structure.*

PROOF. Suppose that X is uniformly homeomorphic to ℓ_p. Then X is an $\mathcal{L}_p$-space (by Theorem 13.3), and so either (a) X has a subspace linearly isomorphic to ℓ_2 or (b) X is linearly isomorphic to ℓ_p (by Theorem 13.6). Option (a) is impossible if $p < 2$ (by Theorem 13.16), and so X is linearly isomorphic to ℓ_p. □

We have addressed the case where $1 < p < 2$, but what about the case where $p > 2$? For that, we will employ a different method.

REMARK 13.21. Suppose that $T : \ell_r \to \ell_p$ is a bounded linear map, where $r > 1$ and $p > 1$. If $r > p$, then T is compact, by Pitt's theorem.[9] If $r < p$, then T is strictly singular. In the case of uniform homeomorphism, the case where $r > p$ was solved using the midpoint technique. For $r < p$, we will employ *asymptotic smoothness techniques*. (We will restrict our attention to the case where $r = 2$, since that is what we need to prove Theorem 13.7.)

Let $\mathbb{N}$ be the set of positive integers, and let $\mathbb{M}$ be a proper infinite subset of $\mathbb{N}$. Let $G_k(\mathbb{M})$ denote the collection of all k-subsets $\{n_1, n_2, \ldots, n_k\}$ of $\mathbb{M}$ such that

$$n_1 < n_2 < n_3 < \cdots < n_k,$$

and when considering the subset $\{n_1, n_2, \ldots, n_k\}$ as an element in $G_k(\mathbb{M})$, we denote it by $(n_1, n_2, \ldots, n_k)$. We define a distance on $G_k(\mathbb{M})$, called the *Hamming distance*, by

$$d\big((m_1, \ldots, m_k), (n_1, \ldots, n_k)\big) = |\{j : m_j \neq n_j\}|,$$

where $m_1 < \cdots < m_k$ and $n_1 < \cdots < n_k$, and both m_j and n_j are in $\mathbb{M}$ for each $j \in \{1, \ldots, k\}$.

COMMENT 13.22. For the sake of brevity, when we say $(m_1, \ldots, m_k)$ is in $G_k(\mathbb{M})$, we will refrain from adding that $m_1, \ldots, m_k$ are in $\mathbb{M}$ with $m_1 < \ldots < m_k$, since that is part of the definition of $G_k(\mathbb{M})$. We will also adopt vector notation, so that we will write $\boldsymbol{m}$ for $(m_1, \ldots, m_k)$.

PROPOSITION 13.23. *Suppose that X is a reflexive Banach space and suppose that for a fixed $p \in (1, \infty)$,*

(13.10) $$\limsup_{n \to \infty} \|x + x_n\|^p \leq \|x\|^p + \limsup_{n \to \infty} \|x_n\|^p$$

for all $x \in X$ whenever $(x_n)_{n=1}^\infty$ is a weakly null sequence in X.[10] *If $\epsilon > 0$ and if $f : G_k(\mathbb{N}) \to X$ is a bounded function, then there exists an infinite subset $\mathbb{M}$ of $\mathbb{N}$ such that*

$$\mathrm{diam}\,\big(f(G_k(\mathbb{M}))\big) < 2\,\mathrm{Lip}(f) k^{1/p} + \epsilon.$$

PROOF. We claim that there exists an infinite subset $\mathbb{M}$ of $\mathbb{N}$ and a point u in X such that

$$\|f(\boldsymbol{m}) - u\| < \mathrm{Lip}(f) k^{1/p} + \frac{\epsilon}{2}$$

for all $\boldsymbol{m}$ in $G_k(\mathbb{M})$. Proposition 13.23 follows from this fact, and so it is this claim that we will prove.

---

[9] See Theorem 2.1.4 in [5] for a proof of Pitt's theorem.

[10] You may wish to compare this to the similar, but also different, Lipschitz weak-star Kadec–Klee condition for embedding into c_0 that can be found in Lemma 10.7.

We will proceed by using a proof by induction on k. For the case $k = 1$, the set $G_k(\mathbb{N})$ is $\mathbb{N}$, and so $f : G_k(\mathbb{N}) \to X$ is an X-valued sequence. Consequently, we can find a subsequence of $(f(n))_{n \in \mathbb{N}}$ that converges weakly to a limit in X. Let u be the weak limit point and let $\mathbb{M}$ be the set for which $(f(m))_{m \in \mathbb{M}}$ is the weakly convergent subsequence. Then, for each $m \in \mathbb{M}$,

$$\|f(m) - u\| \leq \liminf_{n \in \mathbb{M}} \|f(m) - f(n)\| \leq \mathrm{Lip}(f),$$

which gives us the result when $k = 1$.

Now assume that the theorem has been proven for $k - 1$. We will prove it for k. Assume that $f : G_k(\mathbb{N}) \to X$ is a bounded function. We may find an infinite subset $\mathbb{M}_0$ of $\mathbb{N}$ such that

$$\lim_{n_k \in \mathbb{M}_0} f(n_1, n_2, \ldots, n_{k-1}, n_k)$$

exists weakly for each $(n_1, n_2, \ldots, n_{k-1}) \in G_{k-1}(\mathbb{N})$. We now define a new function $\widetilde{f} : G_{k-1}(\mathbb{N}) \to X$ by

$$\widetilde{f}(n_1, n_2, \ldots, n_{k-1}) = \lim_{n_k \in \mathbb{M}_0} f(n_1, n_2, \ldots, n_{k-1}, n_k),$$

where the limit is in the weak topology on X. Note that $\widetilde{f}$ is a Lipschitz function with $\mathrm{Lip}(\widetilde{f}) \leq \mathrm{Lip}(f)$.

By the inductive hypothesis, there exists an infinite subset $\mathbb{M}_1$ of $\mathbb{M}_0$ and a point u in X such that

$$\|\widetilde{f}(\boldsymbol{m}) - u\| < \mathrm{Lip}(\widetilde{f})(k-1)^{1/p} + \frac{\epsilon}{4}$$

for all $\boldsymbol{m}$ in $G_{k-1}(\mathbb{M}_1)$.

We now use the inequality in (13.10) to get[11]

$$\limsup_{n_k \in \mathbb{M}_1} \|f(n_1, n_2, \ldots, n_{k-1}, n_k) - u\|^p \leq \|\widetilde{f}(n_1, n_2, \ldots, n_{k-1}) - u\|^p$$

$$+ \limsup_{n_k \in \mathbb{M}_1} \|f(n_1, n_2, \ldots, n_{k-1}, n_k) - \widetilde{f}(n_1, n_2, \ldots, n_{k-1})\|^p.$$

From what we have shown above, we know this is bounded by

$$\left(\mathrm{Lip}(\widetilde{f})(k-1)^{1/p} + \frac{\epsilon}{4}\right)^p + \mathrm{Lip}(f)^p,$$

which in turn is bounded by

(13.11) $$\left(\mathrm{Lip}(f)(k-1)^{1/p} + \frac{\epsilon}{4}\right)^p + \mathrm{Lip}(f)^p \leq \left(\mathrm{Lip}(f)k^{1/p} + \frac{\epsilon}{4}\right)^p.$$

(See Comment 13.24, for more about this inequality.) Thus, we conclude that

$$\limsup_{n_k \in \mathbb{M}_1} \|f(n_1, n_2, \ldots, n_{k-1}, n_k) - u\| \leq \mathrm{Lip}(f)k^{1/p} + \frac{\epsilon}{4}$$

for fixed $(n_1, \ldots, n_{k-1}) \in G_{k-1}(\mathbb{M}_1)$.

Now, using Ramsey's theorem (see Theorem 11.1.1 in [5], for example), we pass to a further infinite subset $\mathbb{M}$ of $\mathbb{M}_1$ for which

$$\Big|\|f(\boldsymbol{n}) - u\| - \|f(\boldsymbol{m}) - u\|\Big| < \frac{\epsilon}{4}$$

---

[11] Let $x = \widetilde{f}(n_1, n_2, \ldots, n_{k-1}) - u$ and $x_{n_k} = f(n_1, n_2, \ldots, n_{k-1}, n_k) - \widetilde{f}(n_1, n_2, \ldots, n_{k-1})$.

for all $\boldsymbol{n}$ and $\boldsymbol{m}$ in $G_k(\mathbb{M})$. Then

$$\|f(n_1, n_2, \ldots, n_{k-1}, n_k) - u\| \leq \operatorname{Lip}(f) k^{1/p} + \frac{\epsilon}{2},$$

for all $(n_1, n_2, \ldots, n_{k-1}, n_k) \in G_k(\mathbb{M})$, as required. This completes the inductive argument. The conclusion of the proposition follows from this fact, and so the proof is complete. □

COMMENT 13.24. It may not be immediately obvious that the inequality in (13.11) is true, but we claim that

$$\left(\operatorname{Lip}(f)(k-1)^{1/p} + \frac{\epsilon}{4}\right)^p + \operatorname{Lip}(f)^p \leq \left(\operatorname{Lip}(f) k^{1/p} + \frac{\epsilon}{4}\right)^p,$$

where f, k, p, and ϵ are as described in the statement of Proposition 13.23. For notational convenience, we let $c = \frac{\epsilon}{4\operatorname{Lip}(f)}$. Then the above inequality is equivalent to

$$((k-1)^{1/p} + c)^p + 1 \leq (k^{1/p} + c)^p.$$

For each $k \in \mathbb{N}$, define a function $f : [0, \infty) \to \mathbb{R}$ by

$$f(x) = ((k-1)^{1/p} + x)^p + 1 - (k^{1/p} + x)^p.$$

Then $f(0) = 0$ and

$$f'(x) = p((k-1)^{1/p} + x)^{p-1} - p(k^{1/p} + x)^{p-1} \leq 0, \quad x > 0.$$

That means that $f(x) \leq 0$ for all $x \geq 0$, which gives us that

$$((k-1)^{1/p} + x)^p + 1 - (k^{1/p} + x)^p < 0,$$

and from that we get the desired result.

For a slightly different approach to proving the final bound in (13.11), see Theorem 14.5.9 in [5].

If we assume stronger hypotheses on the function $f : G_k(\mathbb{N}) \to X$ than those in Proposition 13.23, then we can get a stronger conclusion with fewer assumptions on the Banach space X.

THEOREM 13.25. *Suppose that X is a Banach space and $f : G_k(\mathbb{N}) \to X$ is a bounded function such that $f(G_k(\mathbb{N}))$ is relatively compact. If $\epsilon > 0$, then there exists an infinite subset $\mathbb{M}$ of $\mathbb{N}$ such that* $\operatorname{diam}\bigl(f(G_k(\mathbb{M}))\bigr) < \epsilon$.

PROOF. We proceed inductively on k. For $k = 1$, we find a Cauchy subsequence $(f(n_j))_{j=1}^\infty$ and let N be a positive integer such that $\|f_{n_i} - f_{n_j}\| < \epsilon$ for all i and j greater than N. Let $\mathbb{M} = \{n_{N+1}, n_{N+2}, \ldots\}$. This completes the case where $k = 1$.

Suppose now that the theorem has been shown to be true for functions on $G_{k-1}(\mathbb{N})$ and assume that $f : G_k(\mathbb{N}) \to X$ has relatively compact image in X. There exists an infinite subset $\mathbb{M}_1$ of $\mathbb{N}$ such that

$$\lim_{n_k \in \mathbb{M}_0} f(n_1, n_2, \ldots, n_{k-1}, n_k)$$

exists for each $(n_1, n_2, \ldots, n_{k-1}) \in G_{k-1}(\mathbb{N})$. Define $\widetilde{f} : G_{k-1}(\mathbb{N}) \to X$ by

$$\widetilde{f}(n_1, n_2, \ldots, n_{k-1}) = \lim_{n_k \in \mathbb{M}_0} f(n_1, n_2, \ldots, n_{k-1}, n_k),$$

where the limit is in the norm topology on X. By the inductive hypothesis, there exists an infinite subset $\mathbb{M}_1$ of $\mathbb{M}_0$ such that $\operatorname{diam}\bigl(\widetilde{f}(G_{k-1}(\mathbb{M}_1))\bigr) < \frac{\epsilon}{4}$.

Pick $m_1, \ldots, m_k$ in $\mathbb{M}_1$ such that
$$\|f(m_1, \ldots, m_{k-1}, m_k) - \widetilde{f}(m_1, \ldots, m_{k-1})\| < \frac{\epsilon}{4}.$$
Next, pick $m_{k+1} \in \mathbb{M}_1$ such that $m_{k+1} > m_k$ and
$$\|f(\boldsymbol{n}, m_{k+1}) - \widetilde{f}(\boldsymbol{n})\| < \frac{\epsilon}{4}$$
for all $\boldsymbol{n} \in G_{k-1}(\{m_1, \ldots, m_k\})$.

Continuing in this way, we construct a set
$$\mathbb{M} = \{m_1, m_2, m_3, \ldots, m_k, m_{k+1}, m_{k+2}, \ldots\}$$
such that $m_j < m_{j+1}$ for all $j \in \mathbb{N}$ and such that
$$\|f(m_{i_1}, \ldots, m_{i_{k-1}}, m_{i_k}) - \widetilde{f}(m_{i_1}, \ldots, m_{i_{k-1}})\| < \frac{\epsilon}{4}$$
when $m_{i_1}, \ldots, m_{i_{k-1}}, m_{i_k}$ are integers in $\mathbb{M}$ with $m_{i_1} < \ldots < m_{i_{k-1}} < m_{i_k}$. Therefore, if $(m_{i_1}, \ldots, m_{i_{k-1}}, m_{i_k})$ and $(n_{j_1}, \ldots, n_{j_{k-1}}, n_{j_k})$ are in $G_k(\mathbb{M})$, then
$$\|f(m_{i_1}, \ldots, m_{i_{k-1}}, m_{i_k}) - f(n_{j_1}, \ldots, n_{j_{k-1}}, n_{j_k})\|$$
$$\leq \|f(m_{i_1}, \ldots, m_{i_{k-1}}, m_{i_k}) - \widetilde{f}(m_{i_1}, \ldots, m_{i_{k-1}})\|$$
$$+ \|\widetilde{f}(m_{i_1}, \ldots, m_{i_{k-1}}) - \widetilde{f}(n_{j_1}, \ldots, n_{j_{k-1}})\|$$
$$+ \|\widetilde{f}(n_{j_1}, \ldots, n_{j_{k-1}}) - f(n_{j_1}, \ldots, n_{j_{k-1}}, n_{j_k})\|$$
$$< \frac{\epsilon}{4} + \frac{\epsilon}{4} + \frac{\epsilon}{4} < \epsilon.$$

This completes the proof. $\square$

For our purposes, we do not need Theorem 13.25 so much as we need a corollary to Theorem 13.25.

COROLLARY 13.26. *Suppose that X is a Banach space and $f : G_k(\mathbb{N}) \to X$ is a function such that $f(G_k(\mathbb{N})) \subset K + \delta B_X$ for some compact set K and some $\delta > 0$. If $\epsilon > 0$, then there exists an infinite subset $\mathbb{M}$ of $\mathbb{N}$ such that*
$$\mathrm{diam}\big(f(G_k(\mathbb{M}))\big) < \epsilon + 2\delta.$$

PROOF. Let $\epsilon > 0$ be given. We start by writing $f = g + h$, where g and h are functions such that $g : G_k(\mathbb{N}) \to K$ and $h : G_k(\mathbb{N}) \to \delta B_X$. Since K is compact, the set $g(G_k(\mathbb{N}))$ is relatively compact, and so by Theorem 13.25, there exists an infinite subset $\mathbb{M}$ of $\mathbb{N}$ such that $\mathrm{diam}\big(g(G_k(\mathbb{M}))\big) < \epsilon$. Since $\mathrm{diam}\big(\delta B_X\big) \leq 2\delta$, the result follows directly. $\square$

We are now ready to establish the second (and final) piece necessary to prove Theorem 13.7. (The first piece being Theorem 13.16.)

THEOREM 13.27. *There is no coarse Lipschitz embedding of ℓ_2 into ℓ_p if $p > 2$.*

PROOF. We start by observing that $X = \ell_p$ satisfies the conditions of Proposition 13.23. (In fact, for $X = \ell_p$, we have equality in (13.10).) Consequently, for a given $\epsilon > 0$, if $g : G_k(\mathbb{N}) \to \ell_p$ is a bounded function, then there exists an infinite subset $\mathbb{M}$ of $\mathbb{N}$ such that

(13.12) $$\mathrm{diam}\big(g(G_k(\mathbb{M}))\big) < 2\,\mathrm{Lip}(g) k^{1/p} + \epsilon.$$

Now we suppose that there exists a coarse Lipschitz embedding $f : \ell_2 \to \ell_p$, where $p > 2$. Then there exist positive constants C and c such that
$$c\|x - y\| \leq \|f(x) - f(y)\| \leq C\|x - y\|$$
whenever $\|x - y\| \geq 1$.

Now define a function $\phi : G_k(\mathbb{M}) \to \ell_2$ by
$$\phi(n_1, \ldots, n_k) = e_{n_1} + \cdots + e_{n_k},$$
where $(e_j)_{j=1}^\infty$ is the standard basis for ℓ_2. Then ϕ is a Lipschitz continuous function with $\text{Lip}(\phi) = 1$. Consequently, the function $f \circ \phi : G_k(\mathbb{M}) \to \ell_p$ is a Lipschitz continuous function with $\text{Lip}(f \circ \phi) \leq C$. (Note that distinct points of the form $\phi(n_1, \ldots, n_k)$ are always at least 1 unit from each other, and so we can apply the "Lipschitz at large distances" condition.)

We now apply (13.12) with $g = f \circ \phi$ and $\epsilon = Ck^{1/p}$ to conclude there exists an infinite subset $\mathbb{M}$ of $\mathbb{N}$ such that

(13.13) $$\text{diam}\big((f \circ \phi)(G_k(\mathbb{M}))\big) < 3Ck^{1/p}.$$

Next, pick positive integers $n_1, \ldots, n_k$ and $m_1, \ldots, m_k$ in $\mathbb{M}$ such that
$$n_1 < \cdots < n_k < m_1 < \cdots < m_k.$$
Then
$$\|\phi(n_1, \ldots, n_k) - \phi(m_1, \ldots, m_k)\| = \sqrt{2k},$$
and so
$$\|(f \circ \phi)(n_1, \ldots, n_k) - (f \circ \phi)(m_1, \ldots, m_k)\| \geq c\sqrt{2k}.$$
Consequently, combining this with (13.13), we have that
$$c\sqrt{2k} < 3Ck^{1/p},$$
which can be rewritten as
$$k^{\frac{1}{2} - \frac{1}{p}} < \frac{3C}{c\sqrt{2}}.$$
This inequality must be true for all $k \in \mathbb{N}$, but that is an impossibility when $p > 2$. Therefore, we have obtained a contradiction, and so there is no coarse Lipschitz embedding of ℓ_2 into ℓ_p if $p > 2$. □

Combining Theorem 13.16 with Theorem 13.27, we obtain the proof of Theorem 13.7 (the Kalton–Randrianarivony theorem), which states that there is no coarse Lipschitz embedding of ℓ_2 into ℓ_p if $p \neq 2$.

This also gives us the second (and final) part of Theorem 13.1 (the theorem of Johnson, Lindenstrauss, and Schechtman). We gave the proof of Theorem 13.1 for $1 < p < 2$ in Corollary 13.20. We now give the proof for $2 < p < \infty$ as another corollary.

COROLLARY 13.28 (J–L–S, $2 < p < \infty$). *If p is a number such that $2 < p < \infty$, then ℓ_p has unique uniform structure.*

PROOF. The proof is essentially the same as the proof of Corollary 13.20. Suppose that X is uniformly homeomorphic to ℓ_p. Then it follows that X is an $\mathcal{L}_p$-space (by Theorem 13.3), and so either (a) X has a subspace linearly isomorphic to ℓ_2 or (b) X is linearly isomorphic to ℓ_p (by Theorem 13.6). Option (a) is impossible if $p > 2$ (by Theorem 13.27), and so X is linearly isomorphic to ℓ_p. □

Together, Corollary 13.20 and Corollary 13.28 complete the proof of Theorem 13.1, and so we conclude that ℓ_p has unique uniform structure.

REMARK 13.29. The proof we have given for Theorem 13.1 is not the same as the one given by Johnson, Lindenstrauss, and Schechtman in [50]. They used the Gorelik principle (Theorem 10.6).

This leads to an important open question.

QUESTION 13.30. *Does $L_p(0,1)$ have unique uniform structure for $1 < p < \infty$?*

For the case $p = 2$, this is already known. (In [33], Enflo proved that every Banach space uniformly homeomorphic to a Hilbert space is isomorphic to it.) It is known that if X is uniformly homeomorphic to $L_p(0,1)$, then X is linearly isomorphic to a complemented subspace of $L_p(0,1)$. (We can prove this similarly to Theorem 13.3.) So, then, a natural question is:

What are the complemented subspaces of $L_p(0,1)$?

It is unknown even how many (nonisomorphic) complemented subspaces $L_p(0,1)$ possesses, though it is known that there are at least $\aleph_1$ of them [20].

So we have established that if X is uniformly homeomorphic to $L_p(0,1)$, then X is linearly isomorphic to a complemented subspace of $L_p(0,1)$. We could then investigate specific complemented subspaces. The simplest option is ℓ_p.

THEOREM 13.31. *$L_p(0,1)$ is not uniformly homeomorphic to ℓ_p for $1 < p < \infty$, unless $p = 2$.*

PROOF. This follows directly from Theorem 13.7, because ℓ_2 is a complemented subspace of $L_p(0,1)$ for all $p \in (1, \infty)$. □

The next candidate for a complemented subspace of $L_p(0,1)$ to which X might be linearly isomorphic is $\ell_p \oplus \ell_2$. So we now wish to address the question:

Is $L_p(0,1)$ uniformly homeomorphic to $\ell_p \oplus \ell_2$?

We will show that the answer to this question is "no" when $p \neq 2$. We will start by considering the case where $1 < p < 2$.

LEMMA 13.32. *Suppose that $1 < p < 2$. Then $L_p(0,1)$ has a subspace isometric to $L_r(0,1)$ for any r such that $p < r \leq 2$.*

PROOF. Since this is a result from the linear theory of Banach spaces, we will not include the complete proof here. The key idea is to find a sequence of independent and identically distributed random variables $f_1, f_2, \ldots$ such that

$$\mathbb{E}(e^{itf_k}) = e^{-|t|^r}$$

for each $k \in \mathbb{N}$. Then f_k is in $L_p(0,1)$ for each k, provided that $p < r \leq 2$. Then it can be shown that $[f_k]_{k=1}^\infty$ is isometric to ℓ_r in $L_p(0,1)$.

We know from the linear theory (Example 8.27) that the Banach space $L_r(0,1)$ is finitely representable in ℓ_r (for all $r \geq 1$), but we have just seen that ℓ_r is isometric to a linear subspace of $L_p(0,1)$ (when $1 < p < r \leq 2$). Since $L_p(0,1)$ is finitely representable in ℓ_p, we conclude that $L_r(0,1)$ is finitely representable in ℓ_p, which means that $L_r(0,1)$ is isometric to a linear subspace of $L_p(0,1)$, when $1 < p < r \leq 2$. This completes the proof.[12] □

---

[12] The construction of the random variables $f_1, f_2, \ldots$ is not a trivial one. The details can be found in [5], in Section 6.4. Lemma 13.32 itself can be found in [5] as Proposition 12.1.10. We

As a consequence of Lemma 13.32, we conclude that if $L_p(0,1)$ is uniformly homeomorphic to $\ell_p \oplus \ell_2$, then ℓ_r coarse Lipschitz embeds into $\ell_p \oplus \ell_2$ for all r such that $p < r \leq 2$. This leads us to consider a natural extension of Theorem 13.1, which states that ℓ_r coarse Lipschitz embeds into $\ell_p \oplus \ell_q$ only when either $r = p$ or $r = q$. In order to prove this theorem, we will need the following proposition, which is a straightforward generalization of Proposition 13.19.

PROPOSITION 13.33. *Suppose $f : \ell_r \to \ell_p \oplus \ell_q$ is a coarse Lipschitz function, where $r > q > p$. Given $t > 0$ and $\delta > 0$, there exists a positive number $\tau > t$, a point $x \in \ell_r$, a subspace E of ℓ_r having the form*

$$E = \{(\xi_j)_{j=1}^\infty \in \ell_r : \xi_1 = 0, \ldots, \xi_N = 0\},$$

and a compact subset K of $\ell_p \oplus \ell_q$ such that

$$f(x + \tau B_E) \subseteq K + \delta \tau B_{\ell_p \oplus \ell_q}.$$

We won't provide the the proof here, but the proof is essentially the same as the proof of Proposition 13.19, observing that Lemma 13.13 can be generalized in a natural way to the direct sum $\ell_p \oplus \ell_q$. (For more details, see Lemma 3.4 and Proposition 3.6 in [**67**].)

We are now ready to prove the extension of Theorem 13.1.[13]

THEOREM 13.34. *Suppose p, q, and r are numbers in the interval $(1, \infty)$ and $p \neq q$. If ℓ_r coarse Lipschitz embeds into $\ell_p \oplus \ell_q$, then either $r = p$ or $r = q$.*

PROOF. We first prove the theorem for the case where $p < r < q$. Suppose that $f : \ell_r \to \ell_p \oplus \ell_q$ is a coarse Lipschitz embedding. Then there exist positive constants C and c such that

$$c\|x - y\| \leq \|f(x) - f(y)\| \leq C\|x - y\|$$

whenever $\|x - y\| \geq 1$. Write

$$f(x) = \big(g(x), h(x)\big)$$

where $g : \ell_r \to \ell_p$ and $h : \ell_r \to \ell_q$ are coarse Lipschitz functions.

Let $k \in \mathbb{N}$ and let $\delta > 0$ be given. By Proposition 13.19, there exists a positive number $\tau > k$, a point $x \in \ell_r$, a subspace E of ℓ_r having the form

$$E = \{(\xi_j)_{j=1}^\infty \in \ell_r : \xi_1 = 0, \ldots, \xi_N = 0\},$$

for some $N \in \mathbb{N}$, and a compact subset K of ℓ_p such that

$$g(x + \tau B_E) \subseteq K + \delta \tau B_{\ell_p}.$$

Let $\mathbb{M} = \{n \in \mathbb{N} : n > N\}$, and define a function $\phi : G_k(\mathbb{M}) \to \ell_r$ by

$$\phi(n_1, \ldots, n_k) = x + \tau k^{-1/r}(e_{n_1} + \cdots + e_{n_k}),$$

where $(e_j)_{j=1}^\infty$ is the standard basis in ℓ_r. Then

$$(g \circ \phi)\big(G_k(\mathbb{M})\big) \subseteq K + \delta \tau B_{\ell_p}.$$

---

will prove a more general result in the next chapter (Corollary 14.25) using positive-definite and negative-definite functions, and so we will be content with an outline of a proof for now.

[13]The theorem we prove here (Theorem 13.34) is a special case of Theorem 5.1 in [**67**], which states that ℓ_r does not coarse Lipschitz embed into the direct sum $\ell_{p_1} = \oplus \cdots \oplus \ell_{p_n}$, where $1 \leq p_1 < p_2 < \cdots < p_n < \infty$ and $r \neq p_k$ for any k.

By Corollary 13.26, there exists an infinite subset $\mathbb{M}_1$ of $\mathbb{M}$ such that
$$\operatorname{diam}\left[(g\circ\phi)\bigl(G_k(\mathbb{M}_1)\bigr)\right] \leq 3\delta\tau.$$
(We apply Corollary 13.26 with $\epsilon = \delta\tau$. Note also that we used the corollary with $\mathbb{M}$ in place of $\mathbb{N}$, which can be done with the obvious changes to the proof.)

Now we turn our attention to $h : \ell_r \to \ell_q$. Since h is Lipschitz at large distances, the composition $h \circ \phi$ is a Lipschitz continuous function, and
$$\operatorname{Lip}(h \circ \phi) \leq C \operatorname{Lip}(\phi).$$
To determine a bound on $\operatorname{Lip}(\phi)$, let $\boldsymbol{n} = (n_1, \ldots, n_k)$ and $\boldsymbol{m} = (m_1, \ldots, m_k)$ both be in $G_k(\mathbb{M})$. Then
$$\begin{aligned}
\|\phi(\boldsymbol{n}) - \phi(\boldsymbol{m})\| &= \tau k^{-1/r}\|(e_{n_1} + \cdots + e_{n_k}) - (e_{m_1} + \cdots + e_{m_k})\| \\
&\leq \tau k^{-1/r} \cdot \bigl|\{n_1,\ldots,n_k\}\Delta\{m_1,\ldots,m_k\}\bigr|^{1/r} \\
&\leq \tau k^{-1/r} \cdot [2 \cdot d(\boldsymbol{n}, \boldsymbol{m})]^{1/r} \\
&\leq 2^{1/r}\tau k^{-1/r} \cdot d(\boldsymbol{n}, \boldsymbol{m}).
\end{aligned}$$

Therefore,
$$\operatorname{Lip}(h \circ \phi) \leq 2^{1/r} C \tau k^{-1/r} \leq 2C\tau k^{-1/r}.$$
Consequently, by Proposition 13.23, there exists an infinite subset $\mathbb{M}_2$ of $\mathbb{M}_1$ such that
$$\operatorname{diam}\left[(h\circ\phi)\bigl(G_k(\mathbb{M}_2)\bigr)\right] \leq 3 \operatorname{Lip}(h\circ\phi) k^{1/q}$$
$$\leq 6C\tau k^{\frac{1}{q} - \frac{1}{r}}.$$
(We chose $\epsilon < \operatorname{Lip}(h\circ\phi) k^{1/q}$ in Proposition 13.23, and we used $\mathbb{M}$ in place of $\mathbb{N}$, which we can do with only minor modifications to the proof.)

Therefore,
$$\operatorname{diam}\left[(f\circ\phi)\bigl(G_k(\mathbb{M}_2)\bigr)\right] \leq \max\left(3\delta\tau,\ 6C\tau k^{\frac{1}{q}-\frac{1}{r}}\right),$$
where we are using the ℓ_∞-sum for $\ell_p \oplus \ell_q$. (That is, we are using the max-norm.) But
$$\operatorname{diam}\left[(f\circ\phi)\bigl(G_k(\mathbb{M}_2)\bigr)\right] \geq 2^{1/r} c\tau,$$
because $\operatorname{diam}\left[\phi\bigl(G_k(\mathbb{M}_2)\bigr)\right] = 2^{1/r}\tau$, and because $\|f(x) - f(y)\| \geq c\|x - y\|$ when $\|x - y\| \geq 1$. Thus,
$$2^{1/r} c\tau \leq \max\left(3\delta\tau,\ 6C\tau k^{\frac{1}{q}-\frac{1}{r}}\right),$$
or
$$2^{1/r} c \leq \max\left(3\delta,\ 6C k^{\frac{1}{q}-\frac{1}{r}}\right).$$

But δ can be made arbitrarily close to 0 and $k^{\frac{1}{q}-\frac{1}{r}} \to 0$ as $k \to \infty$, because $r < q$, by assumption. This leads to a contradiction, and so we conclude that $f : \ell_r \to \ell_p \oplus \ell_q$ is not a coarse Lipschitz embedding if $p < r < q$.

So far, we have proved the theorem for $p < r < q$. The only other case we need to consider is $p < q < r$, and in this case, we require only the midpoint technique. Consequently, suppose that $p < q < r$ and assume that $f : \ell_r \to \ell_p \oplus \ell_q$ is a coarse Lipschitz embedding. As before, there exist positive constants C and c such that
$$c\|x - y\| \leq \|f(x) - f(y)\| \leq C\|x - y\|$$

whenever $\|x - y\| \geq 1$.

Let $k \in \mathbb{N}$ and $\delta > 0$, and let $(e_j)_{j=1}^\infty$ be the standard basis in ℓ_r. Applying Proposition 13.33 there exists a positive number $\tau > k$, a point $x \in \ell_r$, a subspace E of ℓ_r having the form $E = \{(\xi_j)_{j=1}^\infty \in \ell_r : \xi_1 = 0, \ldots, \xi_N = 0\}$ for some $N \in \mathbb{N}$, and a compact subset K of $\ell_p \oplus \ell_q$ such that
$$f(x + \tau B_E) \subseteq K + \delta \tau B_{\ell_p \oplus \ell_q}.$$

Let $\mathbb{M} = \{n \in \mathbb{N} : n > N\}$, and define a function $\phi : G_k(\mathbb{M}) \to \ell_r$ by
$$\phi(n_1, \ldots, n_k) = x + \tau k^{-1/r}(e_{n_1} + \cdots + e_{n_k}).$$

Then
$$(f \circ \phi)\bigl(G_k(\mathbb{M})\bigr) \subseteq K + \delta \tau B_{\ell_p \oplus \ell_q},$$
and so using Corollary 13.26 (with $\epsilon = \delta \tau$ and with $\mathbb{M}$ in place of $\mathbb{N}$), there exists an infinite subset $\mathbb{M}_1$ of $\mathbb{M}$ such that
$$\operatorname{diam}\left[(f \circ \phi)\bigl(G_k(\mathbb{M}_1)\bigr)\right] \leq 3\delta\tau.$$

Since
$$\operatorname{diam}\left[(f \circ \phi)\bigl(G_k(\mathbb{M}_1)\bigr)\right] \geq 2^{1/r} c \tau,$$
it follows that
$$2^{1/r} c\tau \leq 3\delta\tau,$$
or
$$2^{1/r} c \leq 3\delta.$$

But δ can be made arbitrarily close to 0. This leads to a contradiction, and so we conclude that $f : \ell_r \to \ell_p \oplus \ell_q$ is not a coarse Lipschitz embedding if $p < q < r$. □

COROLLARY 13.35. *The direct sum $\ell_p \oplus \ell_2$ is not uniformly homeomorphic to the Banach space $L_p(0,1)$ if $1 < p < 2$.*

PROOF. If $L_p(0,1)$ was uniformly homeomorphic to $\ell_p \oplus \ell_2$, then $L_r(0,1)$, and hence ℓ_r would coarse Lipschitz embed into $\ell_p \oplus \ell_2$ for any r such that $p \leq r \leq 2$, by Lemma 13.32. This is not possible, by Theorem 13.34, and so $L_p(0,1)$ is not uniformly homeomorphic to $\ell_p \oplus \ell_2$ when $1 < p < 2$. □

This answers the question posed earlier when $1 < p < 2$, but what about the case where $p > 2$? It turns out that $\ell_p \oplus \ell_2$ is not uniformly homeomorphic to $L_p(0,1)$ in this case, either. In order to show this, however, we need to talk about the space $\ell_p(\ell_2)$. We let $\ell_p(\ell_2)$ be the set
$$\ell_p(\ell_2) = \Bigl\{(x_k)_{k=1}^\infty : x_k \in \ell_2 \text{ for all } k \in \mathbb{N} \text{ and } \sum_{k=1}^\infty \|x_k\|_{\ell_2}^p < \infty\Bigr\}.$$

We equip $\ell_p(\ell_2)$ with the norm
$$\|(x_k)_{k=1}^\infty\| = \Bigl(\sum_{k=1}^\infty \|x_k\|_{\ell_2}^p\Bigr)^{1/p}.$$

It can be shown that this norm is complete, and so $\ell_p(\ell_2)$ with this norm is a Banach space, but we will omit those details here.

A classical result of Lindenstrauss and Pełczyński tells us that $\ell_p(\ell_2)$ is a complemented subspace of $L_p(0,1)$.[14] Specifically, in Example 8.2 in [**73**], the space

---

[14]Lindenstrauss and Pełczyński use the notation $(\ell_2 \oplus \ell_2 \oplus \cdots)_p$ for what we call $\ell_p(\ell_2)$.

$\ell_p(\ell_2)$ is shown to be an $\mathcal{L}_p$-space, and a separable $\mathcal{L}_p$-space is necessarily isomorphic to a complemented subspace of $L_p(0,1)$. (This last fact follows from Theorem 13.5, which originally comes from Theorem 7.1 in [**73**].)

Consequently, if $L_p(0,1)$ is uniformly homeomorphic to $\ell_p \oplus \ell_2$, then it is necessarily true that $\ell_p(\ell_2)$ coarse Lipschitz embeds into $\ell_p \oplus \ell_2$. We can show that this is impossible using an argument similar to that used in Theorem 13.34, but it requires another generalization of Proposition 13.19.

PROPOSITION 13.36. *Suppose $g : \ell_p(\ell_2) \to \ell_q$ is a coarse Lipschitz function, where $p > q > 1$. Given $t > 0$ and $\delta > 0$, there exists a positive number $\tau > t$, a point $u \in \ell_p(\ell_2)$, a subspace E of $\ell_p(\ell_2)$ having the form*

$$E = \{(\xi_j)_{j=1}^\infty \in \ell_p(\ell_2) : \xi_1 = 0, \ldots, \xi_N = 0\},$$

for some $N \in \mathbb{N}$, and a compact subset K of ℓ_q such that $g(u + \tau B_E) \subseteq K + \delta\tau B_{\ell_q}$.

PROOF. The proof is essentially the same as the proof of Proposition 13.19, and so we won't provide as many details as we did when proving that proposition. Instead, we will sketch the proof and give an indication of what changes are required. For ease of notation, we will denote the norm on $\ell_p(\ell_2)$ by $\|\cdot\|$, but we will use $\|\cdot\|_p$ to denote the norm on ℓ_p in order to minimize ambiguity.

If $\mathrm{Lip}_\infty(g) = 0$, then the proof is like the case where $\mathrm{Lip}_\infty(f) = 0$ in Proposition 13.19, and so we assume that $\mathrm{Lip}_\infty(g) > 0$. By Proposition 13.12, we can find points x and y in $\ell_p(\ell_2)$ with $\|x - y\| \geq T$, with T as large as we like, such that

$$g\bigl(\mathrm{Mid}(x,y,\eta)\bigr) \subseteq \mathrm{Mid}\bigl(g(x), g(y), 2\eta\bigr),$$

where $0 < \eta < 1$ is chosen sufficiently small. (We choose η and T later in the proof.) Furthermore, by Lemma 13.13, there exists a compact set K in ℓ_p such that

$$\mathrm{Mid}\bigl(g(x), g(y), 2\eta\bigr) \subseteq K + 2(2\eta)^{1/q}\left\|\frac{g(x) - g(y)}{2}\right\|_q B_{\ell_q},$$

and so we have

$$g\bigl(\mathrm{Mid}(x,y,\eta)\bigr) \subseteq K + (2\eta)^{1/q}\|g(x) - g(y)\|_q B_{\ell_q}.$$

At this point, we want to show that $\mathrm{Mid}(x,y,\eta)$ contains a suitably "large" subset. When proving Proposition 13.19, we used Lemma 13.15 to provide that suitably "large" subset. We will adapt the proof of Lemma 13.15 to our current case by adjusting the norms that appear in the proof. Start by letting $u = \frac{1}{2}(x+y)$ and $v = \frac{1}{2}(x-y)$. Because $v \in \ell_p(\ell_2)$, we may write $v = (v_k)_{k=1}^\infty$, where $v_k \in \ell_2$ for each $k \in \mathbb{N}$. Let ν be a real number such that

$$0 < \nu < \bigl((1+\eta)^p - 1\bigr)^{1/p}\|v\| - \eta^{1/p}\|v\|.$$

Choose $N \in \mathbb{N}$ so that

$$\sum_{k=N+1}^\infty \|v_k\|_2^p < \nu^p$$

and let

$$E = \{(\xi_j)_{j=1}^\infty \in \ell_p(\ell_2) : \xi_1 = 0, \ldots, \xi_N = 0\}.$$

We claim this E is the subspace of $\ell_p(\ell_2)$ that we asserted exists in the statement Proposition 13.36. To verify this claim, suppose that $z \in E$ and $\|z\| < \eta^{1/p}\|v\|$. Then

$$\|x - (u+z)\|^p = \|v - z\|^p < \|v\|^p + \bigl(\nu + \|z\|\bigr)^p < (1+\eta)^p\|v\|^p$$

and
$$\|y - (u+z)\|^p = \|-v-z\|^p < \|v\|^p + (\nu + \|z\|)^p < (1+\eta)^p \|v\|^p.$$
(Refer to the proof of Lemma 13.15 for the details of these estimates.) It follows that $u + z \in \text{Mid}(x, y, \eta)$, and so we have established that
$$u + \eta^{1/p}\|v\|B_E \subseteq \text{Mid}(x, y, \eta).$$
This gives us the suitably "large" subset of $\text{Mid}(x, y, \eta)$ that we required. Putting our results together, we have that
$$g\big(u + \eta^{1/p}\|v\|B_E\big) \subseteq K + (2\eta)^{1/q}\|g(x) - g(y)\|_q B_{\ell_q}.$$
Now we choose η and T. Let η be a real number in $(0,1)$ chosen so small that
$$8\eta^{1/q - 1/p} \operatorname{Lip}_\infty(g) < \delta.$$
Let $T > 2t\eta^{-1/p}$ be chosen sufficiently large that $\|g(x) - g(y)\|_p \leq 2\operatorname{Lip}_\infty(g)\|x - y\|$, recalling that $\|x - y\| \geq T$. Now let $\tau = \eta^{1/p}\|v\|$. Then for our "small enough" choice of η and our "large enough" choice of T, we have that $\tau > t$ and
$$(2\eta)^{1/q}\|g(x) - g(y)\|_p \leq 4\eta^{1/q}\operatorname{Lip}_\infty(g)\|x - y\| = 8\eta^{1/q - 1/p}\operatorname{Lip}_\infty(g)\tau < \delta\tau,$$
where we used the fact that $2^{1/q} < 2$ and $\|x - y\| = 2\|v\| = 2\tau\eta^{-1/p}$. It follows that
$$g\big(u + \tau B_E\big) \subseteq K + \delta\tau B_{\ell_q},$$
as required. $\square$

What we have given as Proposition 13.36 is not the most general formulation of the result, but it will be sufficient for our purposes. For the more general statement, we refer the reader to Proposition 3.5 in [**67**].

We are now ready to show that $\ell_2 \oplus \ell_p$ is not uniformly homeomorphic to $L_p(0,1)$ if $p \neq 2$.

THEOREM 13.37. *The direct sum $\ell_2 \oplus \ell_p$ is not uniformly homeomorphic to the Banach space $L_p(0,1)$ if $1 < p < 2$ or $2 < p < \infty$.*

PROOF. We already proved this for the case where $1 < p < 2$ in Corollary 13.35. The proof for the case where $2 < p < \infty$ is similar to that used in Theorem 13.34. As we mentioned prior to the statement of Proposition 13.36, the space $\ell_p(\ell_2)$ is a complemented subspace of $L_p(0,1)$,[15] and so it suffices to show that $\ell_p(\ell_2)$ does not coarse Lipschitz embed into $\ell_2 \oplus \ell_p$. Assume to the contrary that $f : \ell_p(\ell_2) \to \ell_2 \oplus \ell_p$ is a coarse Lipschitz embedding. Then there exist positive constants C and c such that
$$c\|x - y\| \leq \|f(x) - f(y)\| \leq C\|x - y\|$$
when $\|x - y\| \geq 1$. Write $f(x) = (g(x), h(x))$ for $x \in \ell_p(\ell_2)$, where $g : \ell_p(\ell_2) \to \ell_2$ and $h : \ell_p(\ell_2) \to \ell_p$.

Let $k \in \mathbb{N}$ and let $\delta > 0$ be given. By Proposition 13.36, there exists a number $\tau > k$, a point $u \in \ell_p(\ell_2)$, a subspace E of $\ell_p(\ell_2)$, and a compact subset K of ℓ_2 such that
$$g\big(u + \tau B_E\big) \subseteq K + \delta\tau B_{\ell_2},$$

---
[15]We remind the reader that this comes from Example 8.2 in [**73**].

where the set E has the form
$$E = \{(\xi_j)_{j=1}^\infty \in \ell_p(\ell_2) : \xi_1 = 0, \ldots, \xi_N = 0\}$$
for some $N \in \mathbb{N}$. Note that ξ_j is a sequence in ℓ_2 for each $j \in \mathbb{N}$. If we let $(e_{i,j})_{j=1}^\infty$ be the standard basis for the copy of ℓ_2 in the ith position, then E is the closed linear span of the set $\{e_{i,j} : i > N, j \in \mathbb{N}\}$.

Define a function $\phi : G_k(\mathbb{N}) \to \ell_p(\ell_2)$ by
$$\phi(n_1, \ldots, n_k) = u + \tau k^{-1/2}(e_{N+1,n_1} + \cdots + e_{N+1,n_k}).$$
Then
$$(g \circ \phi)\bigl(G_k(\mathbb{N})\bigr) \subseteq K + \delta\tau B_{\ell_2},$$
and so (by Corollary 13.26) there exists an infinite subset $\mathbb{M}_1$ of $\mathbb{N}$ such that
$$\operatorname{diam}\Bigl[(g \circ \phi)\bigl(G_k(\mathbb{M}_1)\bigr)\Bigr] \leq 3\delta\tau.$$

The function $h : \ell_p(\ell_2) \to \ell_p$ is Lipschitz at large distances, and thus the composition $h \circ \phi$ is a Lipschitz continuous function with
$$\operatorname{Lip}(h \circ \phi) \leq 2C\tau k^{-1/2}.$$
Now we use Proposition 13.23 to get an infinite subset $\mathbb{M}_2$ of $\mathbb{M}_1$ for which
$$\operatorname{diam}\Bigl[(h \circ \phi)\bigl(G_k(\mathbb{M}_2)\bigr)\Bigr] \leq 6C\tau k^{\frac{1}{p} - \frac{1}{2}}.$$
It follows that
$$2^{1/2} c\tau \leq \operatorname{diam}\Bigl[(f \circ \phi)\bigl(G_k(\mathbb{M}_2)\bigr)\Bigr] \leq \max\Bigl(3\delta\tau,\ 6C\tau k^{\frac{1}{p} - \frac{1}{2}}\Bigr).$$
Thus,
$$2^{1/2} c \leq \max\Bigl(3\delta,\ 6Ck^{\frac{1}{p} - \frac{1}{2}}\Bigr).$$
But $\delta > 0$ and $k \in \mathbb{N}$ are arbitrary, and so we have obtained a contradiction. Therefore, $f : \ell_p(\ell_2) \to \ell_2 \oplus \ell_p$ is not a coarse Lipschitz embedding, and so $L_p(0,1)$ is not uniformly homeomorphic to $\ell_2 \oplus \ell_p$ when $2 < p < \infty$. □

In fact, it can be shown that spaces of the form $\ell_p \oplus \ell_q$, where $2 \notin \{p, q\}$ have unique uniform structure. For completeness, we will state this fact as a theorem, but we will not prove it here.

THEOREM 13.38. *Let p and q be numbers in $(1, \infty)$ so that neither p nor q is equal to 2. If X is uniformly homeomorphic to $\ell_p \oplus \ell_q$, then X is linearly isomorphic to $\ell_p \oplus \ell_q$. That is, the space $\ell_p \oplus \ell_q$ has unique uniform structure.*

The proof of Theorem 13.38 is beyond the scope of this text, but it can be found in [**50**] (for $1 < p < q < 2$ and $2 < p < q < \infty$) and [**67**] (for $1 < p < 2 < q < \infty$).

COMMENT 13.39. In Theorem 13.31, we showed that $L_p(0,1)$ is not uniformly homeomorphic to ℓ_p for $1 < p < \infty$, unless $p = 2$. We also commented at the start of the chapter that Per Enflo showed that $L_1(0,1)$ is not uniformly homeomorphic to ℓ_1 in an unpublished work in 1970, but we did not provide the proof. (Although we mentioned that proofs can be found in [**13**] and [**14**].) We can, however, use the techniques developed in this chapter to prove that $L_1(0,1)$ is not uniformly homeomorphic to ℓ_1.

13. THE UNIQUE UNIFORM STRUCTURE OF SEQUENCE SPACES

THEOREM 13.40 (Enflo). *$L_1(0,1)$ is not uniformly homeomorphic to ℓ_1.*

PROOF. To avoid ambiguity, we will use $\|\cdot\|_{L_1}$ to denote the norm on $L_1(0,1)$ and $\|\cdot\|_{\ell_1}$ to denote the norm on ℓ_1. Suppose that $f : L_1(0,1) \to \ell_1$ is a uniform homeomorphism. Then f and f^{-1} are coarse Lipschitz functions, by Lemma 11.36, and so then there exist constants $C > 0$ and $c > 0$ such that

$$c\|x-y\|_{L_1} \leq \|f(x) - f(y)\|_{\ell_1} \leq C\|x-y\|_{L_1}$$

whenever $\|x-y\|_{L_1} \geq 1$. Let η be an as-yet unspecified real number such that $0 < \eta < \frac{1}{4}$. Choose $s > 1$ so that

$$\operatorname{Lip}_s(f) \leq (1+\eta)\operatorname{Lip}_\infty(f) < 2\operatorname{Lip}_\infty(f),$$

which we can do since $\operatorname{Lip}_s(f)$ is decreasing to $\operatorname{Lip}_\infty(f)$ as $s \to \infty$, and because $\operatorname{Lip}_\infty(f) \geq c > 0$, by Comment 13.17.

By Proposition 13.12, there are points x and y in $L_1(0,1)$ with $\|x-y\|_{L_1} \geq 2s$ such that

$$f\Big(\operatorname{Mid}(x,y,2\eta)\Big) \subseteq \operatorname{Mid}\Big(f(x), f(y), 4\eta\Big).$$

(Let $\epsilon = 1$ and $\delta = 2\eta$ in the statement of Proposition 13.12, noting that $2\eta < 1$.) Then, by Lemma 13.13, there exists a compact set K in ℓ_1 such that

$$\operatorname{Mid}(f(x), f(y), 4\eta) \subseteq K + 8\eta\|f(x) - f(y)\|_{\ell_1} B_{\ell_1}.$$

(We are using $\delta = 4\eta$ when applying Lemma 13.13, noting that $4\eta < 1$, by our original assumption on η.) Consequently, we have that

$$f\Big(\operatorname{Mid}(x,y,2\eta)\Big) \subseteq K + 8\eta\|f(x) - f(y)\|_{\ell_1} B_{\ell_1}.$$

The set of metric midpoints for x and y is the set $\operatorname{Mid}(x,y,0)$, and this is certainly contained in $\operatorname{Mid}(x,y,2\eta)$. Consequently,

$$f\Big(\operatorname{Mid}(x,y,0)\Big) \subseteq K + 8\eta\|f(x) - f(y)\|_{\ell_1} B_{\ell_1}.$$

By Proposition 4.7, there is an infinite sequence $(v_k)_{k=1}^\infty$ of elements in $M(x,y,0)$ for which $\|v_j - v_k\|_{L_1} = \frac{1}{2}\|x-y\|_{L_1} \geq s$ when $j \neq k$. For each $k \in \mathbb{N}$, write $f(v_k) = g(v_k) + h(v_k)$, where $g(v_k) \in K$ and $\|h(v_k)\|_{\ell_1} \leq 8\eta\|f(x) - f(y)\|_{\ell_1}$. Since K is compact, we can cover it in finitely many open balls of radius $\eta\|f(x) - f(y)\|_{\ell_1}$. Since $(v_k)_{k=1}^\infty$ is infinite, we can necessarily find two distinct indices j and k such that v_j and v_k are both in the same open ball of radius $\eta\|f(x) - f(y)\|_{\ell_1}$. Then

$$\|f(v_j) - f(v_k)\|_{\ell_1} \leq \|g(v_j) - g(v_k)\|_{\ell_1} + \|h(v_j)\|_{\ell_1} + \|h(v_k)\|_{\ell_1}$$
$$\leq 2\eta\|f(x) - f(y)\|_{\ell_1} + 16\eta\|f(x) - f(y)\|_{\ell_1}$$
$$= 18\eta\|f(x) - f(y)\|_{\ell_1}.$$

Since $\|v_j - v_k\|_{L_1} = \frac{1}{2}\|x-y\|_{L_1} \geq s > 1$, it follows that

$$\|x-y\|_{L_1} = 2\|v_j - v_k\|_{L_1} \leq 2c\|f(v_j) - f(v_k)\| \leq 2c \cdot 18\eta\|f(x) - f(y)\|_{\ell_1}.$$

But $\|x-y\|_{L_1} \geq s$, and so

$$\|f(x) - f(y)\|_{\ell_1} \leq \operatorname{Lip}_s(f)\|x-y\|_{L_1} \leq 2\operatorname{Lip}_\infty(f)\|x-y\|_{L_1}.$$

Putting all of this together, we have

$$\|x-y\|_{L_1} \leq 72\eta c\operatorname{Lip}_\infty(f)\|x-y\|_{L_1}.$$

This implies that
$$72\eta c \operatorname{Lip}_\infty(f) \geq 1,$$
but if we choose η sufficiently small, this is an impossibility. Consequently, it cannot be true that f is a uniform homeomorphism, which completes the proof. □

CHAPTER 14

Uniform embeddings into a Hilbert space

In this chapter, we wish to address the following general question:
When does a Banach space uniformly embed into a Hilbert space?
Our primary goal is to prove the following theorem.

THEOREM 14.1. *A Banach space uniformly embeds into a Hilbert space if and only if it linearly embeds into $L_p(0,1)$ for every p such that $0 < p < 1$.*

In order to prove Theorem 14.1, we will prove two significant theorems, and then combine them to get Theorem 14.1. The theorems are:
  (a) *Theorem A (Section 14.1).* A Banach space uniformly embeds into a Hilbert space if and only if it *linearly* embeds into $L_0(\Omega, \mathbb{P})$, the set of all $\mathbb{P}$-measurable functions on Ω, for some probability space $(\Omega, \mathbb{P})$.
  (b) *Theorem B (Section 14.2).* Let $(\Omega, \mathbb{P})$ be a probability space. A Banach space linearly embeds into $L_0(\Omega, \mathbb{P})$ if and only if it linearly embeds into $L_p(\Omega, \mathbb{P})$ for every p such that $0 < p < 1$.

REMARK 14.2. Let $(\Omega, \mathbb{P})$ be probability space. We recall that $L_0(\Omega, \mathbb{P})$ is the set of all $\mathbb{P}$-measurable functions on Ω. We can define a metric on $L_0(\Omega, \mathbb{P})$ by

$$d(f,g) = \mathbb{E}\left(1 - e^{-|f-g|}\right) = \int_\Omega \left(1 - e^{-|f(\omega)-g(\omega)|}\right) \mu(d\omega)$$

for $\mathbb{P}$-measurable functions f and g on Ω. This metric induces the topology of convergence in measure (with respect to $\mathbb{P}$). We call this metric *negative-definite* because the integrand is a negative-definite function. (More on that shortly.)
    An alternate choice for metric on $L_0(\Omega, \mathbb{P})$ is given by

$$d(f,g) = \mathbb{E}\left(\frac{|f-g|}{1+|f-g|}\right) = \int_\Omega \frac{|f(\omega)-g(\omega)|}{1+|f(\omega)-g(\omega)|} \mu(d\omega)$$

for $\mathbb{P}$-measurable functions f and g on Ω. This metric also induces the topology of convergence in measure (with respect to $\mathbb{P}$).

We recall that two topological spaces X and Y are *uniformly homeomorphic* if there exists a bijection $f : X \to Y$ such that f and f^{-1} are uniformly continuous, in which case we call f a *uniform homeomorphism* between the two spaces X and Y. If Y is a subspace of Z, then we call f a *uniform embedding* of X into Z.
    If $f : X \to Y$ is a function between normed spaces X and Y, then f is a uniform embedding if there is an increasing function ϕ_f such that

$$\phi_f(\|x-y\|_X) \leq \|f(x)-f(y)\|_Y \leq \omega_f(\|x-y\|_X),$$

with $\phi_f(t) > 0$ for all $t > 0$, and $\lim_{t\to 0^+} \omega_f(t) = 0$, where ω_f is the modulus of continuity of f.

14.1. Theorem A

The first theorem, due to Aharoni, Maurey, and Mityagin [1], identifies the Banach spaces that uniformly embed into a Hilbert space as linear subspaces of spaces of measurable functions.

THEOREM 14.3 (Aharoni, Maurey, and Mityagin). *A Banach space uniformly embeds into a Hilbert space if and only if it linearly embeds into $L_0(\Omega, \mathbb{P})$, the set of all $\mathbb{P}$-measurable functions on Ω, for some probability space $(\Omega, \mathbb{P})$.*

The proof of this theorem makes use of positive-definite and negative-definite functions. We will, therefore, begin by discussing the topics and results related to positive-definite and negative-definite functions.

DEFINITION 14.4. Let $A = (a_{jk})_{j,k=1}^n$ be a matrix with complex entries. We call A *hermitian* if $a_{jk} = \overline{a_{kj}}$ for all j and k in the set $\{1, \ldots, n\}$. A hermitian matrix A is called *positive-definite* if

$$\sum_{j,k=1}^n a_{jk}\, c_j\, \overline{c_k} \geq 0$$

for all complex vectors $(c_1, \ldots, c_n) \in \mathbb{C}^n$.

We remark that the positive-definite condition on A implies that A is hermitian. Lemma 14.5 is straightforward, and so we will not include the proof.

LEMMA 14.5. *If A and B are positive-definite, then so are $A + B$ and AB.*

DEFINITION 14.6. Let X be a set. A function $K : X \times X \to \mathbb{C}$ is called a *hermitian kernel* on X if $K(s,t) = \overline{K(t,s)}$ for all $(s,t) \in X \times X$. A hermitian kernel is called *positive-definite* if

$$\sum_{j,k=1}^n K(s_j, s_k)\, c_j\, \overline{c_k} \geq 0$$

for all $\{s_1, \ldots, s_n\} \subseteq X$ and all $\{c_1, \ldots, c_n\} \subseteq \mathbb{C}$.

Note that Lemma 14.5 applies to positive-definite hermitian kernels, as well.

EXAMPLE 14.7. Let H be a Hilbert space with inner product $\langle \cdot, \cdot \rangle$ and norm $\|\cdot\|$. If $K : H \times H \to \mathbb{C}$ is defined by $K(x,y) = \langle x, y \rangle$ for x and y in H, then K is a positive-definite hermitian kernel on H. This follows from the fact that

$$\sum_{j,k=1}^n \langle x_j, x_k \rangle\, c_j\, \overline{c_k} = \left\| \sum_{j=1}^n c_j x_j \right\|^2 \geq 0$$

for all $\{x_1, \ldots, x_n\} \subseteq H$.

PROPOSITION 14.8. *If K is a positive-definite hermitian kernel, then e^{tK} is a positive-definite hermitian kernel for each $t > 0$.*

PROOF. The Taylor series for e^{tK} is

$$e^{tK} = 1 + tK + \frac{t^2}{2} K^2 + \cdots$$

Consequently, e^{tK} is a positive-definite hermitian kernel as the sum of positive-definite hermitian kernels. □

DEFINITION 14.9. A hermitian kernel N on a set X is called *negative-definite* if
$$\sum_{j,k=1}^{n} N(s_j, s_k) \, c_j \, \overline{c_k} \leq 0$$
whenever $\{s_1, \ldots, s_n\} \subseteq X$ and $\{c_1, \ldots, c_n\} \subseteq \mathbb{C}$ are such that $c_1 + \cdots + c_n = 0$.

PROPOSITION 14.10. *A hermitian kernel N is negative-definite on a set if and only if e^{-tN} is positive-definite on that set for all $t > 0$.*

PROOF. Let X be a set. Assume that e^{-tN} is positive-definite on X for all t. We wish to show that N is negative-definite. Let $\{s_1, \ldots, s_n\} \subseteq X$ and $\{c_1, \ldots, c_n\} \subseteq \mathbb{C}$ be such that $c_1 + \cdots + c_n = 0$. Also let t be a positive real number. Since e^{-tN} is positive-definite (by assumption), we have that
$$\sum_{j,k=1}^{n} e^{-tN(s_j, s_k)} \, c_j \, \overline{c_k} \geq 0.$$
Observe that
$$\sum_{j,k=1}^{n} c_j \, \overline{c_k} = \left| \sum_{j=1}^{n} c_j \right|^2 = 0,$$
and so
$$\sum_{j,k=1}^{n} \left(1 - e^{-tN(s_j, s_k)} \right) c_j \, \overline{c_k} \leq 0.$$
Since t is positive,
$$\sum_{j,k=1}^{n} \left(\frac{1 - e^{-tN(s_j, s_k)}}{t} \right) c_j \, \overline{c_k} \leq 0.$$
Compute the limit as $t \to 0^+$ to deduce that
$$\sum_{j,k=1}^{n} N(s_j, s_k) \, c_j \, \overline{c_k} \leq 0,$$
and so N is negative-definite, as required.

Conversely, assume that N is a negative-definite hermitian kernel on X. Fix a point $x_0 \in X$ and define a function $K : X \times X \to \mathbb{C}$ by
$$K(x, y) = N(x, x_0) + N(x_0, y) - N(x, y) - N(x_0, x_0),$$
for all $(x, y) \in X \times X$. We claim that K is positive-definite. In order to show this, let $\{x_1, \ldots, x_n\} \subseteq X$ and $\{c_1, \ldots, c_n\} \subseteq \mathbb{C}$. Then
$$\sum_{j,k=1}^{n} K(x_j, x_k) \, c_j \, \overline{c_k} = \sum_{j,k=1}^{n} N(x_j, x_0) \, c_j \, \overline{c_k} + \sum_{j,k=1}^{n} N(x_0, x_k) \, c_j \, \overline{c_k}$$
$$- \sum_{j,k=1}^{n} N(x_j, x_k) \, c_j \, \overline{c_k} - \sum_{j,k=1}^{n} N(x_0, x_0) \, c_j \, \overline{c_k}.$$

Let $c_0 = -(c_1 + \cdots + c_n)$, so that $c_0 + c_1 + \cdots + c_n = 0$. Then

$$\sum_{j,k=1}^n K(x_j, x_k) \, c_j \, \overline{c_k} = -\overline{c_0} \sum_{j=1}^n N(x_j, x_0) \, c_j - c_0 \sum_{k=1}^n N(x_0, x_k) \, \overline{c_k}$$
$$- \sum_{j,k=1}^n N(x_j, x_k) \, c_j \, \overline{c_k} - N(x_0, x_0) \sum_{j,k=1}^n c_j \, \overline{c_k}.$$

Note that

$$N(x_0, x_0) \sum_{j,k=1}^n c_j \, \overline{c_k} = N(x_0, x_0) \, c_0 \, \overline{c_0}.$$

Consequently, we can combine all of the sums into one (where j and k start at 0 instead of 1):

$$\sum_{j,k=1}^n K(x_j, x_k) \, c_j \, \overline{c_k} = - \sum_{j,k=0}^n N(x_j, x_k) \, c_j \, \overline{c_k} \geq 0.$$

We observe that the above quantity is nonnegative because N is a negative-definite hermitian kernel, because $c_0 + c_1 + \cdots + c_n = 0$, and because of the negative sign in front of the second term.

Since K is a positive-definite hermitian kernel, so too is e^{tK}, by Proposition 14.8, for each $t > 0$. Recalling the definition of $K(x,y)$, for each $t > 0$ we have:

$$e^{tK(x,y)} = e^{tN(x,x_0) + tN(x_0,y) - tN(x,y) - tN(x_0,x_0)}$$
$$= e^{tN(x,x_0)} \, e^{tN(x_0,y)} \, e^{-tN(x,y)} \, e^{-tN(x_0,x_0)}.$$

Solve for $e^{-tN(x,y)}$ to get

$$e^{-tN(x,y)} = \underbrace{e^{tK(x,y)}}_{\text{pos.-def.}} \underbrace{e^{-tN(x,x_0)} \, e^{-tN(x_0,y)}}_{\text{positive-definite}} \underbrace{e^{-tN(x_0,x_0)}}_{\text{constant}}.$$

The term in the middle is positive-definite because one factor is a function of x and the other is a function of y, and because N is hermitian. To see why, observe that

$$\sum_{j,k=1}^n e^{-tN(x_j,x_0)} \, e^{-tN(x_0,x_k)} \, c_j \, \overline{c_k} = \Bigl(\sum_{j=1}^n e^{-tN(x_j,x_0)} c_j \Bigr) \Bigl(\sum_{k=1}^n e^{-tN(x_0,x_k)} \overline{c_k} \Bigr)$$

$$= \Bigl(\sum_{j=1}^n e^{-tN(x_j,x_0)} c_j \Bigr) \overline{\Bigl(\sum_{k=1}^n e^{-tN(x_k,x_0)} c_k \Bigr)}$$

$$= \Bigl| \sum_{j=1}^n e^{-tN(x_j,x_0)} c_j \Bigr|^2 \geq 0.$$

Therefore, the hermitian kernel $e^{-tN(x,y)}$ is positive-definite, as required. □

DEFINITION 14.11. Let X be a vector space. A function $f : X \to \mathbb{C}$ is called *positive-definite* if $K(x,y) = f(x-y)$ is a positive-definite hermitian kernel. Similarly, the function f is called *negative-definite* if $K(x,y) = f(x-y)$ is a negative-definite hermitian kernel.

EXAMPLE 14.12. Let H be a Hilbert space with inner product $\langle \cdot, \cdot \rangle$. The function $f : H \to \mathbb{R}$ given by $f(x) = \|x\|^2$ is negative-definite. To see this, let $\{x_1, \ldots, x_n\} \subseteq H$ and let $\{c_1, \ldots, c_n\} \subseteq \mathbb{C}$ such that $c_1 + \cdots + c_n = 0$. Then

$$\sum_{j,k=1}^n \|x_j - x_k\|^2 \, c_j \, \overline{c_k} = \sum_{j,k=1}^n c_j \, \overline{c_k} \left(\|x_j\|^2 + \|x_k\|^2 - \langle x_j, x_k \rangle - \langle x_k, x_j \rangle \right).$$

Note that

$$\sum_{j,k=1}^n c_j \, \overline{c_k} \, \|x_j\|^2 = \left(\sum_{j=1}^n c_j \, \|x_j\|^2 \right) \left(\sum_{k=1}^n \overline{c_k} \right) = 0,$$

$$\sum_{j,k=1}^n c_j \, \overline{c_k} \, \|x_k\|^2 = \left(\sum_{j=1}^n c_j \right) \left(\sum_{k=1}^n \overline{c_k} \, \|x_k\|^2 \right) = 0,$$

and so

$$\sum_{j,k=1}^n \|x_j - x_k\|^2 \, c_j \, \overline{c_k} = - \sum_{j,k=1}^n \langle x_j, x_k \rangle c_j \, \overline{c_k} - \sum_{j,k=1}^n \langle x_k, x_j \rangle \overline{c_k \, \overline{c_j}} \leq 0.$$

Therefore, $\|x\|^2$ is negative-definite, as claimed. As a consequence, we also have that $e^{-t\|x\|^2}$ is positive-definite.

Suppose that N is a negative-definite hermitian kernel on a set X such that $N(x, y) \geq 0$ for all $(x, y) \in X \times X$. Then e^{-tN} is positive-definite, and so $1 - e^{-tN}$ is negative-definite. Since this is true for all $t > 0$, it follows that

$$\int_0^\infty t^{-\alpha-1}(1 - e^{-tN}) \, dt$$

is negative-definite whenever $0 < \alpha < 1$. To determine the value of this integral, make the substitution $u = tN(x, y)$. Then $du = N(x, y) \, dt$, and so

$$\int_0^\infty t^{-\alpha-1}(1 - e^{-tN}) \, dt = \int_0^\infty \left(\frac{u}{N} \right)^{-\alpha-1} (1 - e^{-u}) \, \frac{du}{N}$$

$$= \int_0^\infty u^{-\alpha-1} N^{\alpha+1} (1 - e^{-u}) \, \frac{du}{N}$$

$$= N^\alpha \underbrace{\int_0^\infty u^{-\alpha-1} (1 - e^{-u}) \, du}_{\text{constant}}.$$

Therefore,

$$\int_0^\infty t^{-\alpha-1}(1 - e^{-tN}) \, dt = c_\alpha \, N^\alpha$$

for some constant c_α (that depends on α), and so it follows that N^α is negative-definite when $0 < \alpha < 1$. We summarize this result in the following proposition.

PROPOSITION 14.13. *If $N \geq 0$ is a negative-definite hermitian kernel, then so is N^α whenever $0 < \alpha < 1$.*

PROOF. See the discussion preceding the statement of the proposition. □

REMARK 14.14. The restriction on α is imposed because the integral that is used to define α converges when $0 < \alpha < 1$, and diverges otherwise.

EXAMPLE 14.15. On the real line $\mathbb{R}$, the function x^2 is negative-definite, and so $|x|^p$ is negative-definite whenever $0 < p \leq 2$. Consequently, the function $e^{-|x|^p}$ is positive-definite whenever $0 < p \leq 2$.

Theorem 14.16 is a significant result that relates positive-definite kernels and negative-definite kernels to mappings into Hilbert spaces, and is central to our discussion of uniform embeddings into Hilbert spaces.

THEOREM 14.16. *Let X be a vector space.*

(a) *If K is a positive-definite hermitian kernel on X, then there exists a Hilbert space H and a map $T : X \to H$ such that*
$$K(x,y) = \langle T(x), T(y) \rangle, \quad (x,y) \in X \times X.$$

(b) *If N is a real-valued negative-definite kernel on X such that $N(x,x) = 0$ for all $x \in X$, then there exists a Hilbert space H and a map $T : X \to H$ such that*
$$N(x,y) = \|T(x) - T(y)\|^2, \quad (x,y) \in X \times X.$$

PROOF. (a) Let F be the space of finitely-supported functions on X. That is:
$$F = \Big\{ \sum_{i=1}^{n} \lambda_i e_{x_i} : n \in \mathbb{N}, \lambda_i \in \mathbb{C}, x_i \in X \Big\},$$
where
$$e_x(y) = \begin{cases} 0 & \text{if } y \neq x, \\ 1 & \text{if } y = x. \end{cases}$$
We define a semi-inner product on F by
$$\Big\langle \sum_{j=1}^{n} \lambda_j e_{x_j}, \sum_{k=1}^{m} \mu_k e_{y_k} \Big\rangle = \sum_{j=1}^{n} \sum_{k=1}^{m} \lambda_j \overline{\mu_k} K(x_j, y_k).$$
We then let H be the completion of the quotient $F/\{h \in F : \langle h, h \rangle = 0\}$, and define $T : X \to H$ by setting $T(x)$ to be the equivalence class containing e_x, which we write (through an abuse of notation) as $T(x) = e_x$.

(b) Fix $x_0 \in X$ and define a function $K : X \times X \to \mathbb{R}$ by
$$K(x,y) = \frac{1}{2}\big[N(x, x_0) + N(x_0, y) - N(x, y)\big],$$
for all $(x,y) \in X \times X$. Then K is positive-definite. (See the proof of Proposition 14.10. In this case, we have $N(x_0, x_0) = 0$, by our assumption on N. Note also that K is real-valued because N is assumed to be real-valued.)

By part (a), there exists a Hilbert space H with inner product $\langle \cdot, \cdot \rangle$ and norm $\|\cdot\|_H$ such that $\langle T(x), T(y) \rangle = K(x,y)$ for all x and y in X. We claim that $\|T(x) - T(y)\|_H^2 = N(x,y)$ for all x and y in X. Computing directly, we see that
$$\|T(x) - T(y)\|_H^2 = \langle T(x), T(x) \rangle - 2\langle T(x), T(y) \rangle + \langle T(y), T(y) \rangle.$$
Note that
$$\langle T(x), T(x) \rangle = K(x,x) = \frac{1}{2}\big[N(x, x_0) + N(x_0, x) - N(x, x)\big] = N(x, x_0),$$
and
$$\langle T(y), T(y) \rangle = K(y,y) = \frac{1}{2}\big[N(y, x_0) + N(x_0, y) - N(y, y)\big] = N(x_0, y).$$

In the above computations, $N(x, x_0) = N(x_0, x)$ and $N(y, x_0) = N(x_0, y)$ because N is assumed to be real-valued and hermitian. Therefore,
$$\|T(x) - T(y)\|_H^2 = N(x, x_0) - 2K(x, y) + N(y, x_0) = N(x, y),$$
as claimed. $\square$

Note that the converse of Theorem 14.16 is also true. If a function K is defined as in (a), then it is a positive-definite kernel (see Example 14.7), and if a function N is defined as in (b), then it is a negative definite kernel (see Example 14.12).

Using the representation of positive-definite kernels from Theorem 14.16 allows us to easily prove Theorem 14.17.

THEOREM 14.17. *Suppose that X is a real vector space, and suppose that the function $f : X \to \mathbb{R}$ is positive-definite with $f(0) = 1$. Then for all x and y in X, we have:*

(i) $|f(x)| \leq 1$.
(ii) $|f(x) - f(y)|^2 \leq 2(1 - f(x - y))$.
(iii) $|1 - f(nx)| \leq n^2(1 - f(x))$ *for all $n \in \mathbb{N}$.*

PROOF. By definition, since f is a positive-definite function, we have that $f(x - y)$ is a positive-definite hermitian kernel. (In fact, since f is real-valued, the kernel is symmetric.) By Theorem 14.16, there exists a Hilbert space H and a map $T : X \to H$ such that $\langle T(x), T(y) \rangle = f(x - y)$ for all x and y in X, where $\langle \cdot, \cdot \rangle$ is the inner product on H. Let $\|\cdot\|$ denote the norm on H.

(i) Let x be any point in X. Then
$$\|T(x)\|^2 = \langle T(x), T(x) \rangle = f(x - x) = f(0) = 1.$$
Consequently, we have found that $\|T(x)\| = 1$ for all $x \in X$. (We will make frequent use of this fact.) Thus, by the Cauchy–Schwarz inequality,
$$|f(x)| = |\langle T(x), T(0) \rangle| \leq \|T(x)\| \|T(0)\| = 1,$$
for all x in X.

(ii) Computing directly, we see that
$$|f(x) - f(y)|^2 = \big|\langle T(x), T(0) \rangle - \langle T(y), T(0) \rangle\big|^2 = \big|\langle T(x) - T(y), T(0) \rangle\big|^2.$$
Once again using the Cauchy–Schwarz inequality, we have
$$|f(x) - f(y)|^2 \leq \|T(x) - T(y)\|^2 \|T(0)\|^2 = \|T(x) - T(y)\|^2.$$
Since
$$(14.1) \qquad \|T(x) - T(y)\|^2 = \underbrace{\|T(x)\|^2 + \|T(y)\|^2}_{2} - 2\underbrace{\langle T(x), T(y) \rangle}_{f(x-y)},$$
we conclude that
$$|f(x) - f(y)|^2 \leq 2 - 2f(x - y),$$
as required.

(iii) Write
$$1 - f(nx) = \frac{1}{2}\Big(\underbrace{\|T(0)\|^2 + \|T(nx)\|^2}_{2} - \underbrace{2\langle T(0), T(nx) \rangle}_{2f(nx)}\Big).$$

Then
$$1 - f(nx) = \frac{1}{2}\|T(0) - T(nx)\|^2.$$

By repeated application of the triangle inequality,
$$\frac{1}{2}\|T(0) - T(nx)\|^2 \leq \frac{1}{2}\Big(\sum_{k=1}^{n}\|T((k-1)x) - T(kx)\|\Big)^2.$$

At this point, we use (14.1) to observe that for each $k \in \{1, \ldots, n\}$,
$$\|T(kx) - T((k-1)x)\|^2 = 2 - 2f(x),$$

and so
$$1 - f(nx) \leq \frac{1}{2}\Big(\sum_{k=1}^{n}\sqrt{2 - 2f(x)}\Big)^2 = \frac{1}{2}\Big(n\sqrt{2 - 2f(x)}\Big)^2 = n^2(1 - f(x)),$$

as required. $\square$

EXAMPLE 14.18. Let a be a vector in $\mathbb{R}^n$. Then
$$e_a(x) = e^{i\langle x, a\rangle}$$

defines a positive-definite function on $\mathbb{R}^n$. We can easily see this by direct computation. Let $x_1, \ldots, x_m$ be vectors in $\mathbb{R}^n$ and let $\{c_1, \ldots, c_m\} \subseteq \mathbb{C}$. Then for a fixed $a \in \mathbb{R}^n$,

$$\sum_{j,k=1}^{m} e_a(x_j - x_k)\, c_j\, \overline{c_k}$$
$$= \sum_{j,k=1}^{m} e^{i\langle x_j - x_k, a\rangle}\, c_j\, \overline{c_k} = \Big(\sum_{j=1}^{m} e^{i\langle x_j, a\rangle} c_j\Big)\Big(\sum_{k=1}^{m} e^{-i\langle x_k, a\rangle} \overline{c_k}\Big)$$
$$= \Big(\sum_{j=1}^{m} e^{i\langle x_j, a\rangle} c_j\Big)\overline{\Big(\sum_{k=1}^{m} e^{i\langle x_k, a\rangle} c_k\Big)} = \Big|\sum_{j=1}^{m} e^{i\langle x_j, a\rangle} c_j\Big|^2 \geq 0.$$

Therefore, the function $x \mapsto e_a(x)$ is positive-definite, as claimed.

If the argument above looks familiar, that is because it is very similar to the proof of Proposition 14.10, and in fact we could use Proposition 14.10 in this case, because the function $x \mapsto i\langle x, a\rangle$ is *negative-definite*. This is also easy to show. Let $x_1, \ldots, x_m$ be vectors in $\mathbb{R}^n$ and let $\{c_1, \ldots, c_m\} \subseteq \mathbb{C}$ be such that $c_1 + \cdots + c_m = 0$. Then for a fixed $a \in \mathbb{R}^n$,

$$\sum_{j,k=1}^{m} i\langle x_j - x_k, a\rangle\, c_j\, \overline{c_k}$$
$$= \sum_{j,k=1}^{m} i\langle x_j, a\rangle\, c_j\, \overline{c_k} - \sum_{j,k=1}^{m} i\langle x_k, a\rangle\, c_j\, \overline{c_k}$$
$$= \Big(\sum_{j=1}^{m} i\langle x_j, a\rangle c_j\Big)\Big(\underbrace{\sum_{k=1}^{m} \overline{c_k}}_{0}\Big) - \Big(\sum_{k=1}^{m} i\langle x_k, a\rangle \overline{c_k}\Big)\Big(\underbrace{\sum_{j=1}^{m} c_j}_{0}\Big) = 0.$$

Therefore, the function $x \mapsto i\langle x, a\rangle$ is negative-definite, and so $x \mapsto e^{-i\langle x,a\rangle}$ is positive-definite, by Proposition 14.10. Since this is true for every $a \in \mathbb{R}^n$, we can replace a with $-a$ to get that $x \mapsto e^{i\langle x,a\rangle}$ is positive-definite, as required.

Alternately, we could use the same argument to show that $x \mapsto -i\langle x, a\rangle$ is a negative-definite function, and then drawn the desired conclusion immediately from Proposition 14.10.

REMARK 14.19. In Example 14.18, we showed that $x \mapsto e^{-i\langle x,a\rangle}$ is positive-definite for every $a \in \mathbb{R}^n$. Furthermore, if μ is a finite Borel measure on $\mathbb{R}^n$, then the function

$$x \mapsto \int_{\mathbb{R}^n} e^{-i\langle x,y\rangle} \mu(dy)$$

is positive-definite, since it is the limit of a sum of positive-definite functions. That is, the Fourier transform of a finite Borel measure on $\mathbb{R}^n$ is positive-definite. Theorem 14.20 tells us that the converse is also true.

THEOREM 14.20 (Bochner's Theorem). *If $f : \mathbb{R}^n \to \mathbb{R}$ is a continuous positive-definite function with $f(\mathbf{0}) = 1$, then there exists a probability measure μ on $\mathbb{R}^n$ such that*

$$\widehat{\mu}(x) = \int_{\mathbb{R}^n} e^{-i\langle x,y\rangle} \mu(dy) = f(x)$$

for all $x \in \mathbb{R}^n$.

A proof of Bochner's theorem[1] would take us too far afield, but we can use it to prove the following theorem.

PROPOSITION 14.21. *Let X be a real linear metric space. If $f : X \to \mathbb{R}$ is a continuous positive-definite function with $f(0) = 1$, then there exists a probability space $(\Omega, \mathbb{P})$ and a continuous linear map $U : X \to L_0(\Omega, \mathbb{P})$ such that*

$$\mathbb{E}\left[e^{itU(x)}\right] = f(tx), \quad t \in \mathbb{R},\, x \in X.$$

PROOF. Suppose that X is finite-dimensional. Since X is finite-dimensional, we can use Bochner's theorem (Theorem 14.20) to find a probability measure μ on X^* such that

$$\int_{X^*} e^{-ix^*(x)} \mu(dx^*) = f(x).$$

So, in this (finite-dimensional) case, let $\Omega = X^*$ and $\mathbb{P} = \mu$, and define the measurable function $U(x) \in L_0(\Omega, \mathbb{P})$ by $U(x)(x^*) = -x^*(x)$. Then

$$\mathbb{E}\left[e^{itU(x)}\right] = \int_\Omega e^{itU(x)}\, d\mathbb{P} = \int_{X^*} e^{-itx^*(x)} \mu(dx^*) = f(tx).$$

This proves the theorem when X is finite-dimensional.

In order to extend the theorem to infinite-dimensional spaces, use Kolmogorov's consistency theorem.[2] □

---

[1] Due to the significance of Bochner's theorem, proofs are abundant in the literature. A good example can be found in [**69**].

[2] A proof of Kolmogorov's consistency theorem, also known as Kolmogorov's extension theorem, is beyond the scope of this text, but it is an important result, and so can be found in many texts, such as [**30**].

In other words, if $f : X \to \mathbb{R}$ is a continuous positive-definite function with $f(0) = 1$, then there exists a probability space and a continuous linear map U from X into the set of random variables on that probability space such that the characteristic function of $U(x)$ is $f(xt)$ for each $x \in X$.

EXAMPLE 14.22. If $0 < p \leq 2$, then the map $f \mapsto \|f\|_p^p$ is negative-definite on $L_p(0,1)$, and so $f \mapsto e^{-\|f\|_p^p}$ is positive-definite on that space, by Proposition 14.10. Therefore, by Proposition 14.21, there exists a probability space $(\Omega, \mathbb{P})$ and a continuous linear map $U : L_p(0,1) \to L_0(\Omega, \mathbb{P})$ such that
$$\mathbb{E}\big[e^{itU(f)}\big] = e^{-\|tf\|_p^p} = e^{-|t|^p\|f\|_p^p},$$
for all $t \in \mathbb{R}$ and all $f \in L_p(0,1)$. In particular, this means that the random variable $U(f)$ has characteristic function $e^{-|t|^p\|f\|_p^p}$.

DEFINITION 14.23. If a random variable has characteristic function $e^{-c|t|^p}$, where c is a positive constant, then the random variable is called *p-stable*.

THEOREM 14.24. *Let (Ω, μ) be a probability space and suppose that $0 < r < 2$. If g is a r-stable random variable on (Ω, μ), then*

(a) $g \in L_p(\Omega, \mu)$ *for all p such that $0 < p < r$, and*
(b) $g \notin L_r(\Omega, \mu)$.

PROOF. Suppose that g is an r-stable random variable on (Ω, μ). Then
$$\mathbb{E}\big[e^{itg}\big] = e^{-c|t|^r},$$
for some $c > 0$. Let μ_g be the probability distribution of g on $\mathbb{R}$, so that
$$\widehat{\mu}_g(-t) = e^{-c|t|^r}$$
for all $t \in \mathbb{R}$. Note that $\widehat{\mu}_g(-t) = \widehat{\mu}_g(t)$, so that g is a symmetric random variable, because that implies that g and $-g$ have the same characteristic function. In particular, this means that
$$\mathbb{E}\big[\cos(tg)\big] = e^{-c|t|^r},$$
for each $t \in \mathbb{R}$.

Let p be a real number such that $0 < p < r$. Then
$$\mathbb{E}\big[|g|^p\big] = \int_\Omega |g(\omega)|^p \, \mu(d\omega) = \int_\mathbb{R} |x|^p \, \mu_g(dx).$$
We will use the following identity:
$$|x|^p = \frac{1}{k_p} \int_0^\infty t^{-p-1} \big[1 - \cos(tx)\big] \, dt,$$
where k_p is a positive constant that depends on p. To verify this identity (which is valid for $0 < p < 2$) make the substitution $u = t|x|$ in the integral. Then $du = |x| \, dt$, and so
$$\int_0^\infty t^{-p-1}\big[1-\cos(tx)\big]\, dt = \int_0^\infty \left(\frac{u}{|x|}\right)^{-p-1}\big[1-\cos(u)\big]\frac{du}{|x|}$$
$$= |x|^p \underbrace{\int_0^\infty u^{-p-1}\big[1-\cos(u)\big]\, du}_{k_p},$$

as required. We comment that the integral defining k_p converges when $0 < p < 2$, and diverges otherwise. We also point out that $\cos(t|x|) = \cos(tx)$, because cosine is an even function.

We now use the identity to compute $\|g\|_p^p$:

$$\mathbb{E}[|g|^p] = \int_\mathbb{R} |x|^p \mu_g(dx) = \int_\mathbb{R} \left(\frac{1}{k_p} \int_0^\infty \frac{1 - \cos(tx)}{t^{p+1}} dt \right) \mu_g(dx).$$

We may change the order of integration, and so

$$\mathbb{E}[|g|^p] = \frac{1}{k_p} \int_0^\infty \frac{1}{t^{p+1}} \left(\int_\mathbb{R} [1 - \cos(tx)] \mu_g(dx) \right) dt$$

$$= \frac{1}{k_p} \int_0^\infty \frac{1}{t^{p+1}} \left(1 - e^{-ct^r} \right) dt$$

This integral converges if $0 < p < r$ and diverges if $p = r$, which completes the proof. $\square$

COROLLARY 14.25. *If $0 < p < r \leq 2$, then $L_r(0,1)$ is isometric to a subspace of $L_p(0,1)$.*[3]

PROOF. Since $L_2(0,1)$ is a separable Hilbert space, we already know that it is isometrically isomorphic to a subspace of $L_p(0,1)$ for any p. Consequently, we may assume that $r < 2$, and so assume that $0 < p < r < 2$.

The map $f \mapsto \|f\|_r^r$ is negative-definite on $L_r(0,1)$ if $0 < r < 2$. It follows that $f \mapsto e^{-\|f\|_r^r}$ is positive-definite. By Proposition 14.21, there exists a probability space $(\Omega, \mathbb{P})$ and a continuous linear map $U : L_r(0,1) \to L_0(\Omega, \mathbb{P})$ such that

$$\mathbb{E}[e^{itU(f)}] = e^{-\|tf\|_r^r} = e^{-|t|^r \|f\|_r^r},$$

for all $t \in \mathbb{R}$ and all $f \in L_r(0,1)$. We know that $U(f)$ is a $\mathbb{P}$-measurable function. Indeed, we have that $U(f) \in L_p(\Omega, \mathbb{P})$ because $p < r$ and $U(f)$ is r-stable. Consequently, the map U is an isomorphism from $L_r(0,1)$ into $L_p(\Omega, \mathbb{P})$.

Next, we observe that

$$\|U(f)\|_p^p = \frac{1}{k_p} \int_0^\infty \frac{1}{t^{p+1}} \left(1 - e^{-ct^r} \right) dt,$$

where $c = \|f\|_r^r$ and k_p is a constant that depends on p. (See the proof of Theorem 14.24 for the derivation of this equation.) Let $u = ct^r$. Then $du = crt^{r-1} dt$, and so

$$\|U(f)\|_p^p = \frac{1}{k_p} \int_0^\infty \frac{1}{t^{p+1}} \underbrace{\left(1 - e^{-ct^r} \right)}_{1 - e^{-u}} \cdot \frac{1}{crt^{r-1}} \cdot \underbrace{crt^{r-1} dt}_{du}$$

$$= \frac{1}{k_p cr} \int_0^\infty \frac{c^{\frac{p}{r}+1}}{u^{\frac{p}{r}+1}} \left(1 - e^{-u} \right) du,$$

noting that

$$\frac{1}{t^{(p+1)+(r-1)}} = \frac{1}{t^{p+r}} = \frac{1}{(u/c)^{(p+r)/r}} = \frac{c^{\frac{p}{r}+1}}{u^{\frac{p}{r}+1}}.$$

---

[3] We have seen this result before, in Lemma 13.32, but in that case it was for $1 < p < r \leq 2$.

Therefore,
$$\|U(f)\|_p^p = c^{\frac{p}{r}} \cdot \underbrace{\frac{1}{k_p r} \int_0^\infty u^{-\frac{p}{r}-1}\left(1-e^{-u}\right) du}_{\text{constant}},$$

and so
$$\|U(f)\|_p = c^{\frac{1}{r}} \cdot \underbrace{\left(\frac{1}{k_p r}\int_0^\infty u^{-\frac{p}{r}-1}\left(1-e^{-u}\right) du\right)^{1/p}}_{\text{constant}}.$$

Now we recall that $c = \|f\|_r^r$, and so if we let the above constant be called $C_{p,r}$ (because it depends on p and r), then we have that
$$\|U(f)\|_p = C_{p,r}\|f\|_r.$$

It follows that U is a constant multiple of an isometry from $L_r(0,1)$ into $L_p(\Omega, \mathbb{P})$. Therefore, we can normalize U to get an isometric isomorphism into $L_p(\Omega, \mathbb{P})$. The conclusion follows. $\square$

THEOREM 14.26. *Let X be a real Banach space and let $1 \le p \le 2$. The Banach space X is linearly isomorphic to a subspace of $L_p(0,1)$ if and only if $x \mapsto e^{-\|x\|^p}$ is positive-definite (or $x \mapsto \|x\|^p$ is negative-definite).*

The proof of Theorem 14.26 uses some of the same ideas as the proof of Corollary 14.25, but is more technical, and so we will omit the details[4].

We wish to show that $L_0(\Omega, \mathbb{P})$, the set of $\mathbb{P}$-measurable functions on Ω, where $\mathbb{P}$ is a probability measure on Ω, uniformly embeds into a Hilbert space. Before we can do that, however, we need a lemma.

LEMMA 14.27. *Let X be a linear metric space, and let $f: X \to \mathbb{R}$ be a continuous positive-definite function such that $f(0) = 1$. If for sequences $(x_n)_{n=1}^\infty$ in X the following statement is true:*

(14.2) $$\lim_{n \to \infty} f(x_n) = 1 \iff \lim_{n \to \infty} \|x_n\| = 0,$$

then X uniformly embeds into the unit sphere of a Hilbert space.

PROOF. Since f is assumed to be positive-definite, it follows (by definition) that the function given by the formula $K(x,y) = f(x-y)$ is a positive-definite hermitian kernel. Consequently, by Theorem 14.16, there exists a Hilbert space H and a map $T: X \to H$ such that
$$K(x,y) = \langle T(x), T(y) \rangle$$
for all $(x,y) \in X \times X$. We claim that T is a uniform homeomorphism.

Recall from the proof of Theorem 14.17 that $\|T(x)\| = 1$ for all $x \in X$. Therefore, T is a map from X into the set $\partial B_H = \{u \in H : \|u\| = 1\}$. Also from the proof of Theorem 14.17, we learned that
$$\|T(x) - T(y)\|^2 = 2(1 - f(x-y))$$
for all $(x,y) \in X \times X$. This condition together with (14.2) implies that T is a uniform homeomorphism into ∂B_H (using the sequential characterization of uniform continuity). $\square$

---

[4]For a proof, see Theorem 8.9 in [**13**]. (Although their Theorem 8.9 is more general.)

COROLLARY 14.28. *A Hilbert space is uniformly homeomorphic to a subset of its unit sphere.*

PROOF. Let H be a Hilbert space. Apply Lemma 14.27 to $x \mapsto e^{-\|x\|^2}$, where $x \in H$. We get that H is uniformly homeomorphic to a subset of the unit sphere of a Hilbert space H'. The result follows from the fact that H' (or a subspace of it) is isometrically isomorphic to H. □

COROLLARY 14.29. *The function space $L_p(0,1)$ is uniformly homeomorphic to a subset of a Hilbert space if $0 < p \leq 2$.*

PROOF. Apply Lemma 14.27 to $x \mapsto e^{-\|x\|^p}$, where $x \in L_p(0,1)$. This is positive-definite provided that $0 < p \leq 2$. □

We are now prepared to prove Theorem 14.30.

THEOREM 14.30. *If $(\Omega, \mathbb{P})$ is a probability space, then $L_0(\Omega, \mathbb{P})$ uniformly embeds into a Hilbert space.*

PROOF. The function given by $e^{-|t|}$ is positive-definite on $\mathbb{R}$, which means that $1 - e^{-|t|}$ is negative-definite on $\mathbb{R}$. If we define
$$N(x) = \int_\Omega \left(1 - e^{-|x(\omega)|}\right) \mathbb{P}(d\omega),$$
then $N(x)$ is negative-definite on $L_0(\Omega, \mathbb{P})$ (because the property of being negative-definite is preserved by taking limits and linear combinations). That means that $f(x) = e^{-N(x)}$ is positive-definite on $L_0(\Omega, \mathbb{P})$. The conclusion will follow from Lemma 14.27 if we can verify (14.2).

Note that
$$f(x) = \exp\left[-\int_\Omega \left(1 - e^{-|x(\omega)|}\right)\mathbb{P}(d\omega)\right].$$

If $(x_n)_{n=1}^\infty$ is a sequence in $L_0(\Omega, \mathbb{P})$ that converges to 0 in measure, then $N(x_n) \to 0$ as $n \to \infty$, and so $f(x_n) \to 1$ as $n \to \infty$.

Conversely, if $(x_n)_{n=1}^\infty$ is a sequence in $L_0(\Omega, \mathbb{P})$ such that $f(x_n) \to 1$ as $n \to \infty$, then $N(x_n) \to 0$ as $n \to \infty$, and so $(x_n)_{n=1}^\infty$ converges to 0 in measure. The result follows from Lemma 14.27. □

REMARK 14.31. At the end of the proof of Theorem 14.30, we used the fact that the sequence $(x_n)_{n=1}^\infty$ in $L_0(\Omega, \mathbb{P})$ converges to 0 in measure if and only if $N(x_n) \to 0$ as $n \to \infty$. This follows from the fact that the metric
$$d(f, g) = \int_\Omega \left(1 - e^{-|f(\omega) - g(\omega)|}\right)\mu(d\omega)$$
induces the topology of convergence in measure on $L_0(\Omega, \mathbb{P})$, and $N(x) = d(x, 0)$.

We are now ready to prove the theorem of Aharoni, Maurey, and Mityagin (Theorem 14.3).

PROOF OF THEOREM 14.3. We wish to show that a real Banach space uniformly embeds into a Hilbert space if and only if it linearly embeds into $L_0(\Omega, \mathbb{P})$, for some probability space $(\Omega, \mathbb{P})$.

We know from Theorem 14.30 that $L_0(\Omega, \mathbb{P})$ uniformly embeds into a Hilbert space for each probability space $(\Omega, \mathbb{P})$. Consequently, it remains to show that a

Banach space which uniformly embeds into a Hilbert space also linearly embeds into $L_0(\Omega, \mathbb{P})$ for some probability space $(\Omega, \mathbb{P})$.

Without loss of generality, we may assume that $H = L_2(0,1)$. Let X be a Banach space and assume there exists a uniform embedding $\phi : X \to \partial B_{L_2(0,1)}$. We may assume ϕ takes values in the unit sphere of $L_2(0,1)$ because of Corollary 14.28. Define a function on $X \times X$ by

$$K(x,y) = \langle \phi(x), \phi(y) \rangle, \quad (x,y) \in X \times X.$$

Then K is a positive-definite kernel on X (see Example 14.7) and $K(x,x) = 1$ for all $x \in X$.

Let $\mathcal{M} : \ell_\infty(X) \to \mathbb{R}$ be an invariant mean on X (see Definition 9.12) and define a function $f : X \to \mathbb{R}$ by

$$f(x) = \mathcal{M}\Big(\big(K(x+y,y)\big)_{y \in X} \Big),$$

for all $x \in X$. We claim that f is a continuous positive-definite function with $f(0) = 1$. Once we verify this, we will use Proposition 14.21 to identify a probability space $(\Omega, \mathbb{P})$ and a continuous linear map $U : X \to L_0(\Omega, \mathbb{P})$, and we will then show that this U has a continuous inverse.

The continuity of f follows from the continuity of ϕ. Furthermore, because $K(x,x) = 1$ for all $x \in X$, we have that

$$f(0) = \mathcal{M}\Big(\big(K(y,y)\big)_{y \in X} \Big) = \mathcal{M}\Big((1)_{y \in X} \Big) = 1,$$

since an invariant mean takes the identity in $\ell_\infty(X)$ to 1. Now it only remains to show that f is positive-definite. This will follow from the fact that an invariant mean is nonnegative (meaning that $\mathcal{M}(g) \geq 0$ whenever $g \geq 0$), as well as the other properties of invariant means. Suppose $\{x_1, \ldots, x_n\} \subseteq X$ and $\{c_1, \ldots, c_n\} \subseteq \mathbb{C}$. Then

$$\sum_{j,k=1}^n f(x_j - x_k) c_j \overline{c_k} = \sum_{j,k=1}^n \mathcal{M}\Big(\big(K(x_j - x_k + y, y)\big)_{y \in X} \Big) c_j \overline{c_k}.$$

By the translation invariance of $\mathcal{M}$, for each fixed j and k, we have that

$$\mathcal{M}\Big(\big(K(x_j - x_k + y, y)\big)_{y \in X} \Big) = \mathcal{M}\Big(\big(K(x_j + y, x_k + y)\big)_{y \in X} \Big).$$

(The translation is $y \mapsto x_k + y$ for each y.) Now we use linearity of $\mathcal{M}$ to get

$$\sum_{j,k=1}^n f(x_j - x_k) c_j \overline{c_k} = \sum_{j,k=1}^n \mathcal{M}\Big(\big(K(x_j + y, y + x_k)\big)_{y \in X} \Big) c_j \overline{c_k}$$

$$= \mathcal{M}\Big[\Big(\sum_{j,k=1}^n K(x_j + y, y + x_k) c_j \overline{c_k} \Big)_{y \in X} \Big].$$

Since K is a positive-definite kernel,

$$\sum_{j,k=1}^n K(x_j + y, y + x_k) c_j \overline{c_k} \geq 0$$

for each $y \in X$, and so

$$\mathcal{M}\Big[\Big(\sum_{j,k=1}^n K(x_j + y, y + x_k) c_j \overline{c_k} \Big)_{y \in X} \Big] \geq 0$$

because $\mathcal{M}$ is nonnegative. Consequently, we have shown that f is a positive-definite function on X.

We now use Proposition 14.21 to conclude that there is a probability space $(\Omega, \mathbb{P})$ and a continuous linear map $U : X \to L_0(\Omega, \mathbb{P})$ such that
$$\mathbb{E}\big[e^{itU(x)}\big] = f(tx)$$
for all $t \in \mathbb{R}$ and $x \in X$. In particular, if we let $t = 1$, then the characteristic function of $U(x)$ is $f(x)$.

Now suppose that $(x_n)_{n=1}^\infty$ is a sequence in X such that $\lim_{n\to\infty} U(x_n) = 0$ in $L_0(\Omega, \mathbb{P})$. That is, $U(x_n) \to 0$ in measure as $n \to \infty$. Then $f(x_n) \to 1$ as $n \to \infty$. But
$$f(x_n) = \mathcal{M}\Big(\big(K(x_n + y, y)\big)_{y \in X}\Big),$$
and
$$K(x_n + y, y) = \langle \phi(x_n + y), \phi(y) \rangle = 1 - \frac{1}{2}\|\phi(x_n + y) - \phi(y)\|^2,$$
for each $y \in X$. Consequently, if $f(x_n) \to 1$ as $n \to \infty$, then
$$\lim_{n \to \infty} \mathcal{M}\Big(\big(\|\phi(x_n + y) - \phi(y)\|^2\big)_{y \in X}\Big) = 0.$$
This implies that
$$\lim_{n \to \infty} \inf_{y \in X} \|\phi(x_n + y) - \phi(y)\|^2 = 0,$$
and so for each $n \in \mathbb{N}$, there exists some $y_n \in X$ such that
$$\lim_{n \to \infty} \|\phi(x_n + y_n) - \phi(y_n)\|^2 = 0.$$
Since ϕ^{-1} is uniformly continuous (by assumption), it follows that $\lim_{n\to\infty} x_n = 0$ in the norm topology of X. Therefore, the continuous linear function U has a inverse U^{-1} that is also continuous, and so X linearly embeds into $L_0(\Omega, \mathbb{P})$, as required. □

In Theorem 14.3, we showed that a real Banach space uniformly embeds into a Hilbert space if and only if it linearly embeds into $L_0(\Omega, \mathbb{P})$, for some probability space $(\Omega, \mathbb{P})$. The theorem remains true for if we replace uniform embeddings with coarse embeddings.

We recall that a function $f : X \to Y$ between normed spaces X and Y is a *coarse embedding* if there is an increasing function ϕ_f such that
$$\phi_f(\|x - y\|_X) \leq \|f(x) - f(y)\|_Y \leq \omega_f(\|x - y\|_X),$$
with $\lim_{t \to \infty} \phi_f(t) > 0$ and $\omega_f(t) < \infty$ for all $t > 0$, where ω_f is the modulus of continuity of f.[5] With that in mind, we provide the following companion to Theorem 14.3 (without proof).

THEOREM 14.32 (Randrianarivony [99]). *A real Banach space coarsely embeds into a Hilbert space if and only if it linearly embeds into $L_0(\Omega, \mathbb{P})$, for some probability space $(\Omega, \mathbb{P})$.*[6]

It is worth noting that the above theorem of Randrianarivony and Theorem 14.3 together imply that a real Banach space uniformly embeds into a Hilbert space if and only if it coarsely embeds into a Hilbert space.

---

[5]Note for a uniform embedding, we instead require $\phi_f(t) > 0$ for all $t > 0$ and $\lim_{t \to 0^+} \omega_f(t) = 0$.

[6]In fact, Randrianarivony proved the theorem for quasi-Banach spaces in [99].

14.2. Theorem B

We have proven the first of the two theorems that we need to prove Theorem 14.1. The second theorem is due to Nikišin [89].[7] We will follow the development of the proof in [56]. Without loss of generality, we will take our probability space $(\Omega, \mathbb{P})$ to be the closed unit interval $[0,1]$ with Lebesgue measure λ.

THEOREM 14.33 (Nikišin's Theorem). *A Banach space linearly embeds into $L_0(0,1)$ if and only if it linearly embeds into $L_p(0,1)$ for every p such that $0 < p < 1$.*

COMMENT 14.34. When $0 < p < 1$, the collection of (equivalence classes of) p-integrable functions $L_p(0,1)$ is not a Banach space, because

$$\|f\|_p = \left(\int_0^1 |f(x)|^p \, dx\right)^{1/p}$$

does not determine a norm on $L_p(0,1)$. The reason for this is that $\|\cdot\|_p$ does not satisfy the triangle inequality when $0 < p < 1$. Instead, when $0 < p < 1$, we have a weakened form of the triangle inequality that leads to something called a *quasi-norm*.

DEFINITION 14.35. Let X be a real vector space. A real-valued map $x \mapsto \|x\|$ defined for $x \in X$ is called a *quasi-norm* if
  (i) $\|x\| \geq 0$ for all $x \in X$ and $\|x\| = 0$ if and only if $x = 0$.
  (ii) $\|tx\| = |t|\|x\|$ for all $x \in X$ and $t \in \mathbb{R}$.
  (iii) There exists a constant $C \geq 1$ such that
$$\|x + y\| \leq C(\|x\| + \|y\|)$$
  for all x and y in X.

The value of C must be independent of choice of vectors x and y in X. The smallest possible value for C (that works for all x and y in X) is called the *modulus of concavity* for the quasi-norm. If the modulus of concavity is 1, then the quasi-norm is a *norm* on X. In particular, a norm is a quasi-norm.

A complete quasi-normed space is called a *quasi-Banach space*.

The spaces $L_p(0,1)$ are not Banach spaces when $0 < p < 1$, but they are quasi-Banach spaces. When $0 < p < 1$, the modulus of concavity for $\|\cdot\|_p$ is

$$C_p = 2^{\frac{1}{p}-1}.$$

Of course, when $p \geq 1$, the spaces $L_p(0,1)$ are Banach spaces, so in these cases the modulus of concavity is $C_p = 1$.

In order to prove Nikišin's theorem (Theorem 14.33), we will make use of the *weak L_1 space* $L_{1,\infty}(0,1)$.

DEFINITION 14.36. A measurable function f on a measure space (Ω, μ) is said to be *weak L_1* on (Ω, μ) if there exists a constant K such that

(14.3) $$t\mu(\{\omega \in \Omega : |f(\omega)| > t\}) \leq K, \quad t > 0.$$

---

[7]Nikišin proves a more general result in [89]. (An earlier version can also be found in [88].)

The set of all (equivalence classes of) weak L_1 functions on (Ω, μ) is denoted $L_{1,\infty}(\Omega, \mu)$. For a given function f in $L_{1,\infty}(\Omega, \mu)$, the smallest constant K satisfying (14.3) is denoted $\|f\|_{1,\infty}$.

We will adopt the practice of writing $L_{1,\infty}(0,1)$ when the measure space (Ω, μ) is the closed unit interval $[0,1]$ with Lebesgue measure λ.

It is natural at this point to wonder how we will use the weak L_p spaces to prove Nikišin's theorem. Our goal is to prove that a Banach space linearly embeds into $L_0(0,1)$ if and only if it linearly embeds into $L_p(0,1)$ for every p such that $0 < p < 1$. One direction of this theorem we can already prove using results from Section 14.1. The spaces $L_p(0,1)$, where $0 < p < 1$, uniformly embed into a Hilbert space (Corollary 14.29), and so if X is a Banach space that linearly embeds into $L_p(0,1)$ for some p such that $0 < p < 1$, then X uniformly embeds into a Hilbert space. But that implies that X linearly embeds into $L_0(0,1)$, by Theorem 14.3 (the theorem of Aharoni, Maurey, and Mityagin).

Nikišin showed in [88] that if a Banach space X can be uniformly embedded in $L_0(0,1)$, then it can be linearly embedded into $L_{1,\infty}(0,1)$. This follows from Theorem 14.38, which we will spend the majority of this section proving. This is the key result because $L_{1,\infty}(0,1)$ is a linear subspace of $L_p(0,1)$ for each p such that $0 < p < 1$. We state this last fact as a theorem now.

THEOREM 14.37. *If p is any number such that $0 < p < 1$, then $L_1(0,1)$ is a linear subspace of $L_{1,\infty}(0,1)$ and $L_{1,\infty}(0,1)$ is a linear subspace of $L_p(0,1)$. That is,*
$$L_1(0,1) \subseteq L_{1,\infty}(0,1) \subseteq L_p(0,1),$$
where the subset symbol here is used to represent inclusion as a linear subspace.

PROOF. The inclusion $L_1(0,1) \subseteq L_{1,\infty}(0,1)$ follows immediately from Chebyshev's inequality[8], which says that
$$\lambda\big(\{\omega \in [0,1] : |f(\omega)| \geq t\}\big) \leq \frac{1}{t} \int_{\{|f|>t\}} |f(\omega)|\, \lambda(d\omega).$$

To be more precise, let $f \in L_1(0,1)$. Then Chebyshev's inequality gives us that
$$t\lambda\big(\{\omega \in [0,1] : |f(\omega)| \geq t\}\big) \leq \int_{\{|f|>t\}} |f(\omega)|\, \lambda(d\omega) \leq \|f\|_1.$$

Taking the supremum over $t > 0$, we get
$$\|f\|_{1,\infty} \leq \|f\|_1,$$
which is the desired result.

In order to show that $L_{1,\infty}(0,1) \subseteq L_p(0,1)$ for each $p \in (0,1)$, we use the following identity:[9]
$$\int_{[0,1]} |f|^p\, d\lambda = \int_0^\infty p t^{p-1} \lambda\big(|f| > t\big)\, dt.$$

---

[8] Chebyshev's inequality, which is a special case of Markov's inequality, is a standard result in probability theory, and can be found in many texts, such as [30]. Chebyshev's inequality is stated explicitly in Theorem 14.56, in the comments at the end of this chapter, but the proof is not included there.

[9] This identity, which is sometimes called the layer cake representation, is an immediate consequence of the Fubini–Tonelli theorem. A proof can be found in the comments at the end of the chapter. (See Lemma 14.58.)

To see explicitly how this gives us the desired result, let $f \in L_{1,\infty}(0,1)$, and assume that $\|f\|_{1,\infty} \neq 0$. Then, since $\|f\|_{1,\infty}$ is finite, we can compute the above integral as follows:
$$\int_{[0,1]} |f|^p \, d\lambda = \int_0^{\|f\|_{1,\infty}} pt^{p-1} \lambda(|f|>t) \, dt + \int_{\|f\|_{1,\infty}}^\infty pt^{p-1} \lambda(|f|>t) \, dt.$$
For the first of these integrals, we use the bound $\lambda(|f|>t) \leq 1$, which is true because λ is a probability measure on $[0,1]$, and for the second, we use
$$\lambda(|f|>t) \leq \frac{\|f\|_{1,\infty}}{t},$$
which is true by the definition of the weak L_1 norm. Consequently,
$$\int_{[0,1]} |f|^p \, d\lambda \leq \int_0^{\|f\|_{1,\infty}} pt^{p-1} \, dt + \int_{\|f\|_{1,\infty}}^\infty pt^{p-2} \|f\|_{1,\infty} \, dt$$
$$= \left(t^p\right)\Big|_0^{\|f\|_{1,\infty}} + \left(\frac{p}{p-1}\right)\|f\|_{1,\infty} \left(t^{p-1}\right)\Big|_{\|f\|_{1,\infty}}^\infty$$
$$= \|f\|_{1,\infty}^p - \frac{p}{p-1}\|f\|_{1,\infty}^p.$$
Therefore,
$$\int_{[0,1]} |f|^p \, d\lambda \leq \frac{1}{1-p}\|f\|_{1,\infty}^p,$$
and so
$$\|f\|_p \leq \left(\frac{1}{1-p}\right)^{1/p} \|f\|_{1,\infty}.$$
From this it follows that $L_{1,\infty}(0,1) \subseteq L_p(0,1)$ for each $p \in (0,1)$, as required. $\square$

Once we show that X can be linearly embedded into $L_{1,\infty}(0,1)$ (which follows from Theorem 14.38), we can use Theorem 14.37 to conclude that X can be linearly embedded into $L_p(0,1)$ for each p such that $0 < p < 1$.

THEOREM 14.38 (Nikišin). *Let X be a Banach space and let $T : X \to L_0(0,1)$ be a continuous linear map. If $\epsilon > 0$ is given, then there is a measurable subset E of $[0,1]$ with $\lambda(E) \geq 1 - \epsilon$ such that $\chi_E T$ is a bounded linear operator from X into $L_{1,\infty}(0,1)$, where*
$$\chi_E(x) = \begin{cases} 1 & \text{if } x \in E \\ 0 & \text{if } x \notin E \end{cases}$$
is the indicator function for E.

Before proving Theorem 14.38, we will need to lay some groundwork. The space $L_0(0,1)$ has the topology given by convergence in measure—the measure in this case being Lebesgue measure λ on $[0,1]$. Consequently, a typical open neighborhood of zero in $L_0(0,1)$ has the form
$$V_\epsilon = \left\{f : \lambda(|f| > \epsilon) < \epsilon\right\}$$
for $\epsilon > 0$.

If T is a continuous linear operator from a Banach space X into $L_0(0,1)$, then it maps bounded sets to bounded sets. That means that $T(B_X)$ is a bounded set in $L_0(0,1)$, and so for each $\epsilon > 0$, there exists some $r > 0$ such that $T(B_X) \subseteq rV_\epsilon$. What does the set rV_ϵ look like? If $f \in rV_\epsilon$, where $r > 0$, then $\frac{f}{r} \in V_\epsilon$. That means that $\lambda(|\frac{f}{r}| > \epsilon) < \epsilon$, or (in other words) $\lambda(|f| > r\epsilon) < \epsilon$. Consequently, if we let $M_\epsilon = r\epsilon$, then for any $x \in B_X$, we have that

$$T(x) \in \Big\{ f : \lambda(|f| > M_\epsilon) < \epsilon \Big\}.$$

Note that M_ϵ depends only on ϵ, since r depends only on ϵ.

Similarly, if T is a linear *embedding*, then we can bound ∂B_X away from zero. That means that there exists some $\delta > 0$ such that $T(x) \notin V_\delta$ for any $x \in X$ with $\|x\| = 1$. That is, there exists a $\delta > 0$ such that

$$T(x) \in \Big\{ f : \lambda(|f| > \delta) \geq \delta \Big\}$$

for all $x \in \partial B_X$. Otherwise, if there were no such $\delta > 0$, we would be able to find a sequence $(T(x_n))_{n=1}^\infty$, with $\|x_n\| = 1$ for each n, which goes to zero in $L_0(0,1)$, an impossibility, since $x_n \in \partial B_X$ for each $n \in \mathbb{N}$.

Now we will proceed with a series of lemmas that will ultimately allow us to prove Theorem 14.38, which will then enable us to reach our final objective, a proof of Theorem 14.33 (Nikišin's theorem).

LEMMA 14.39. *Suppose that $\epsilon_1, \ldots, \epsilon_n$ are independent and identically distributed Rademacher random variables on a probability space $(\Omega, \mathbb{P})$. If $a_1, \ldots, a_n$ are real numbers with $a_1^2 + \cdots + a_n^2 = 1$, then*

$$(a) \ \Big(\mathbb{E} \Big| \sum_{k=1}^n \epsilon_k a_k \Big|^2 \Big)^{1/2} = 1 \quad and \quad (b) \ \Big(\mathbb{E} \Big| \sum_{k=1}^n \epsilon_k a_k \Big|^4 \Big)^{1/4} \leq 3^{1/4}.$$

COMMENT 14.40. We recall that a *Rademacher random variable* is a random variable that is uniformly distributed over the two-point set $\{-1, 1\}$.

PROOF OF LEMMA 14.39. Note that $\mathbb{E}[\epsilon_j \epsilon_k] = \delta_{jk}$, where

$$\delta_{jk} = \begin{cases} 1 & \text{if } j = k \\ 0 & \text{if } j \neq k \end{cases}$$

is the the *Kronecker delta*. Thus,

$$\mathbb{E} \Big| \sum_{k=1}^n \epsilon_k a_k \Big|^2 = \mathbb{E} \Big[\sum_{j,k=1}^n \epsilon_j \epsilon_k a_j a_k \Big] = \sum_{j=1}^n a_j^2 = 1,$$

and then taking square gives us (a). For (b), we compute:

$$\mathbb{E} \Big| \sum_{k=1}^n \epsilon_k a_k \Big|^4 = \mathbb{E} \Big[\sum_{j_1, j_2, j_3, j_4 = 1}^n \epsilon_{j_1} \epsilon_{j_2} \epsilon_{j_3} \epsilon_{j_4} a_{j_1} a_{j_2} a_{j_3} a_{j_4} \Big]$$

$$= \sum_{j=1}^n a_j^4 + \binom{4}{2} \sum_{j=1}^n \sum_{k=j+1}^n a_j^2 a_k^2 \leq 3 \Big(\sum_{j=1}^n a_j^2 \Big)^2 = 3.$$

Now take the fourth root of both sides to get (b). □

COMMENT 14.41. In Lemma 14.39, the equation in (a) and the inequality in (b) are special case of Khintchine's inequalities.

THEOREM 14.42 (Khintchine's inequalities). *If $\epsilon_1, \ldots, \epsilon_n$ are independent Rademacher functions on a probability space $(\Omega, \mathbb{P})$, then there exist positive constants A_p and B_p (that depend only on p) such that*

$$A_p \Big(\sum_{k=1}^n |a_n|^2\Big)^{1/2} \le \Big(\mathbb{E}\Big|\sum_{k=1}^n \epsilon_k a_k\Big|^p\Big)^{1/p} \le B_p \Big(\sum_{k=1}^n |a_n|^2\Big)^{1/2}.$$

(See Section 6.2 in [5] for a proof and discussion of Khintchine's inequalities.)

In 1981, Haagerup found the best constants for each value of p. In particular, the best constants for $p = 2$ are $A_2 = 1$ and $B_2 = 1$, which gives us equality in the case $p = 2$, and that in turn gives us the equality in (a) from Lemma 14.39. The inequality in (b) from Lemma 14.39 follows from the (nontrivial) fact that the best possible value for B_4 is

$$B_4 = \sqrt{2} \left(\frac{\Gamma(\frac{5}{2})}{\sqrt{\pi}}\right)^{1/4} = 3^{1/4},$$

where Γ is the Gamma function, and $\Gamma(\frac{5}{2}) = \frac{3\sqrt{\pi}}{4}$.

LEMMA 14.43. *There exists a constant $c > 0$ such that*

$$\mathbb{P}\left(\Big|\sum_{k=1}^n \epsilon_k a_k\Big| \ge \frac{1}{2}\Big(\sum_{k=1}^n a_k^2\Big)^{1/2}\right) \ge c,$$

whenever $\epsilon_1, \ldots, \epsilon_n$ are independent and identically distributed Rademacher random variables on a probability space $(\Omega, \mathbb{P})$, and $a_1, \ldots, a_n$ are real numbers, $n \in \mathbb{N}$.

PROOF. Let $\epsilon_1, \ldots, \epsilon_n$ be independent Rademacher random variables on a probability space $(\Omega, \mathbb{P})$, and let $a_1, \ldots, a_n$ be real numbers. Without loss of generality, we assume that $a_1^2 + \cdots + a_n^2 = 1$. Define a subset A of Ω by

$$A = \Big\{\Big|\sum_{k=1}^n \epsilon_k a_k\Big| \ge \frac{1}{2}\Big\}.$$

Then

$$\mathbb{E}\Big|\sum_{k=1}^n \epsilon_k a_k\Big|^2 = \int_\Omega \Big|\sum_{k=1}^n \epsilon_k a_k\Big|^2 d\mathbb{P}$$

$$= \int_A \Big|\sum_{k=1}^n \epsilon_k a_k\Big|^2 d\mathbb{P} + \underbrace{\int_{\Omega \setminus A} \Big|\sum_{k=1}^n \epsilon_k a_k\Big|^2 d\mathbb{P}}_{\le 1/4}.$$

On the set $\Omega \setminus A$, we have (by definition) that

$$\Big|\sum_{k=1}^n \epsilon_k a_k\Big| < \frac{1}{2},$$

and for that reason, the second integral above is bounded above by $\frac{1}{4}$. Consequently,

$$\mathbb{E}\Big|\sum_{k=1}^{n}\epsilon_k a_k\Big|^2 \leq \int_A \Big|\sum_{k=1}^{n}\epsilon_k a_k\Big|^2 d\mathbb{P} + \frac{1}{4}$$

$$= \int_\Omega \Big(\chi_A \cdot \Big|\sum_{k=1}^{n}\epsilon_k a_k\Big|^2\Big) d\mathbb{P} + \frac{1}{4},$$

where χ_A is the indicator function for the measurable set A. Now we apply Hölder's inequality on the last integral above (with $p = 2$) to get

$$\mathbb{E}\Big|\sum_{k=1}^{n}\epsilon_k a_k\Big|^2 \leq \underbrace{\Big(\int_\Omega \chi_A^2 \, d\mathbb{P}\Big)^{1/2}}_{\mathbb{P}(A)} \underbrace{\Big(\int_\Omega \Big|\sum_{k=1}^{n}\epsilon_k a_k\Big|^4 d\mathbb{P}\Big)^{1/2}}_{\leq 3} + \frac{1}{4}.$$

The second integral in the above product (not including the square root that surrounds it) is bounded by 3 by (b) from Lemma 14.39. Therefore, we have found that

$$\mathbb{E}\Big|\sum_{k=1}^{n}\epsilon_k a_k\Big|^2 \leq \mathbb{P}(A)^{1/2}\sqrt{3} + \frac{1}{4}.$$

The left side of this inequality is equal to 1, by (a) from Lemma 14.39. Consequently,

$$1 \leq \sqrt{3}\,\mathbb{P}(A)^{1/2} + \frac{1}{4},$$

which means that $\mathbb{P}(A) \geq \frac{3}{16}$. Therefore, the proof is complete with $c = \frac{3}{16}$. $\square$

REMARK 14.44. In the proof of Lemma 14.43, instead of defining the set A the way we did, we could have chosen a set A_λ to be the set

$$A_\lambda = \Big\{\Big|\sum_{k=1}^{n}\epsilon_k a_k\Big| \geq \lambda\Big\}.$$

We can repeat the same argument with A_λ to show that

$$\mathbb{P}(A_\lambda) \geq \frac{(1-\lambda^2)^2}{3}.$$

With this notation, the set A we chose in the proof of Lemma 14.43 is $A_{1/2}$.

COROLLARY 14.45. *Suppose $f_1, \ldots, f_n$ are measurable functions in $L_0(0,1)$. Then for all $t \in \mathbb{R}$,*

$$\mathbb{P}\Big[\Big|\sum_{k=1}^{n}\epsilon_k f_k(t)\Big| \geq \frac{1}{2}\Big(\sum_{k=1}^{n}|f_k(t)|^2\Big)^{1/2}\Big] \geq \frac{3}{16},$$

where $\epsilon_1, \ldots, \epsilon_n$ are independent and identically distributed Rademacher random variables on a probability space $(\Omega, \mathbb{P})$.

PROOF. This follows immediately from Lemma 14.43, using the value of the constant $c = \frac{3}{16}$ found in the proof. $\square$

LEMMA 14.46. *Let X be a Banach space and let $T : X \to L_0(0,1)$ be a continuous linear operator. For any $\delta > 0$, there exists a number $M_\delta > 0$ (that depends only on δ) such that*

$$\lambda\left[\Big(\sum_{j=1}^n |T(x_j)|^2\Big)^{1/2} > 2M_\delta\right] \leq \frac{\delta}{c},$$

whenever $x_1, \ldots, x_n$ are points in X such that $\sum_{j=1}^n \|x_j\| \leq 1$, where λ is Lebesgue measure on $[0,1]$ and $c = \frac{3}{16}$.

PROOF. Let $\delta > 0$ be given. By assumption, the map $T : X \to L_0(0,1)$ is a continuous linear operator, and so $T(B_X)$ is a bounded set. Thus, there exists a positive number M_δ (that depends only on δ) such that

$$\lambda(|T(x)| > M_\delta) \leq \delta$$

for all $x \in B_X$. (See the comments following the statement of Theorem 14.38.)

Now let $x_1, \ldots, x_n$ be in X and suppose that $\sum_{j=1}^n \|x_j\| \leq 1$. Let $\epsilon_1, \ldots, \epsilon_n$ be independent Rademacher random variables on a probability space $(\Omega, \mathbb{P})$. Observe that for each $\omega \in \Omega$,

$$\|\epsilon_1(\omega)x_1 + \cdots + \epsilon_n(\omega)x_n\| \leq 1.$$

Consequently, for each $\omega \in \Omega$, we have that

$$\lambda\Big(\Big|T\Big(\epsilon_1(\omega)x_1 + \cdots + \epsilon_n(\omega)x_n\Big)\Big| > M_\delta\Big) \leq \delta.$$

Using the linearity of T we can rewrite the above inequality as

$$\lambda\Big(\Big|\sum_{j=1}^n \epsilon_j(\omega)T(x_j)\Big| > M_\delta\Big) \leq \delta.$$

For ease of notation, let $f_j = T(x_j)$ for each $j \in \{1, \ldots, n\}$. Then each f_j is a measurable function on $[0,1]$. Using this notation, we can write the above inequality (no longer suppressing the variable) as

$$\lambda\Big(\Big\{t : \Big|\sum_{j=1}^n \epsilon_j(\omega)f_j(t)\Big| > M_\delta\Big\}\Big) \leq \delta.$$

Since this is true for every $\omega \in \Omega$, we can then conclude that

(14.4) $$(\lambda \times \mathbb{P})\Big(\Big\{(t,\omega) : \Big|\sum_{j=1}^n \epsilon_j(\omega)f_j(t)\Big| > M_\delta\Big\}\Big) \leq \delta,$$

where $\lambda \times \mathbb{P}$ is the product measure on $[0,1] \times \Omega$.

We are now going to consider the probability of a set smaller than the one appearing on the left side of (14.4). Let

$$A = \Big\{t : \Big(\sum_{j=1}^n |f_j(t)|^2\Big)^{1/2} > 2M_\delta\Big\}.$$

Let B denote the set appearing on the left side of (14.4). Then

$$\underbrace{(\lambda \times \mathbb{P})(A \times \Omega)}_{\lambda(A)} \cdot \underbrace{(\lambda \times \mathbb{P})(B | \{t \in A\})}_{\text{we will bound this below}} \leq (\lambda \times \mathbb{P})(B) \leq \delta.$$

We will show that $(\lambda \times \mathbb{P})(B|\{t \in A\}) \geq c$. To that end, suppose that $t \in A$. Then
$$\frac{1}{2}\Big(\sum_{j=1}^n |f_j(t)|^2\Big)^{1/2} > M_\delta,$$
and so, by Corollary 14.45, we have that
$$\mathbb{P}\left[\Big|\sum_{j=1}^n f_j(t)\epsilon_j\Big| \geq M_\delta\right] \geq \mathbb{P}\left[\Big|\sum_{j=1}^n f_j(t)\epsilon_j\Big| \geq \frac{1}{2}\Big(\sum_{j=1}^n |f_j(t)|^2\Big)^{1/2}\right] \geq c.$$

Therefore,
$$\lambda(A) \cdot c \leq \delta, \quad \text{and so} \quad \lambda(A) \leq \frac{\delta}{c},$$
and this is the inequality that we wanted to prove. □

COROLLARY 14.47. *Let X be a Banach space and let $T : X \to L_0(0,1)$ be a continuous linear operator. For any $\delta > 0$, there exists a number $M_\delta > 0$ (that depends only on δ) such that*
$$\lambda\left(\max_{1 \leq j \leq n} |T(x_j)| > 2M_\delta\right) \leq \frac{\delta}{c},$$
whenever $x_1, \ldots, x_n$ are points in X such that $\sum_{j=1}^n \|x_j\| \leq 1$, where λ is Lebesgue measure on $[0,1]$ and $c = \frac{3}{16}$.

PROOF. This follows immediately from Lemma 14.46. □

We are now ready to prove Theorem 14.38.

PROOF OF THEOREM 14.38. Let us recall what we wish to prove. We have a Banach space X and a continuous linear operator $T : X \to L_0(0,1)$. Our objective is to show that if $\epsilon > 0$ is given, then there is a measurable subset E of $[0,1]$ with $\lambda(E) \geq 1 - \epsilon$ such that $\chi_E T$ is a bounded linear operator from X into $L_{1,\infty}(0,1)$, where χ_E is the indicator function for E and λ is Lebesgue measure on $[0,1]$.

By Corollary 14.47, there exists a number $M_\epsilon > 0$ such that
$$\lambda\left(\Big\{\omega \in [0,1] : \max_{1 \leq j \leq n} |T(x_j)(\omega)| > 2M_\epsilon\Big\}\right) < \epsilon, \tag{14.5}$$
whenever $x_1, \ldots, x_n$ are points in X such that $\sum_{j=1}^n \|x_j\| \leq 1$, for each $n \in \mathbb{N}$. Let K be a positive number such that $K > 2M_\epsilon$.

We first ask ourselves "Is T bounded into $L_{1,\infty}(0,1)$ with constant K?" If so, then we let $E = [0,1]$ and we are done. Therefore, suppose that T is not bounded into $L_{1,\infty}(0,1)$ with constant K.

If T *is* bounded into $L_{1,\infty}(0,1)$ with constant K, then for every $x \in \partial B_X$, we have that
$$\lambda\Big(\{\omega \in [0,1] : |T(x)(\omega)| > t\}\Big) \leq \frac{K}{t},$$
for all $t > 0$. Since we are assuming that T is *not* bounded into $L_{1,\infty}(0,1)$ with constant K, it follows that there is some $x \in X$ with $\|x\| = 1$ such that
$$\lambda\Big(\{\omega \in [0,1] : |T(x)(\omega)| > t\}\Big) > \frac{K}{t}, \tag{14.6}$$

for *some* $t > 0$. Let τ_1 be the infimum over all t such that (14.6) is true for some $x \in B_X$. That is, let

$$\tau_1 = \inf\left\{t > 0 : \exists x \in \partial B_X \text{ for which } \lambda\big(|T(x)| > t\big) > \frac{K}{t}\right\}.$$

(From here on, we will suppress the $\omega \in [0,1]$ for ease of notation.) In particular, we know that $\tau_1 < \infty$, because we know there is some pair (x,t) for which (14.6) is true. We also know that $\tau_1 > 0$, because $t \geq K$ (or else we would have a contradiction in (14.6)).

Choose $x_1 \in B_X$ and $t_1 > 0$ such that

$$\lambda\big(|T(x_1)| > t_1\big) > \frac{K}{t_1}, \quad t_1 \leq 2\tau_1.$$

Let A_1 be the subset of $[0,1]$ given by $A_1 = \big\{|T(x_1)| > t_1\big\}$, which we know is nonempty, and let $A_1^c = [0,1] \setminus A_1$ be the complement of A_1 in $[0,1]$. Note that $t_1 \geq K$, and so $\lambda(A_1) < \epsilon$ by (14.5), which means that $\lambda(A_1^c) > 1 - \epsilon$. In particular, we have that A_1^c is nonempty and satisfies the requirement on the measure of the set E from the statement of the theorem.

If $\chi_{A_1^c} T$ is bounded into $L_{1,\infty}(0,1)$ with constant K, then let $E = A_1^c$ and we are done. Otherwise, we repeat the procedure. In such a case, we let

$$\tau_2 = \inf\left\{t > 0 : \exists x \in \partial B_X \text{ for which } \lambda\big(|\chi_{A_1^c} T(x)| > t\big) > \frac{K}{t}\right\},$$

and then we choose $x_2 \in \partial B_X$ and $t_2 > 0$ such that

$$\lambda\big(|\chi_{A_1^c} T(x_2)| > t_2\big) > \frac{K}{t_2}, \quad t_2 \leq 2\tau_2.$$

Let $A_2 = A_1 \cup \big\{|\chi_{A_1^c} T(x_2)| > t_2\big\}$, which (we will see shortly) has measure $\lambda(A_2) < \epsilon$ by (14.5), and consequently we have that $\lambda(A_2^c) \geq 1 - \epsilon$. If $\chi_{A_2^c} T$ is bounded into $L_{1,\infty}(0,1)$ with constant K, then let $E = A_2^c$ and we are done, otherwise we continue (again).

We proceed inductively. Suppose that we have found a sequence $(x_j)_{j=1}^N$ in ∂B_X, a sequence $(t_j)_{j=1}^N$ of positive real numbers, and a sequence $(A_j)_{j=1}^N$ of measurable subsets of $[0,1]$ with measure $\lambda(A_{j-1}) < \epsilon$ such that

$$\lambda\big(|\chi_{A_{j-1}^c} T(x_j)| > t_j\big) > \frac{K}{t_j}, \quad t_j \leq 2\tau_j,$$

where

$$\tau_j = \inf\left\{t > 0 : \exists x \in \partial B_X \text{ for which } \lambda\big(|\chi_{A_{j-1}^c} T(x)| > t\big) > \frac{K}{t}\right\}$$

and

$$A_j = A_{j-1} \cup \big\{|\chi_{A_{j-1}^c} T(x_j)| > t_j\big\},$$

for all $j \in \{1, \ldots, N\}$.

We claim that $\lambda(A_N) < \epsilon$. Once we have shown that, then either $\chi_{A_N^c} T$ is bounded into $L_{1,\infty}(0,1)$ with constant K, in which case we let $E = A_N^c$ (and we are done), or we continue with the inductive procedure.

In order to show that $\lambda(A_N) < \epsilon$, we let $\omega \in A_N$. Then $|T(x_j)(\omega)| > t_j$ for at least one integer j in the set $\{1, \ldots, N\}$. Thus, for each $\omega \in A_N$,

$$\max_{1 \leq j \leq N} \left\{ \frac{1}{t_j} \cdot T(x_j)(\omega) \right\} > 1.$$

Multiply by $2M_\epsilon$ to get

$$\max_{1 \leq j \leq N} \left\{ \frac{2M_\epsilon}{t_j} \cdot T(x_j)(\omega) \right\} > 2M_\epsilon,$$

which (by linearity of T) can be written as

$$\max_{1 \leq j \leq N} \left\{ T\left(\frac{2M_\epsilon}{t_j} \cdot x_j \right)(\omega) \right\} > 2M_\epsilon.$$

Consequently, we will be able to conclude that $\lambda(A_N) < \epsilon$ from (14.5), provided we can show that $\sum_{j=1}^{N} \left\| \frac{2M_\epsilon}{t_j} \cdot x_j \right\| \leq 1$.

Suppose to the contrary that $\sum_{j=1}^{N} \frac{2M_\epsilon}{t_j} \|x_j\| > 1$. Then, since x_j has norm 1 for each j, we have that

$$\sum_{j=1}^{N} \frac{2M_\epsilon}{t_j} > 1.$$

However,

$$\lambda(A_N) = \lambda\Big(A_1 \cup \{|\chi_{A_1^c} T(x_2)| > t_2\} \cup \cdots \cup \{|\chi_{A_{N-1}^c} T(x_N)| > t_N\} \Big)$$

$$= \lambda(A_1) + \lambda\Big(\{|\chi_{A_1^c} T(x_2)| > t_2\} \Big) + \cdots + \lambda\Big(\{|\chi_{A_{N-1}^c} T(x_N)| > t_N\} \Big),$$

and so (because $K > 2M_\epsilon$)

$$\lambda(A_N) \geq \sum_{j=1}^{N} \frac{K}{t_j} > \sum_{j=1}^{N} \frac{2M_\epsilon}{t_j} > 1.$$

This is a contradiction, and so we conclude that $\sum_{j=1}^{N} \left\| \frac{2M_\epsilon}{t_j} x_j \right\| \leq 1$, which in turn implies that $\lambda(A_N) < \epsilon$ by (14.5).

We continue inductively until we find a set E that satisfies the requirements of the theorem. If there is no such set E, then we continue this procedure indefinitely, until we have constructed a sequence $(x_j)_{j=1}^{\infty}$ in ∂B_X, a sequence $(t_j)_{j=1}^{\infty}$ of positive real numbers, and a sequence $(A_j)_{j=1}^{\infty}$ of measurable subsets of $[0,1]$ such that

$$\lambda\big(|\chi_{A_{j-1}^c} T(x_j)| > t_j\big) > \frac{K}{t_j}, \quad t_j \leq 2\tau_j,$$

where

$$\tau_j = \inf \left\{ t > 0 : \exists x \in B_X \text{ for which } \lambda\big(|\chi_{A_{j-1}^c} T(x)| > t\big) > \frac{K}{t} \right\}$$

and $A_j = A_{j-1} \cup \{|\chi_{A_{j-1}^c} T(x_j)| > t_j\}$ with $\lambda(A_j) < \epsilon$ for all $j \in \mathbb{N}$. Let

$$A = \bigcup_{j=1}^{\infty} A_j.$$

Note that $\lambda(A) \leq \epsilon$ because A is the union of nested sets, all of which have measure less than ϵ. Thus, the complement A^c has measure $\lambda(A^c) \geq 1 - \epsilon$.

We claim that $\chi_{A^c}T$ is bounded into $L_{1,\infty}(0,1)$ with constant K. Suppose to the contrary that $\chi_{A^c}T$ is not bounded with constant K. Then there exists $x \in \partial B_X$ and a $t > 0$ such that
$$\lambda\Big(\big|\chi_{A^c}T(x)\big| > t\Big) > \frac{K}{t}.$$
It must be that $t \geq \tau_n$ for each $n \in \mathbb{N}$, and so it follows that $t_n \leq 2t$ for each $n \in \mathbb{N}$. Thus, for each $N \in \mathbb{N}$, we have that
$$\lambda(A_N) \geq \sum_{j=1}^{N} \frac{K}{t_j} > \sum_{j=1}^{N} \frac{K}{2t} = \frac{NK}{2t},$$
an impossibility. Consequently, the operator $\chi_{A^c}T$ is bounded into $L_{1,\infty}(0,1)$ with constant K, and so in this case we let $E = A^c$.

Therefore, for $\epsilon > 0$, there exists a measurable set E with $\lambda(E) \geq 1 - \epsilon$ such that $\chi_E T$ is bounded into $L_{1,\infty}(0,1)$ with constant K for any $K \leq 2M_\epsilon$, which means that $\|\chi_E T\| \leq 2M_\epsilon$. This completes the proof. $\square$

Finally, we are ready to prove Theorem 14.33 (Nikišin's theorem), which we restate here.

THEOREM 14.48 (Nikišin's theorem, restated). *A Banach space linearly embeds into $L_0(0,1)$ if and only if it linearly embeds into $L_p(0,1)$ for every p such that $0 < p < 1$.*

PROOF. One direction of this theorem follows from the results of Section 14.1. Suppose X is a Banach space that linearly embeds into $L_p(0,1)$ for every p such that $0 < p < 1$. By Corollary 14.29, the spaces $L_p(0,1)$, where $0 < p < 1$, uniformly embed into a Hilbert space. That means that X uniformly embeds into a Hilbert space, and so X linearly embeds into $L_0(0,1)$, by Theorem 14.3.

Conversely, assume that X is a Banach space and that $T : X \to L_0(0,1)$ is a linear embedding. For each $n \in \mathbb{N}$, there exists a measurable set E_n such that $\lambda(E_n) \geq 1 - \frac{1}{n}$ and such that $\chi_{E_n} T$ is bounded into $L_{1,\infty}(0,1)$, by Theorem 14.38.

Define a positive real-valued function ϕ on $[0,1]$ by
$$\phi = \sum_{n=1}^{\infty} \frac{2^{-n}}{\|\chi_{E_n}T\|} \chi_{E_n \setminus (E_1 \cup \cdots \cup E_{n-1})}.$$

Next, define M_ϕ by $M_\phi(f) = \phi f$ for each $f \in L_0(0,1)$. Then $M_\phi T$ is a bounded linear operator from X into $L_{1,\infty}(0,1)$, by construction. Furthermore, since ϕ is positive, the map $M_\phi T$ is an isomorphism into $L_{1,\infty}(0,1)$. Therefore, it is also an isomorphism into $L_p(0,1)$ for all $p \in (0,1)$, by Theorem 14.37. This completes the proof. $\square$

It is possible to prove a more general version of Theorem 14.38. Before stating the theorem, let us introduce a definition.

DEFINITION 14.49. A Banach space X is said to have *Rademacher type p* if there exists a constant $c \geq 1$ such that
$$\Big(\mathbb{E}\Big\|\sum_{i=1}^{n} \epsilon_i x_i\Big\|^p\Big)^{1/p} \leq c\Big(\sum_{i=1}^{n} \|x_i\|^p\Big)^{1/p},$$

for any finite sequence $(x_i)_{i=1}^n$ of points in X, where $(\epsilon_i)_{i=1}^n$ is a finite sequence of independent and identically distributed random variables on a probability space $(\Omega, \mathbb{P})$ for which $\mathbb{P}(\epsilon_i = +1) = \frac{1}{2}$ and $\mathbb{P}(\epsilon_i = -1) = \frac{1}{2}$ for each $i \in \{1, \ldots, n\}$.

COMMENT 14.50. We recall that a random variable ϵ that is uniformly distributed over the two-point set $\{-1, 1\}$ is known as a *Rademacher random variable*, or simply as a *Rademacher*.

EXAMPLE 14.51. The function space $L_p(0,1)$ has Rademacher type p if $p \in [1,2]$ and has Rademacher type 2 if $p \geq 2$.

THEOREM 14.52. *Let X be a Banach space that has Rademacher type p, and let $T : X \to L_0(0,1)$ be a continuous linear map. If $\epsilon > 0$ is given, then there is a measurable subset E of $[0,1]$ with $\lambda(E) \geq 1 - \epsilon$ such that $\chi_E T$ is a bounded linear operator from X into $L_{p,\infty}(0,1)$ (and so into $L_q(0,1)$ for all $q < p$).*

We won't prove Theorem 14.52, because it is similar to the proof of Theorem 14.38.

COMMENT 14.53. Theorem 14.52 is about embeddings into $L_{p,\infty}(0,1)$. In Definition 14.36, we defined weak L_1, but weak L_p is defined analogously when $1 < p < \infty$.

DEFINITION 14.54. A measurable function f on a measure space (Ω, μ) is said to be *weak L_p* on (Ω, μ) if there exists a constant K such that

$$(14.7) \qquad t^p \mu(\{\omega \in \Omega : |f(\omega)| > t\}) \leq K^p, \quad t > 0.$$

The set of all (equivalence classes of) weak L_p functions on (Ω, μ) is denoted $L_{p,\infty}(\Omega, \mu)$. For a given function f in $L_{p,\infty}(\Omega, \mu)$, the smallest constant K satisfying (14.7) is denoted $\|f\|_{p,\infty}$. The set *weak L_∞* is defined to be the same as the set of μ-essentially bounded functions on Ω, so that $L_{\infty,\infty}(\Omega, \mu)$ is the same as $L_\infty(\Omega, \mu)$.

We also have an analogue of Theorem 14.37 for weak L_p, $1 < p < \infty$. We will state the analogue as two separate theorems: Theorem 14.55 and Theorem 14.57.

THEOREM 14.55. *Let (Ω, μ) be a measure space and let p be a positive number such that $0 < p < \infty$. If $f \in L_p(\Omega, \mu)$, then*

$$\|f\|_{p,\infty} \leq \|f\|_p,$$

and so $L_p(\Omega, \mu) \subseteq L_{p,\infty}(\Omega, \mu)$.

Theorem 14.55 follows from Chebyshev's inequality, in the same way that it did when $p = 1$. The more general form of Chebyshev's inequality is as follows.

THEOREM 14.56 (Chebyshev's inequality). *If $t > 0$ and $p \in (0, \infty)$, then*

$$\mu(\{\omega \in \Omega : |f(\omega)| \geq t\}) \leq \frac{1}{t^p} \int_{\{|f| > t\}} |f(\omega)|^p \, d\mu.$$

We won't prove Chebyshev's inequality here, but we will use it to prove Theorem 14.55. As an aside, we remark that sometimes the name Chebyshev's inequality is reserved for the case $p = 2$, in which case the more general theorem is called Markov's inequality, although Markov's inequality is sometimes reserved for an even more general theorem. (See [**30**], for example.)

PROOF OF THEOREM 14.55. Since the right side of Chebyshev's inequality (as it is given above) is bounded by $\frac{1}{t^p}\|f\|_p^p$, we immediately have that

$$t^p \mu(\{\omega \in \Omega : |f(\omega)| \geq t\}) \leq \|f\|_p^p,$$

and from here we get the desired inequality. □

THEOREM 14.57. *Let (Ω, μ) be a finite measure space. If p and q are positive numbers such that $0 < q < p$, then $L_q(\Omega, \mu) \subseteq L_{p,\infty}(\Omega, \mu)$ and*

$$\|f\|_q \leq \left(\frac{p}{p-q}\right)^{1/q} \mu(\Omega)^{\frac{1}{q}-\frac{1}{p}} \|f\|_{p,\infty}.$$

In order to prove this, we use the following useful lemma.

LEMMA 14.58. *Let (Ω, μ) be a measure space. If f is a measurable function and $q > 0$, then*

$$\int_\Omega |f|^q \, d\mu = \int_0^\infty q t^{q-1} \mu(|f| > t) \, dt.$$

PROOF. This is a straightforward application of the Fubini–Tonelli theorem:

$$\int_0^\infty q t^{q-1} \mu(|f| > t) \, dt = \int_0^\infty q t^{q-1} \left(\int_\Omega \chi_{\{|f|>t\}} \, d\mu\right) dt$$

$$= \int_\Omega \left(\int_0^\infty q t^{q-1} \chi_{\{|f|>t\}} \, dt\right) d\mu.$$

Here, we recall that $\{|f| > t\}$ is shorthand for $\{\omega \in \Omega : |f(\omega)| > t\}$, and so for a given $\omega \in \Omega$, the inner integral is over the set $(0, |f(\omega)|)$. Therefore,

$$\int_0^\infty q t^{q-1} \mu(|f| > t) \, dt = \int_\Omega \left(\int_0^{|f(\omega)|} q t^{q-1} \, dt\right) d\mu$$

$$= \int_\Omega |f(\omega)|^q \, d\mu.$$

And that proves Lemma 14.58. □

Now we can prove Theorem 14.57.

PROOF OF THEOREM 14.57. For this theorem, (Ω, μ) is a finite measure space and p and q are numbers such that $0 < q < p < \infty$. Assume that $f \in L_{p,\infty}(\Omega, \mu)$, and assume $\|f\|_{p,\infty} \neq 0$. Then, by Lemma 14.58,

$$\int_\Omega |f|^q \, d\mu = \int_0^\infty q t^{q-1} \mu(|f| > t) \, dt.$$

We will evaluate this integral in a similar way to the analogous integral in the proof of Theorem 14.37, by splitting it into the sum of two integrals. In this case, we will split the integral at

$$T = \frac{\|f\|_{p,\infty}}{\mu(\Omega)^{1/p}},$$

and we will call it T for ease of notation. Consequently, we have

$$\int_\Omega |f|^q \, d\mu = \int_0^T qt^{q-1} \mu(|f| > t) \, dt + \int_T^\infty qt^{q-1} \mu(|f| > t) \, dt.$$

For the first of these two integrals, we use the bound $\mu(|f| > t) \leq \mu(\Omega)$, which is finite, by assumption. For the second of the two integrals, we use

$$\mu(|f| > t) \leq \frac{\|f\|_{p,\infty}^p}{t^p},$$

which is true by the definition of the weak L_p norm. Thus,

$$\int_\Omega |f|^q \, d\mu \leq \int_0^T qt^{q-1} \mu(\Omega) \, dt + \int_T^\infty qt^{q-p-1} \|f\|_{p,\infty}^p \, dt$$

$$= \mu(\Omega) \left(t^q\right)\Big|_0^T + \left(\frac{q}{q-p}\right) \|f\|_{p,\infty}^p \left(t^{q-p}\right)\Big|_T^\infty$$

$$= \mu(\Omega) \, T^q - \left(\frac{q}{q-p}\right) \|f\|_{p,\infty}^p \, T^{q-p}.$$

Recalling the definition of T, we now have

$$\int_\Omega |f|^q \, d\mu \leq \mu(\Omega) \left(\frac{\|f\|_{p,\infty}}{\mu(\Omega)^{1/p}}\right)^q - \left(\frac{q}{q-p}\right) \|f\|_{p,\infty}^p \left(\frac{\|f\|_{p,\infty}}{\mu(\Omega)^{1/p}}\right)^{q-p}$$

$$= \mu(\Omega)^{1-\frac{q}{p}} \|f\|_{p,\infty}^q - \left(\frac{q}{q-p}\right) \mu(\Omega)^{1-\frac{q}{p}} \|f\|_{p,\infty}^q.$$

Therefore,

$$\int_\Omega |f|^q \, d\mu \leq \left(\frac{p}{p-q}\right) \mu(\Omega)^{1-\frac{q}{p}} \|f\|_{p,\infty}^q.$$

The desired inequality is obtained by computing the qth root of both sides. □

14.3. Conclusion

We can now combine the results of Section 14.1 and Section 14.2 to prove Theorem 14.1, which we restate below.

THEOREM 14.59 (Theorem 14.1, restated). *A Banach space uniformly embeds into a Hilbert space if and only if it linearly embeds into $L_p(0,1)$ for every p such that $0 < p < 1$.*

PROOF. By Theorem 14.3, a Banach space uniformly embeds into a Hilbert space if and only if it linearly embeds into $L_0(\Omega, \mathbb{P})$, for some probability space $(\Omega, \mathbb{P})$. By Theorem 14.33, a Banach space linearly embeds into $L_0(\Omega, \mathbb{P})$ if and only if it linearly embeds into $L_p(\Omega, \mathbb{P})$ for every p such that $0 < p < 1$. That completes the proof. □

In Theorem 14.1, we have shown that if a Banach space uniformly embeds into a Hilbert space, then it must linearly embed into $L_p(0,1)$ for each p such that $0 < p < 1$. This leads to an open question.

QUESTION 14.60 (Kwapień [**71**]). *If a Banach space uniformly embeds into a Hilbert space, does it linearly embed into $L_1(0,1)$?*

Alternately, we could ask if a closed Banach subspace of $L_p(0,1)$ for $0 < p < 1$ must be isomorphic to a closed subspace of $L_1(0,1)$. The strongest known result in this direction is due to Kalton [**58**].

THEOREM 14.61. *A Banach space X is isomorphic to a closed subspace of $L_1(0,1)$ if and only if $\ell_1(X)$ is linearly isomorphic to a closed subspace of $L_0(0,1)$, and hence it is linearly isomorphic to a closed subspace of $L_p(0,1)$ for $0 < p < 1$.*

CHAPTER 15

Uniform embeddings into reflexive spaces

In Chapter 14, we addressed the question of what types of spaces uniformly embed into a Hilbert space. In this chapter, we wish to address a more general question:

> Under what conditions does a Banach space uniformly embed into a reflexive Banach space?

This is a much broader question that the one we looked at in the last chapter, and one we are not able to fully answer, and so instead we will answer a more specific question:

> Does every separable Banach space uniformly embed into a reflexive Banach space?

We will show that the answer to this question is "No." We will determine that answer by proving the following theorem, due to Kalton [**60**].[1]

THEOREM 15.1. *The space c_0 of real-valued null sequences does not uniformly embed into a reflexive Banach space.*

In order to prove Theorem 15.1, we will make use of methods similar to the asymptotic smoothness techniques we used in Chapter 13.

For $r \in \mathbb{N}$, let $\mathcal{P}_r(\mathbb{N})$ be the collection of r-subsets $\{m_1, m_2, \ldots, m_r\}$ of r members of $\mathbb{N}$, which we write in increasing order, so that

$$m_1 < m_2 < \cdots < m_{r-1} < m_r.$$

For convenience, we will adopt vector notation, so that $\{m_1, m_2, \ldots, m_r\}$ in $\mathcal{P}_r(\mathbb{N})$ can be written as $\boldsymbol{m} = (m_1, m_2, \ldots, m_r)$, a vector having length r, such that $m_1, \ldots, m_r$ are in $\mathbb{N}$ with

$$m_1 < m_2 < \cdots < m_{r-1} < m_r.$$

If $\boldsymbol{m} = (m_1, \ldots, m_r)$ and $\boldsymbol{n} = (n_1, \ldots, n_r)$ are both in $\mathcal{P}_r(\mathbb{N})$, then we call them *adjacent* if either

$$m_1 \leq n_1 \leq m_2 \leq n_2 \cdots \leq m_r \leq n_r$$

or

$$n_1 \leq m_1 \leq n_2 \leq m_2 \cdots \leq n_r \leq m_r.$$

In this case, we say that $\boldsymbol{m}$ and $\boldsymbol{n}$ are *interlaced*. We write $\boldsymbol{m} < \boldsymbol{n}$ if $m_r < n_1$.

Defining adjacent vectors in this way allows us to make $\mathcal{P}_r(\mathbb{N})$ into a graph, and gives us an associated metric d, the *shortest path metric*, where $d(\boldsymbol{m}, \boldsymbol{n})$ is the length of the shortest path from $\boldsymbol{m}$ to $\boldsymbol{n}$. We observe that with this metric,

$$\operatorname{diam}\left(\mathcal{P}_r(\mathbb{N})\right) = r,$$

---

[1] In fact, in [**60**], Kalton shows that c_0 does not embed uniformly or coarsely into a reflexive Banach space.

and $d(\boldsymbol{m}, \boldsymbol{n}) = r$ if and only if $\boldsymbol{m} < \boldsymbol{n}$ or $\boldsymbol{n} < \boldsymbol{m}$.

If $\mathbb{A}$ is a subset of $\mathbb{N}$, then we define $\mathcal{P}_r(\mathbb{A})$ in the natural way. Furthermore, we let $\mathcal{P}_0(\mathbb{N})$ denote the set $\{\emptyset\}$.

REMARK 15.2. In Chapter 13, we used a different notation. In that chapter's notation, the collection of r-subsets $\{m_1, m_2, \ldots, m_r\}$ of r members of $\mathbb{A} \subseteq \mathbb{N}$, written in increasing order, would be denoted $G_r(\mathbb{A})$ instead of $\mathcal{P}_r(\mathbb{A})$. We justify the change in notation because in Chapter 13, the set $G_r(\mathbb{A})$ was equipped with the Hamming distance, while here we are viewing $\mathcal{P}_r(\mathbb{A})$ as a *partially ordered set* with the shortest path metric.[2]

THEOREM 15.3. *Suppose X is a reflexive Banach space and let $f : \mathcal{P}_r(\mathbb{N}) \to X$ be a bounded map, where $r \in \mathbb{N}$. If $\epsilon > 0$, then there exists an infinite subset $\mathbb{A}$ of $\mathbb{N}$ such that*
$$\operatorname{diam}\left(\mathcal{P}_r(\mathbb{A})\right) < 2\operatorname{Lip}(f) + \epsilon.$$

(Compare to Proposition 13.23.) Before proving Theorem 15.3, we will introduce some notation. Let $\mathcal{U}$ be a non-principal ultrafilter. If $f : \mathcal{P}_r(\mathbb{N}) \to X$ is a bounded map, where $r \in \mathbb{N}$, we define a map
$$\partial_{\mathcal{U}} f : \mathcal{P}_{r-1}(\mathbb{N}) \to X$$
by
$$\partial_{\mathcal{U}} f(m_1, \ldots, m_{r-1}) = \lim_{m_r \in \mathcal{U}} f(m_1, \ldots, m_{r-1}, m_r),$$
where the limit is in the weak topology on X. This limit exists in the weak topology because X is reflexive. (If X was not reflexive, then we would conclude that the limit existed in X^{**} in the weak* topology on X^{**}.)

We can iterate this procedure to generate a map
$$\partial_{\mathcal{U}}^k f : \mathcal{P}_{r-k}(\mathbb{N}) \to X$$
for each k in $\{1, \ldots, r\}$, where $\partial_{\mathcal{U}}^1 = \partial_{\mathcal{U}}$. In each case, the map goes to the second dual, but X is assumed to be reflexive, so we can view $\partial_{\mathcal{U}}^k f$ as a map into X for each k in $\{1, \ldots, r\}$. Note also that this interpretation means that $\partial_{\mathcal{U}}^r f \in X$.

We can define $\partial_{\mathcal{U}} f$ when X is not reflexive. In such a case, we have that
$$\partial_{\mathcal{U}}^k f : \mathcal{P}_{r-k}(\mathbb{N}) \to X^{(2k)},$$
where $X^{(2k)}$ is the $2k$th dual of X. Of course, when X is reflexive, we can identify $X^{(2k)}$ with X for each k.

LEMMA 15.4. *Let $h : \mathcal{P}_r(\mathbb{N}) \to \mathbb{R}$ be a bounded function, where $r \in \mathbb{N}$. If $\epsilon > 0$, then there exists an infinite subset $\mathbb{A}$ of $\mathbb{N}$ such that*
$$|h(\sigma) - \partial_{\mathcal{U}}^r h| < \epsilon, \quad \sigma \in \mathcal{P}_r(\mathbb{A}).$$

PROOF. We will construct the set $\mathbb{A} = \{m_1, m_2, \ldots\}$ inductively. We start by choosing m_1 so that
$$|\partial_{\mathcal{U}}^{r-1} h(m_1) - \partial_{\mathcal{U}}^r h| < \epsilon,$$

---

[2] During the actual lectures, Kalton used $G_k(\mathbb{M})$ for the topic of Chapter 13 and $\mathcal{P}_r(\mathbb{A})$ for the topic in this chapter. This is consistent with the published literature. The presentation in Chapter 13 followed that of Kalton and Randrianarivony in [**67**], where $G_k(\mathbb{M})$ was used. The presentation in this chapter follows that of Kalton in [**60**], where $\mathcal{P}_r(\mathbb{A})$ is the notation used.

which we can do because $\partial_{\mathcal{U}}^r h$ is defined so that
$$\partial_{\mathcal{U}}^r h = \lim_{m \in \mathcal{U}} \partial_{\mathcal{U}}^{r-1} h(m),$$
where the limit is taken in the weak topology on X.

Now suppose that we have chosen $\{m_1, m_2, \ldots, m_k\}$ with $m_1 < m_2 < \cdots < m_k$, so that if σ is a s-subset of $\{m_1, m_2, \ldots, m_k\}$, where s is any number such that $1 \leq s \leq \min\{k, r\}$, then
$$|\partial_{\mathcal{U}}^{r-s} h(\sigma) - \partial_{\mathcal{U}}^r h| < \frac{\epsilon}{2}.$$
We need to find an integer m_{k+1} to add to out set $\{m_1, m_2, \ldots, m_k\}$ that is larger than the current members and satisfies the desired inequality.

Let σ be a s-subset of $\{m_1, m_2, \ldots, m_k\}$, where s is any number such that $1 \leq s \leq \min\{k, r-1\}$. By definition,
$$\partial_{\mathcal{U}}^{r-s} h(\sigma) = \lim_{m \in \mathcal{U}} \partial_{\mathcal{U}}^{r-s-1} h(\sigma, m),$$
and so there exists an infinite set $\mathbb{A}_\sigma \in \mathcal{U}$ such that
$$|\partial_{\mathcal{U}}^{r-s} h(\sigma) - \partial_{\mathcal{U}}^{r-s-1} h(\sigma, m)| < \frac{\epsilon}{2}$$
whenever $m \in \mathbb{A}_\sigma$. We may assume that $m > m_k$ for every $m \in \mathbb{A}_\sigma$.

Since $\mathbb{A}_\sigma \in \mathcal{U}$ for each $\sigma \subseteq \{m_1, \ldots, m_k\}$, the intersection of all such sets is also in $\mathcal{U}$. That is,
$$\mathbb{A} = \bigcap_{\sigma \subseteq \{m_1, \ldots, m_k\}} \mathbb{A}_\sigma \in \mathcal{U},$$
because $\mathcal{U}$ is a filter, and so is closed under finite intersections. (See Definition 11.1.) If $m \in \mathbb{A}$, then
$$|\partial_{\mathcal{U}}^{r-s-1} h(\sigma, m) - \partial_{\mathcal{U}}^r h| \leq |\partial_{\mathcal{U}}^{r-s-1} h(\sigma, m) - \partial_{\mathcal{U}}^{r-s} h(\sigma)| + |\partial_{\mathcal{U}}^{r-s} h(\sigma) - \partial_{\mathcal{U}}^r h| < \epsilon.$$
Therefore, if we pick $m_{k+1} \in \mathbb{A}$, then we have verified the inductive step. This completes the proof. $\square$

Now suppose that X is a Banach space and assume that $f : \mathcal{P}_r(\mathbb{N}) \to X$ and $g : \mathcal{P}_r(\mathbb{N}) \to X^*$ are both bounded functions. Define a new function
$$f \otimes g : \mathcal{P}_{2r}(\mathbb{N}) \to \mathbb{R}$$
by the rule
$$(f \otimes g)(m_1, \ldots, m_{2r}) = \Big\langle f(m_2, m_4, \ldots, m_{2r}), g(m_1, m_3, \ldots, m_{2r-1}) \Big\rangle,$$
where $\langle \cdot, \cdot \rangle$ denotes the dual action of X^* on X and $m_1 < m_2 < \cdots < m_{2r-1} < m_{2r}$.

LEMMA 15.5. *If X is a Banach space and $f : \mathcal{P}_r(\mathbb{N}) \to X$ and $g : \mathcal{P}_r(\mathbb{N}) \to X^*$ are bounded functions, then*
$$\partial_{\mathcal{U}}^2 (f \otimes g) = \partial_{\mathcal{U}} f \otimes \partial_{\mathcal{U}} g.$$

PROOF. First, we compute $\partial_{\mathcal{U}}(f \otimes g)$ by taking the limit with respect to m_{2r} through $\mathcal{U}$. That gives us
$$\partial_{\mathcal{U}}(f \otimes g)(m_1, \ldots, m_{2r-1}) = \lim_{m_{2r} \in \mathcal{U}} \Big\langle f(m_2, \ldots, m_{2r}), g(m_1, \ldots, m_{2r-1}) \Big\rangle$$
$$= \Big\langle \partial_{\mathcal{U}} f(m_2, \ldots, m_{2r-2}), g(m_1, \ldots, m_{2r-1}) \Big\rangle.$$

Next, in order to get $\partial_{\mathcal{U}}^2(f \otimes g)$, we take the limit with respect to m_{2r-1} through $\mathcal{U}$. That gives us

$$\partial_{\mathcal{U}}^2(f \otimes g)(m_1, \ldots, m_{2r-2}) = \lim_{m_{2r-1} \in \mathcal{U}} \Big\langle \partial_{\mathcal{U}} f(m_2, \ldots, m_{2r-2}), g(m_1, \ldots, m_{2r-1}) \Big\rangle$$

$$= \Big\langle \partial_{\mathcal{U}} f(m_2, \ldots, m_{2r-2}), \partial_{\mathcal{U}} g(m_1, \ldots, m_{2r-3}) \Big\rangle.$$

Therefore,
$$\partial_{\mathcal{U}}^2(f \otimes g) = \partial_{\mathcal{U}} f \otimes \partial_{\mathcal{U}} g,$$
as required. □

COROLLARY 15.6. *If X is a reflexive Banach space and $f : \mathcal{P}_r(\mathbb{N}) \to X$ and $g : \mathcal{P}_r(\mathbb{N}) \to X^*$ are bounded functions, then*

$$\partial_{\mathcal{U}}^{2r}(f \otimes g) = \Big\langle \partial_{\mathcal{U}}^r f, \partial_{\mathcal{U}}^r g \Big\rangle,$$

where $\langle \cdot, \cdot \rangle$ denotes the dual action of X^* on X.

PROOF. This follows from repeated application of Lemma 15.5. □

LEMMA 15.7. *Suppose X is a reflexive Banach space and let $f : \mathcal{P}_r(\mathbb{N}) \to X$ be a bounded map, where $r \in \mathbb{N}$. If $\epsilon > 0$, then there exists an infinite subset $\mathbb{A}$ of $\mathbb{N}$ such that*

$$\|f(\sigma)\| < \|\partial_{\mathcal{U}}^r f\| + \mathrm{Lip}(f) + \epsilon, \quad \sigma \in \mathcal{P}_r(\mathbb{A}).$$

PROOF. Let $\epsilon > 0$ be given. By the Hahn–Banach theorem, we can find a function $g : \mathcal{P}_r(\mathbb{N}) \to X^*$ such that $\|g(\sigma)\| = 1$ and

$$\langle f(\sigma), g(\sigma) \rangle = \|f(\sigma)\|$$

for each $\sigma \in \mathcal{P}_r(\mathbb{N})$. By Corollary 15.5, we have that

(15.1) $$\partial_{\mathcal{U}}^{2r}(f \otimes g) = \Big\langle \partial_{\mathcal{U}}^r f, \partial_{\mathcal{U}}^r g \Big\rangle.$$

The function $f \otimes g : \mathcal{P}_{2r}(\mathbb{N}) \to \mathbb{R}$ is bounded, and so (by Lemma 15.4), there exists an infinite subset $\mathbb{M}$ of $\mathbb{N}$ such that

(15.2) $$\left| (f \otimes g)(\sigma) - \partial_{\mathcal{U}}^{2r}(f \otimes g) \right| < \epsilon$$

for all $\sigma \in \mathcal{P}_{2r}(\mathbb{M})$. Write $\mathbb{M}$ as

$$\mathbb{M} = \{m_1, n_1, m_2, n_2, \ldots, m_j, n_j, \ldots\},$$

where the terms are written in increasing order, so that

$$m_1 < n_1 < m_2 < n_2 < \ldots < m_j < n_j < \ldots.$$

Now define $\mathbb{A}$ by
$$\mathbb{A} = \{m_1, m_2, \ldots, m_j \ldots\},$$
so that $\mathbb{A}$ is composed of every other member from $\mathbb{M}$, starting with the first.

We claim that
$$\|f(\boldsymbol{m})\| < \|\partial_{\mathcal{U}}^r f\| + \mathrm{Lip}(f) + \epsilon$$
for all $\boldsymbol{m} \in \mathcal{P}_r(\mathbb{A})$, where $\mathbb{A}$ is the infinite subset of $\mathbb{N}$ that we constructed in the preceding paragraph. In order to verify this claim, let $\boldsymbol{m} \in \mathcal{P}_r(\mathbb{A})$. Write

$$\boldsymbol{m} = (m_{j_1}, m_{j_2}, \ldots, m_{j_r})$$

where $j_1 < j_2 < \cdots < j_r$ (so that $m_{j_1} < m_{j_2} < \cdots < m_{j_r}$). Now let
$$\boldsymbol{n} = (n_{j_1}, n_{j_2}, \ldots, n_{j_r}).$$
Then the vectors $\boldsymbol{m}$ and $\boldsymbol{n}$ are adjacent (because of how $\mathbb{M}$ was constructed), and so (in particular) the distance between them is one. That is,
$$d(\boldsymbol{m}, \boldsymbol{n}) = 1.$$
Therefore,
$$\|f(\boldsymbol{m}) - f(\boldsymbol{n})\| \leq \mathrm{Lip}(f)\, d(\boldsymbol{m}, \boldsymbol{n}) = \mathrm{Lip}(f),$$
and so
$$\|f(\boldsymbol{m})\| = \langle f(\boldsymbol{m}), g(\boldsymbol{m})\rangle = \langle f(\boldsymbol{m}) - f(\boldsymbol{n}), g(\boldsymbol{m})\rangle + \langle f(\boldsymbol{n}), g(\boldsymbol{m})\rangle$$
$$\leq \mathrm{Lip}(f)\|g(\boldsymbol{m})\| + \langle f(\boldsymbol{n}), g(\boldsymbol{m})\rangle$$
(15.3)
$$\leq \mathrm{Lip}(f) + \langle f(\boldsymbol{n}), g(\boldsymbol{m})\rangle.$$

Our next step is to find a bound for $\langle f(\boldsymbol{n}), g(\boldsymbol{m})\rangle$. To that end, let σ be the result of interlacing $\boldsymbol{m}$ and $\boldsymbol{n}$, by which we mean let
$$\sigma = (m_{j_1}, n_{j_1}, m_{j_2}, n_{j_2}, \ldots, m_{j_r}, n_{j_r}).$$
Then $\sigma \in \mathcal{P}_{2r}(\mathbb{M})$ and $(f \otimes g)(\sigma) = \langle f(\boldsymbol{n}), g(\boldsymbol{m})\rangle$. Consequently, by (15.2),
$$\left|\langle f(\boldsymbol{n}), g(\boldsymbol{m})\rangle\right| < \left|\partial_\mathcal{U}^{2r}(f \otimes g)\right| + \epsilon.$$
Thus, by (15.1), we have that
$$\left|\langle f(\boldsymbol{n}), g(\boldsymbol{m})\rangle\right| < \left|\langle \partial_\mathcal{U}^r f, \partial_\mathcal{U}^r g\rangle\right| + \epsilon < \left\|\partial_\mathcal{U}^r f\right\| \left\|\partial_\mathcal{U}^r g\right\| + \epsilon.$$
Since $\left\|\partial_\mathcal{U}^r g\right\| \leq 1$, we have that
$$\left|\langle f(\boldsymbol{n}), g(\boldsymbol{m})\rangle\right| < \left\|\partial_\mathcal{U}^r f\right\| + \epsilon.$$
combining this bound with the inequality in (15.3), we conclude that
$$\|f(\boldsymbol{m})\| < \mathrm{Lip}(f) + \left\|\partial_\mathcal{U}^r f\right\| + \epsilon,$$
which is the desired result. $\square$

We are now ready to prove Theorem 15.3.

PROOF OF THEOREM 15.3. We recall our assumptions: We suppose that X is a reflexive Banach space and that $f : \mathcal{P}_r(\mathbb{N}) \to X$ is a bounded map, where $r \in \mathbb{N}$. Let $\epsilon > 0$ be given. We wish to show that there exists an infinite subset $\mathbb{A}$ of $\mathbb{N}$ such that
$$\mathrm{diam}\left(\mathcal{P}_r(\mathbb{A})\right) < 2\,\mathrm{Lip}(f) + \epsilon.$$
Define a map $\widehat{f} : \mathcal{P}_r(\mathbb{N}) \to X$ by
$$\widehat{f}(\sigma) = f(\sigma) - \partial_\mathcal{U}^r f,$$
for all $\sigma \in \mathcal{P}_r(\mathbb{N})$. By Lemma 15.7 (applied to $\widehat{f}$), there is an infinite subset $\mathbb{A}$ of $\mathbb{N}$ such that
$$\|\widehat{f}(\sigma)\| < \|\partial_\mathcal{U}^r \widehat{f}\| + \mathrm{Lip}(\widehat{f}) + \frac{\epsilon}{2}$$
for all $\sigma \in \mathcal{P}_r(\mathbb{A})$. But $\partial_\mathcal{U}^r \widehat{f} = 0$ and $\mathrm{Lip}(\widehat{f}) = \mathrm{Lip}(f)$, and so
$$\|\widehat{f}(\sigma)\| < \mathrm{Lip}(f) + \frac{\epsilon}{2}$$

for all $\sigma \in \mathcal{P}_r(\mathbb{A})$.

Now let σ_1 and σ_2 be in $\mathcal{P}_r(\mathbb{N})$. Then

$$\|f(\sigma_1) - f(\sigma_2)\| \leq \|\widehat{f}(\sigma_1)\| + \|\widehat{f}(\sigma_2)\| < 2\operatorname{Lip}(f) + \epsilon.$$

The desired result follows. $\square$

DEFINITION 15.8. Let M be a metric space and let $\epsilon > 0$ and $\delta \geq 0$. We say that M has *property $Q(\epsilon, \delta)$* if the following condition is satisfied: For every $r \in \mathbb{N}$, if $f : \mathcal{P}_r(\mathbb{N}) \to M$ has $\operatorname{Lip}(f) \leq \delta$, then there exists an infinite subset $\mathbb{M}$ of $\mathbb{N}$ such that $d(f(\sigma), f(\tau)) \leq \epsilon$ for all σ and τ in $\mathcal{P}_r(\mathbb{N})$.

Furthermore, for each $\epsilon > 0$, we define $\Delta_M(\epsilon)$ to be the supremum of all $\delta \geq 0$ for which M has property $Q(\epsilon, \delta)$. That is,

$$\Delta_M(\epsilon) = \sup\left\{\delta \geq 0 : M \text{ has property } Q(\epsilon, \delta)\right\}.$$

THEOREM 15.9. *Let M be a metric space. If M uniformly embeds into a reflexive Banach space, then $\Delta_M(\epsilon) > 0$ for every $\epsilon > 0$.*

PROOF. Let $\epsilon > 0$ be given. Assume that $h : M \to X$ is a uniform embedding of M into a reflexive Banach space X. Let $r \in \mathbb{N}$ and let $f : \mathcal{P}_r(\mathbb{N}) \to M$ have $\operatorname{Lip}(f) \leq \delta$. Then $h \circ f : \mathcal{P}_r(\mathbb{N}) \to X$ is a map such that

$$\operatorname{Lip}(h \circ f) \leq \omega_h(\delta),$$

where ω_h is the modulus of continuity of h. Therefore, by Theorem 15.3, for any $\epsilon' > 0$, there exists an infinite subset $\mathbb{A}$ of $\mathbb{N}$ such that

$$\operatorname{diam}\left(\mathcal{P}_r(\mathbb{A})\right) < 2\operatorname{Lip}(h \circ f) + \epsilon'.$$

Pick $\epsilon' < \operatorname{Lip}(h \circ f)$. Then there exists an infinite subset $\mathbb{A}$ of $\mathbb{N}$ such that

$$\operatorname{diam}\left(\mathcal{P}_r(\mathbb{A})\right) < 3\operatorname{Lip}(h \circ f) \leq 3\omega_h(\delta).$$

Consequently, if σ and τ are in $\mathcal{P}_r(\mathbb{A})$, then

$$\|(h \circ f)(\sigma) - (h \circ f)(\tau)\| \leq 3\omega_h(\delta),$$

and so

$$d(f(\sigma), f(\tau)) \leq \omega_{h^{-1}}(3\omega_h(\delta)).$$

Since h is a uniform embedding, by choosing $\delta > 0$ sufficiently small, we can ensure that

$$d(f(\sigma), f(\tau)) \leq \omega_{h^{-1}}(3\omega_h(\delta)) < \epsilon,$$

which means that there exists some $\delta > 0$ such that M has property $Q(\epsilon, \delta)$.

We have established that for each $\epsilon > 0$, we can find a $\delta > 0$ such that M has property $Q(\epsilon, \delta)$, and therefore $\Delta_M(\epsilon) > 0$ for each $\epsilon > 0$, as required. $\square$

If M is a Banach space, then $\Delta_M(\epsilon) = k\epsilon$ for some constant k that depends on M. We let Q_M denote this constant k. That is, when M is a Banach space, the real number Q_M is the constant for which $\Delta_M(\epsilon) = Q_M \epsilon$.

THEOREM 15.10. $Q_{c_0} = 0$.

PROOF. Let $(e_j)_{j=1}^\infty$ be the standard basis for c_0. For each $n \in \mathbb{N}$, let

$$s_n = e_1 + \cdots + e_n = \sum_{j=1}^n e_j.$$

For each $r \in \mathbb{N}$ and $\delta > 0$, define $f_{r,\delta} : \mathcal{P}_r(\mathbb{N}) \to c_0$ by

$$f_{r,\delta}(m_1, \ldots, m_r) = \delta(s_{m_1} + \cdots + s_{m_r}) = \delta \sum_{j=1}^r s_{m_j}.$$

Then for any $\delta > 0$, we have that $\text{Lip}(f_{r,\delta}) \leq \delta$, but for any infinite subset $\mathbb{A}$ of $\mathbb{N}$, the image of $\mathcal{P}_r(\mathbb{A})$ has diameter δr, which exceeds any $\epsilon > 0$ for r sufficiently large. Consequently, c_0 fails to have property $Q(\epsilon, \delta)$ for each $\epsilon > 0$, and so $\Delta_M(\epsilon) = 0$ for all $\epsilon > 0$. □

We now get Theorem 15.1 (our goal for this chapter) as an immediate corollary.

PROOF OF THEOREM 15.1. We wish to show that the space c_0 does not uniformly embed into a reflexive Banach space. Combine Theorem 15.9 and Theorem 15.10 to get this result. □

We can say something even stronger than Theorem 15.1.

PROPOSITION 15.11. *The space c_0 does not uniformly embed into a Banach space X if all of the dual spaces for X are separable.*

Proposition 15.11 can be proven using methods similar to those that we used when proving Theorem 15.10, and so we will not include the proof here.

An example of a Banach space having separable dual spaces is the James space. We recall that the *James space* is the Banach space $\mathcal{J}$ consisting of all null real sequences having finite *second variation*. That is, the space $\mathcal{J}$ consists of all real sequences $\xi = (\xi_n)_{n=1}^\infty$ such that $\lim_{n \to \infty} \xi_n = 0$ and such that

$$\|\xi\|_\mathcal{J} = \sup_{i_0 < i_1 < \cdots < i_n} \left[\sum_{j=1}^n (\xi_{i_j} - \xi_{i_{j-1}})^2 \right]^{1/2} < \infty.$$

The James space is a separable Banach space with separable dual spaces. The James space is not reflexive, but in a sense is almost reflexive, because the codimension of $\mathcal{J}$ in $\mathcal{J}^{**}$ is one. The following proposition, which we will not prove, tells us that the James space does not uniformly embed into a reflexive Banach space, despite being almost reflexive itself.

PROPOSITION 15.12. *If $\mathcal{J}$ denotes the James space, then $Q_\mathcal{J} = 0$, and so the James space does not uniformly embed into a reflexive Banach space.*

Being "close" to reflexive does not guarantee that a Banach space will uniformly embed into a reflexive space. Indeed, the spaces ℓ_1 and $L_1(0,1)$ are known to uniformly embed into $L_2(0,1)$, but ℓ_1 and $L_1(0,1)$ are (in a sense) far from being reflexive.

The James space is an example (and indeed it was the first example) of what is known as a *quasi-reflexive* space.

DEFINITION 15.13. A Banach space X is called *quasi-reflexive* if X has finite codimension in X^{**}.

There are quasi-reflexive Banach spaces X that have $Q_X > 0$. The possibility of a quasi-reflexive Banach space uniformly embedding into a reflexive space is closely tied to something called the ω-dual of the Banach space.

DEFINITION 15.14. Let X be a Banach space and let $X^{(2n)}$ denote the $2n$th dual of X. For each $n \in \mathbb{N}$, view $X^{(2n-2)}$ as a subspace of $X^{(2n)}$ through the canonical embedding. The ω-dual of X is the completion of the union
$$\bigcup_{n=0}^{\infty} X^{(2n)},$$
where $X^{(0)} = X$. We denote the ω-dual of X by $X^{(\omega)}$.

THEOREM 15.15. *Suppose that X is a quasi-reflexive Banach space with codimension 1 in X^{**}. If X uniformly embeds into a reflexive Banach space, then $X^{(\omega)}$ is linearly isomorphic to $X \oplus \ell_1$.*

The proof of this theorem is beyond the scope of this book, but we will comment that in 1982, Bellenot [11] constructed a quasi-reflexive Banach space X for which $X^{(\omega)}$ is linearly isomorphic to $X \oplus \ell_1$. Consequently, it is possible that Bellenot's space embeds into a reflexive space, but it is not currently known if this is the case.

We did not prove Proposition 15.11, Proposition 15.12, or Theorem 15.15, but proofs for all three of these results can be found in Kalton's 2007 paper [60].

Some comments on unit balls

A general question of interest is the following:

When is B_X uniformly homeomorphic to B_Y?

It has been known since Mazur's work in 1929 [80] that $B_{L_p(0,1)}$ is uniformly homeomorphic to $B_{L_2(0,1)}$ for $1 \leq p < \infty$ via the Mazur map. If $f \in L_p(0,1)$, then we define a function $M_{p,2}(f)$ on $(0,1)$ by
$$M_{p,2}(f)(s) = \|f\|_p^{1-\frac{p}{2}} |f(s)|^{\frac{p}{2}} \operatorname{sign}(f(s)),$$
for all $s \in (0,1)$, with the convention that $M_{p,2}(f)(s) = 0$ whenever $f(s) = 0$. The function $M_{p,2}(f)$ is in $L_2(0,1)$ for all $L_p(0,1)$, where $1 \leq p < \infty$. We call the function $M_{p,2} : L_p(0,1) \to L_2(0,1)$ the *Mazur map* from $L_p(0,1)$ to $L_2(0,1)$, and it gives a uniform homeomorphism of $B_{L_p(0,1)}$ onto $B_{L_2(0,1)}$ for $1 \leq p < \infty$.

The Mazur map can be modified in the natural way to give a map between the sequence spaces ℓ_p and ℓ_2, and so we can see that there is a uniform homeomorphism of B_{ℓ_p} onto B_{ℓ_2} for $1 \leq p < \infty$.

The case of c_0 is quite different from the case of ℓ_p for $1 \leq p < \infty$. The result of Theorem 15.1 tells us that c_0 cannot be uniformly embedded into a reflexive Banach space. However, as a result of Aharoni's work [2], we know that c_0 can be uniformly embedded into B_{c_0}. It follows that B_{c_0} cannot be uniformly embedded into a reflexive Banach space. In particular, we know that B_{c_0} is not uniformly homeomorphic to B_{ℓ_2}.

More generally, we have Theorem 15.16.

THEOREM 15.16. *If X is a Banach space that fails to have non-trivial cotype, then B_X cannot be embedded in ℓ_2 (or any uniformly convex space).*

We recall the definition of cotype (or Rademacher cotype).

DEFINITION 15.17. A Banach space X is said to have *Rademacher cotype* (or *cotype* for short) p for some $p \geq 2$ if there exists a constant $c \geq 1$ such that

$$\Big(\mathbb{E}\Big\|\sum_{i=1}^{n}\epsilon_i x_i\Big\|^p\Big)^{1/p} \geq c\Big(\sum_{i=1}^{n}\|x_i\|^p\Big)^{1/p},$$

for any finite sequence $(x_i)_{i=1}^n$ of points in X, where $(\epsilon_i)_{i=1}^n$ is a finite sequence of independent Rademacher random variables on a probability space $(\Omega, \mathbb{P})$.

When we say that a Banach space fails to have non-trivial cotype, we mean that there is no number p with $2 \leq p < \infty$ such that the Banach space has Rademacher cotype p.[3]

In 1994, Odell and Schlumprecht [90] found a way to extend the Mazur map to a more general setting, which led to the following generalization of our earlier results.

THEOREM 15.18 (Odell–Schlumprecht). *If X has an unconditional basis and non-trivial cotype, then B_X is uniformly homeomorphic to B_{ℓ_2}.*

We recall the definition of an unconditional basis.

DEFINITION 15.19. Let X be a Banach space. A basis $(e_n)_{n=1}^\infty$ of X is called an *unconditional basis* if for each $x \in X$, the series

$$\sum_{n=1}^{\infty} e_n^*(x) e_n$$

converges unconditionally, where $(e_n^*)_{n=1}^\infty$ are the biorthogonal functionals.

An alternate interpretation of the Odell–Schlumprecht theorem leads to Theorem 15.20 in terms of super-reflexive Banach spaces.

THEOREM 15.20. *If X is a super-reflexive Banach space with unconditional basis (or if X is a super-reflexive Banach lattice), and E is a closed infinite-dimensional subspace of X, then B_E is uniformly homeomorphic to B_{ℓ_2}.*

We won't present the proof of Theorem 15.20 here, but we do wish to single out a key ingredient in the proof of Theorem 15.20.

PROPOSITION 15.21. *Suppose that X is a super-reflexive Banach space, and suppose that E is a closed subspace of X. If $Q : X \to X/E$ is the quotient map, then there is a uniformly continuous map $\psi : B_{X/E} \to X$ for which $Q \circ \psi = \mathrm{Id}|_{B_{X/E}}$.*

The map ψ in the statement of Proposition 15.21 is called a *selection*. The existence of the uniformly continuous selection implies that B_X is uniformly homeomorphic to $B_E \oplus B_{X/E}$. (The details can be found in Chapter 9 of [13].[4])

This leads to a natural question.

QUESTION 15.22. *If X and Y are Banach spaces such that $Q : X \to Y$ is a quotient map, under what conditions is there a uniformly continuous selection $\psi : B_Y \to X$ such that $Q \circ \psi = \mathrm{Id}|_{B_Y}$?*

---

[3]It is not possible for a Banach space to have cotype p for $p < 2$. Similarly, it is not possible for a Banach space to have Rademacher type p for $p > 2$. Every Banach space has Rademacher type 1, and so we call that trivial type. We can think of trivial cotype as being cotype ∞.

[4]In particular, our Theorem 15.20 and Proposition 15.21 are Corollary 9.11 and Lemma 9.10 in Benyamini and Lindenstrauss' text [13], respectively.

If such a uniformly continuous selection exists, then it follows that B_X is uniformly homeomorphic to $B_Y \oplus B_{\ker(Q)}$. Proposition 15.21 tells us that we have such a uniformly continuous selection when X is a super-reflexive Banach space. Such a selection also exists for a quotient map of ℓ_1 onto $L_1(0,1)$. (See [**59**].)

The following are open questions related to this topic.

QUESTION 15.23. *If X is a super-reflexive Banach space, is B_X uniformly homeomorphic to B_{ℓ_2}?*

QUESTION 15.24. *If X is a closed subspace of $L_1(0,1)$, is B_X uniformly homeomorphic to B_{ℓ_2}?*

Finally, we ask one more question.

QUESTION 15.25. *If X is a Banach space and B_X uniformly embeds into a reflexive space, is X weakly sequentially complete? That is, if $(x_n)_{n=1}^\infty$ is a weakly Cauchy sequence, must it converge weakly?*

CHAPTER 16

Exercises

The following exercises are based on the two problem sets that were given during Nigel Kalton's last course on nonlinear functional analysis.

16.1. Examples I

1. Let S be an arbitrary set and let C be a nonempty closed subset of $\ell_\infty(S)$ such that the following conditions are satisfied:
   (i) The set C is convex.
   (ii) If A is a subset of C for which $\sup(A)$ exists in $\ell_\infty(S)$, then $\sup(A) \in C$.
   Show that C is an 8-absolute Lipschitz retract.

2. Let p be a real number such that $1 \leq p \leq 2$. Show that
$$\frac{1}{2}\Big(|1+t|^p + |1-t|^p\Big) \leq 1 + |t|^p$$
for all $t \in \mathbb{R}$. Deduce that
$$\frac{1}{2}\Big(\|x+y\|^p + \|x-y\|^p\Big) \leq \|x\|^p + \|y\|^p$$
for all x and y in $L_p(0,1)$ (or ℓ_p).

3. Let p be a real number such that $1 \leq p \leq 2$. Let Ω be a finite set with $\mathbb{P}$ normalized counting measure on Ω. Suppose that $\sigma_1 : \Omega \to \Omega$ and $\sigma_2 : \Omega \to \Omega$ are two commuting permutations with $\sigma_1^2 = \text{Id}\,|_\Omega$ and $\sigma_2^2 = \text{Id}\,|_\Omega$, where $\text{Id}\,|_\Omega$ is the identity on Ω. Show for any map $f : \Omega \to L_p(0,1)$,
$$\mathbb{E}\|f - f \circ \sigma_1 \circ \sigma_2\|_p^p \leq \mathbb{E}\|f - f \circ \sigma_1\|_p^p + \mathbb{E}\|f - f \circ \sigma_2\|_p^p.$$
Hint: Replace f by $g = \frac{1}{2}(f - f \circ \sigma_1 \circ \sigma_2)$ and use Exercise 2.

4. (Continuation of Exercise 3.) Deduce that if $\sigma_1, \ldots, \sigma_n$ are commuting permutations on Ω with $\sigma_j^2 = \text{Id}\,|_\Omega$, then
$$\mathbb{E}\|f - f \circ \sigma_1 \circ \cdots \circ \sigma_n\|_p^p \leq \sum_{j=1}^n \mathbb{E}\|f - f \circ \sigma_j\|_p^p.$$

5. Let $\mathbb{D}_n = \{-1,1\}^n$ with $\mathbb{P}$ normalized counting measure on $\mathbb{D}_n$. For each j in the set $\{1, \ldots, n\}$, let $\sigma_j : \mathbb{D}_n \to \mathbb{D}_n$ be the permutation which reverses the sign of the jth element. A metric space (M, d) is said to be of *Enflo*

type p with constant C if

$$\mathbb{E}\Big[d(f, f\circ\sigma_1\circ\cdots\circ\sigma_n)^p\Big] \leq C^p \sum_{j=1}^n \mathbb{E}\Big[d(f, f\circ\sigma_j)^p\Big],$$

whenever $f : \mathbb{D}_n \to M$ is a map from $\mathbb{D}_n$ into M. Show that for $p \in [1,2]$, the function space $L_p(0,1)$ is of Enflo type p with constant 1, but not of Enflo type q for any $q > p$.

6. Suppose $f : X \to \ell_p$ (or into $L_p(0,1)$) is a uniformly continuous map, where $1 < p \leq 2$. Show that there exists a constant C so that whenever $x_1, \ldots, x_n$ are points in ∂B_X, there exist $\epsilon_1, \ldots, \epsilon_n$ with $\epsilon_j \in \{-1, 1\}$ for each j so that

$$\Big\| f\Big(\sum_{j=1}^n \epsilon_j x_j\Big) - f\Big(-\sum_{j=1}^n \epsilon_j x_j\Big)\Big\|_p \leq C n^{1/p}.$$

Hint: Uniformly continuous maps on a Banach space are Lipschitz at large distances!

7. Show that ℓ_p and ℓ_q are not uniformly homeomorphic if $1 \leq p < q \leq 2$.

COMMENT 16.1. Enflo type was developed by Enflo in [32] (although not with that name) as a nonlinear analogue to Rademacher type, using only the metric structure of the underlying metric space (where Rademacher type requires a linear structure). If a Banach space X has Enflo type p, then it must have Rademacher type p, which can be seen by considering the function

$$f(\epsilon_1, \ldots, \epsilon_n) = \epsilon_1 x_1 + \cdots + \epsilon_n x_n,$$

where $x_1, \ldots, x_n$ are in X and $\epsilon_j \in \{-1, 1\}$ for each $j \in \{1, \ldots, n\}$. It was asked by Enflo in [35] if a Banach space having Rademacher type p must necessarily also have Enflo type p. This remained an open question for a long time. It was not until 2020 that this question was answered in the positive by Ivanisvili, et al. in [49] (over forty years after the question was posed by Enflo).

16.2. Examples II

1. Show that there is a uniform embedding of c_0 into the unit ball of c_0.

2. Let c_0^+ be the positive cone in c_0. Show that $\mathbb{R}$ cannot be λ-embedded into c_0^+ unless $\lambda \geq 2$.

3. Show that ℓ_1 cannot be λ-embedded into c_0^+ unless $\lambda \geq 3$.

4. Let (M,d) be a separable metric space and suppose $0 < p < 1$. Show that there is an embedding $f : M \to c_0$ such that
$$d(x,y)^p < \|f(x) - f(y)\|_{c_0} \leq 2^p d(x,y)^p, \quad x \neq y.$$
   Hint: Consider (M, d^p).

5. A metric space (M,d) is called an *ultrametric space* if it satisfies the inequality
$$d(x,y) \leq \max\Big(d(x,z), d(y,z)\Big),$$
whenever x, y, and z are in M. Show that if $d(x,z) \neq d(y,z)$, then
$$d(x,y) = \max\Big(d(x,z), d(y,z)\Big).$$

6. Let M be a separable ultrametric space. For each $x \in M$, show that the set
$$\Gamma_x = \big\{d(x,y) : y \in M\big\}$$
is countable.

7. Let M be a separable ultrametric space. Show $\Gamma = \big\{d(x,y) : x \in M, y \in M\big\}$ is a countable set.

8. Let M be a separable ultrametric space, and let $\Gamma = \big\{d(x,y) : x \in M, y \in M\big\}$. Let $(a_j)_{j=1}^\infty$ be a dense sequence in the separable metric space M. Let $\mathcal{D}$ denote the set of finite sequences with terms in Γ. That is, let $\mathcal{D}$ be the collection of sequences of the form $(r_1, \ldots, r_n)$, where $n \in \mathbb{N}$, and $r_j \in \Gamma$ for each $j \in \mathbb{N}$. For each $(r_1, \ldots, r_n)$ in $\mathcal{D}$, define
$$f_{r_1,\ldots,r_n}(x) = \begin{cases} \min(r_1, \ldots, r_n) & \text{if } d(x, a_j) = r_j \text{ for } 1 \leq j \leq n, \\ 0 & \text{otherwise.} \end{cases}$$
Show that each function $f_{r_1,\ldots,r_n}$ is Lipschitz continuous with Lipschitz constant 1, and
$$f(x) = \big(f_{r_1,\ldots,r_n}(x)\big)_{r_1,\ldots,r_n} \in c_0(\mathcal{D})$$
for each $x \in M$.

9. Show that the map $f : M \to c_0(\mathcal{D})$ from Exercise 8 is an isometry.

COMMENT 16.2. Exercises 5–9 come from [66].

Afterword (where to from here)

Nonlinear functional analysis is a vast and growing field, and many of the results in this text were quite new when Nigel Kalton gave his course that formed the basis for this text. Where should a reader who wishes to learn more go next? A standard reference text for many of the topics in this book is Benyamini and Lindenstrauss' significant book *Geometric nonlinear functional analysis Vol. 1* [13], a massive tome that is frequently cited in the current slim volume. However, the subject has evolved a lot in the more than twenty years since Benyamini and Lindenstrauss produced their highly influential text. Indeed, some of the results in the current book first appeared after [13] was published.

The forthcoming book *Nonlinear Geometry of Banach Spaces* by Gilles Lancien and Florent Baudier[1] will provide a comprehensive update to [13], and should be a good source for those interested in continuing their study of this active and interesting subject.

Additionally, many of the results in this book appeared in articles that can be found in the literature. We gave reference throughout the text, during our discussions, and anyone who made it through this book should be ready to read those articles now. Of particular interest may be Nigel Kalton's articles [63] and [64], which have some of Kalton's final valuable contributions to the subject.

Also of note is the fine work *The non-linear geometry of Banach spaces after Nigel Kalton* by Godefroy, Lancien, and Zizler [40], which gives an overview on developments beyond what can be found in this book.

---

[1]To appear in the series "Cours Spécialisés" by the Société Mathématique de France.

APPENDIX A

Vector integration

The purpose of this appendix is to introduce the Bochner integral and present some of its key properties. The Bochner integral is a type of integral that is used for integrating vector-valued functions. In our case, we will consider functions that take values in a Banach space. This will be only a brief introduction. For details and proofs, see the classic text by Diestel and Uhl [**28**] or the comprehensive work on geometric nonlinear functional analysis by Benyamini and Lindenstrauss [**13**].

Suppose that (Ω, μ) is a measure space where μ is a countably additive measure. We will assume that μ is real-valued and nonnegative. Suppose that X is a real Banach space with norm $\|\cdot\|$. Let $f : \Omega \to X$ be a function that takes values in X.

We say that f is a *simple function* if there are finitely many points $x_1, \ldots, x_n$ in X and pairwise-disjoint measurable sets $E_1, \ldots, E_n$ such that

$$f(\omega) = \sum_{j=1}^{n} x_j \, \chi_{E_j}(\omega),$$

where χ_E is the *characteristic function* or *indicator function* for the set E:

$$\chi_E(\omega) = \begin{cases} 1 & \text{if } \omega \in E, \\ 0 & \text{if } \omega \notin E. \end{cases}$$

We define the *integral* of this simple function f with respect to μ according to the rule

$$\int_\Omega f \, d\mu = \sum_{j=1}^{n} x_j \mu(E_j).$$

If E is a measurable subset of Ω, then we can integrate over E by considering intersections with E:

$$\int_E f \, d\mu = \sum_{j=1}^{n} x_j \mu(E_j \cap E).$$

A function $f : \Omega \to X$ is called *measurable*, or *strongly measurable*, if there exists a sequence $(f_j)_{j=1}^{\infty}$ of simple functions such that

$$\lim_{j \to \infty} \|f_j - f\| = 0 \quad \text{a.e.}(\mu).$$

Furthermore, we call f *Bochner integrable* if there exists such a sequence of simple functions $(f_j)_{j=1}^{\infty}$ for which

$$\lim_{j \to \infty} \int_\Omega \|f_j - f\| \, d\mu = 0,$$

and in this case, we define the *Bochner integral* of f over a measurable subset E of Ω to be
$$\int_E f\,d\mu = \lim_{j \to \infty} \int_E f_j\,d\mu.$$
The value of $\int_E f\,d\mu$ is independent of the sequence of simple functions used in computing it. (We will not show that here.) When we wish to explicitly include the variable ω in the integral, we will follow the usual convention of writing
$$\int_\Omega f\,d\mu = \int_\Omega f(\omega)\,\mu(d\omega).$$
The Bochner integral shares many properties with the Lebesgue integral. The three theorems below are analogous to theorems in the classical theory of Lebesgue integration of scalar-valued functions.

THEOREM A.1. *Let (Ω, μ) be a measure space and let X be a Banach space. A strongly measurable function $f : \Omega \to X$ is Bochner integrable if and only if*
$$\int_\Omega \|f\|\,d\mu < \infty.$$

THEOREM A.2. *Let (Ω, μ) be a measure space and let X be a Banach space. If f and g are strongly measurable functions from Ω into X such that*
$$\int_E f\,d\mu = \int_E g\,d\mu$$
for each measurable set E, the $f = g$ almost everywhere with respect to μ.

THEOREM A.3. *Suppose that X is a Banach space. If $f : [0,1] \to X$ is Bochner integrable with respect to Lebesgue measure on $[0,1]$, then*
$$\lim_{h \to 0} \frac{1}{h} \int_x^{x+h} \|f(t) - f(x)\|\,dt = 0$$
for almost every x in $[0, 1]$. Consequently,
$$\lim_{h \to 0} \frac{1}{h} \int_x^{x+h} f(t)\,dt = f(x)$$
for almost every x in $[0, 1]$.

The following theorem of Hille and Phillips (although it is usually called Hille's theorem) gives an important property of the Bochner integral that does not have a natural analogue in the scalar-valued case.

THEOREM A.4 (Hille's theorem). *Let X and Y be two Banach spaces and let (Ω, μ) be a measure space. Suppose that $T : X \to Y$ is a bounded linear operator between Banach spaces. If $f : \Omega \to X$ is Bochner integrable, then $Tf : \Omega \to Y$ is also Bochner integrable and*
$$\int_E T(f)\,d\mu = T\left(\int_E f\,d\mu\right)$$
for each measurable set E in Ω.

This is not the most general formulation of Hille's theorem. The conclusion holds for closed linear operators, so long as you know that Tf is Bochner integrable. If T is continuous (i.e., bounded), then it is necessarily true that Tf is Bochner integrable whenever f is Bochner integrable.

An important special case of the Hille's theorem is when T is a bounded linear functional into R.

COROLLARY A.5. *Let X be a Banach space and let (Ω, μ) be a measure space, and suppose that $x^* \in X^*$. If $f : \Omega \to X$ is Bochner integrable, then $x^*f : \Omega \to \mathbb{R}$ is an integrable scalar-valued function (in the Lebesgue sense) and*

$$\int_E x^* f \, d\mu = x^* \left(\int_E f \, d\mu \right)$$

for each measurable set E.

DEFINITION A.6. Let (Ω, μ) be a measure space and let X be a Banach space with dual space X^*. A function $f : \Omega \to X$ is said to be *weakly measurable* if for each $x^* \in X^*$ the scalar valued function $x^*f : \Omega \to \mathbb{R}$ is a measurable function.

A strongly measurable function is necessarily weakly measurable, but there are functions that are weakly measurable and not strongly measurable. The following theorem of Pettis tells us when a weakly measurable function is also strongly measurable.

THEOREM A.7 (Pettis measurability theorem). *Let X be a Banach space and let (Ω, μ) be a measure space. A function $f : \Omega \to X$ is strongly measurable if and only if it is weakly measurable and has essentially separable range.*

When we say that the function f has "essentially separable range", we mean that there is a set E in Ω such that $\mu(E) = 0$ and such that $f(\Omega \setminus E)$ is a separable subset of X (in the norm topology). In particular, when X is a separable Banach space, weak measurability is the same as strong measurability.

One nice consequence of the Pettis measurability theorem is that any strongly measurable function is almost separably valued, and so we have the following useful approximation result for strongly measurable functions.

COROLLARY A.8. *Let X be a Banach space and let (Ω, μ) be a measure space. If $f : \Omega \to X$ is strongly measurable, then for every $\epsilon > 0$, there exists a countably valued function $g : \Omega \to X$ such that $\|f - g\| < \epsilon$ almost everywhere with respect to the measure μ.*

In particular, if f is a strongly measurable function, then there exists a sequence of functions $(g_n)_{n=1}^\infty$ such that g_n is countably valued for each $n \in \mathbb{N}$ and such that $g_n \to f$ uniformly outside of a set of μ-measure zero.

It is possible to develop a theory of integration for weakly measurable functions. Suppose that $f : (\Omega, \mu) \to X$ is a weakly measurable function. A theorem of Dunford states that if x^*f is in $L_1(\mu)$ (that is, if x^*f is Lebesgue integrable) for each $x^* \in X^*$, then for each measurable set E, there exists an element x_E^{**} in the second dual X^{**} such that

$$x_E^{**}(x^*) = \int_E x^* f \, d\mu$$

for all $x^* \in X^*$. The element x_E^{**} of the second dual is called the *Dunford integral* and is usually denoted by
$$x_E^{**} = \int_E f\,d\mu,$$
provided there is no confusion about what kind of integral it is. Note that the Dunford integral is defined as a linear functional on X^*, and consequently is an element of the second dual X^{**}.

The Banach space X can be naturally identified with a subspace of X^{**}. Consequently, it is possible that the linear functional x_E^{**} can actually be found in X (under this identification with a subspace of X^{**}). If $x_E^{**} \in X$ for each measurable set E, then f is called *Pettis integrable* and $\int_E f\,d\mu$ is called the *Pettis integral* of f. Note that if X is a reflexive space, then the Dunford integral and the Pettis integral coincide. Note also that if f is Bochner integrable, then it is also Pettis integrable and the Bochner integral equals the Pettis integral, by Corollary A.5.

APPENDIX B

The Radon–Nikodym property

In Chapter 7, we gave the following definition for the Radon–Nikodym property. (See Definition 7.1.)

DEFINITION B.1. A Banach space X has the *Radon–Nikodym property (RNP)* if any Lipschitz function $f : [0, 1] \to X$ is necessarily differentiable almost everywhere with respect to Lebesgue measure on $\mathbb{R}$.

This definition was not the original formulation of the Radon–Nikodym property given by Diestel and Uhl [27]. The original formulation of RNP was given in the context of vector measure theory. We wish now to give a brief description of the history of the Radon–Nikodym property, but in order to do so, we must define a vector measure, and revisit the topics of Appendix A.

DEFINITION B.2. Let $(\Omega, \mathcal{A})$ be a measurable space with $\mathcal{A}$ a σ-algebra of subsets of Ω. Let X be a Banach space, and suppose that $\nu : \mathcal{A} \to X$ is an X-valued set function defined on $\mathcal{A}$. If

$$\nu\Big(\bigcup_{n=1}^{\infty} E_n\Big) = \sum_{n=1}^{\infty} \nu(E_n)$$

whenever $(E_n)_{n=1}^{\infty}$ is a sequence of pairwise disjoint sets in $\mathcal{A}$, where convergence is in the norm topology on X, then ν is called a *countably additive vector measure*, or simply a *vector measure*.

If we require only that $\nu(E_1 \cup E_2) = \nu(E_1) + \nu(E_2)$ whenever E_1 and E_2 are disjoint sets in $\mathcal{A}$, then ν is called a *finitely additive vector measure*. For our purposes, we will consider only countably additive vector measures, but much of the theory of vector integration can be carried through using only the requirement that a vector measure be finitely additive.

DEFINITION B.3. Let $(\Omega, \mathcal{A})$ be a measurable space with $\mathcal{A}$ a σ-algebra of subsets of Ω, and let X be a Banach space. Suppose that $\nu : \mathcal{A} \to X$ is an X-valued countably additive vector measure. The *total variation* of ν—or the *total variation measure* of ν—is the nonnegative (possibly infinite) real-valued set function $|\nu| : \mathcal{A} \to [0, \infty]$ defined by

$$|\nu|(E) = \sup\Big\{\sum_{k=1}^{N} \|\nu(E_k)\| : (E_k)_{k=1}^{N} \text{ is a finite partition of } E\Big\}.$$

When we say that $(E_k)_{k=1}^{N}$ is a finite partition of E, we mean that the sets in $(E_k)_{k=1}^{N}$ are pairwise disjoin and their union is E:

$$E = \bigcup_{k=1}^{N} E_k.$$

If $|\nu|(\Omega) < \infty$, then the measure is said to have *bounded variation*, or be of bounded variation. The total variation $|\nu|$ of a vector measure ν of bounded variation is also a measure, and in such a case $|\nu|$ is countably additive if and only if the original measure ν is countably additive.

An easy way to construct vector measures is via Bochner integration.

THEOREM B.4. *Let X be a Banach space and let $(\Omega, \mathcal{A}, \mu)$ be a finite measure space with μ a nonnegative countably additive measure. If $f : \Omega \to X$ is Bochner integrable, then the set function $\nu : \mathcal{A} \to X$ defined by*

$$\nu(E) = \int_E f \, d\mu$$

for $E \in \mathcal{A}$ is a countably additive vector measure of bounded variation and

$$|\nu|(E) = \int_E \|f\| \, d\mu$$

for all measurable sets E in $\mathcal{A}$.

In the above theorem, the vector-valued function f is called the *Radon–Nikodym derivative* of ν with respect to μ, and is usually denoted by $\frac{d\nu}{d\mu}$. A natural question is the following: Given a countably additive vector measure ν, under what circumstances does it have a Radon–Nikodym derivative with respect to a finite nonnegative countably additive measure μ?

In the classical case, where ν is scalar-valued, the answer to this question is given by the Radon–Nikodym theorem, which says that a Radon–Nikodym derivative of ν with respect to μ exists provided that ν is absolutely continuous with respect to μ. We remind the reader that a measure ν is *absolutely continuous* with respect to a measure μ (defined on the same σ-algebra), written $\nu \ll \mu$, provided that $\nu(E) = 0$ whenever E is a measurable set for which $\mu(E) = 0$.

The situation is more complicated when ν is a vector measure. It is possible to find a Banach space X and a vector measure ν taking values in X such that ν is absolutely continuous with respect to a nonnegative measure μ, but does not admit a Radon-Nikodym derivative with respect to μ. (In this context, ν being absolutely continuous with respect to μ means $\|\nu(E)\|_X = 0$ whenever $\mu(E) = 0$.) Banach spaces where absolute continuity is sufficient for the existence of a derivative are said to have the *Radon–Nikodym property*.

In [**27**], Diestel and Uhl stated that Banach spaces possessing the Radon–Nikodym property are

> ...those Banach spaces X with the property ... that given a countably additive X-valued map F defined on a sigma-algebra possessing finite variation $|F|$ then there exists a Bochner $|F|$-integrable function f such that $F(A) = \int_A f \, d|F|$ for each A in F's domain.

So in the original formulation, a Banach space had the Radon–Nikodym property when it possessed a Radon–Nikodym type of theorem.

Turning our attention to Lipschitz functions, a classical result of Rademacher (dating back to 1919) states that a Lipschitz function $f : [0,1]^n \to \mathbb{R}^m$ is differentiable almost everywhere (with respect to Lebesgue measure). However, there are Banach spaces X and Banach space-valued functions $f : [0,1] \to X$ that are

Lipschitz continuous, but not differentiable almost everywhere. For example, the function $f : [0,1] \to L_1(0,1)$ defined by the formula
$$f(t) = \chi_{[0,t]}, \quad 0 \le t \le 1$$
is not differentiable at any t in the interval $[0,1]$. (This can be found in [**78**].)

In 1938, Gelfand showed that every Lipschitz function from $[0,1]$ into a separable dual space had a derivative almost everywhere. For that reason, Banach spaces X with this property (all Lipschitz functions $f : [0,1] \to X$ have a derivative almost everywhere) came to be known as Gelfand spaces or Gelfand–Fréchet spaces. (See for example the classic paper of Mankiewicz [**78**].)

It is now known that a space is a Gelfand–Fréchet space if and only if it has the Radon–Nikodym property. An accessible proof of this equivalence can be found in Theorem 5.21 in Benyamini and Lindenstrauss [**13**]. (It is not clear when this equivalence was first identified, but Benyamini and Lindenstrauss describe it as "folklore" in the Chapter 5 notes in [**13**].)

APPENDIX C

Gaussian measures

We will give only some basic facts about Gaussian Measures. For a more detailed treatment, see Chapter 6 of [13] or the comprehensive volume *Gaussian Measures* by Bogachev [16].

DEFINITION C.1. Let X be a Banach space. A probability measure μ on X is called a *Gaussian measure* if for each $x^* \in X^*$, the measure μ_{x^*} on $\mathbb{R}$ given by the formula

$$\mu_{x^*}(A) = \mu\Big(\{x \in X : x^*(x) \in A\}\Big)$$

has a Gaussian distribution on $\mathbb{R}$. We say that μ is *nondegenerate* if the μ_{x^*} is nondegenerate for each $x^* \in X^*$.

Gaussian measures are plentiful on a separable Hilbert space, where they are intimately related to trace-class operators. Let us recall the definition of a trace-class operator.

DEFINITION C.2. Let H be a separable Hilbert space with inner product $\langle \cdot, \cdot \rangle$. A compact operator $T : H \to H$ is said to be *trace-class* if

$$\text{Tr}(T) = \sum_{n=1}^{\infty} \langle T(e_n), e_n \rangle$$

is finite and independent of choice of orthonormal basis $(e_n)_{n=1}^{\infty}$. The value $\text{Tr}(T)$ is called the *trace* of T.

We will continue to use H to denote a separable Hilbert space and we will use $\langle \cdot, \cdot \rangle$ to denote the inner product on H.

PROPOSITION C.3. *Let H be a separable Hilbert space. A probability measure μ on H is a Gaussian measure if and only if there is a $v \in H$ and a nonnegative trace-class operator $T : H \to H$ such that*

$$\int_H e^{i\langle x, y \rangle} \mu(dy) = \exp\left[i\langle x, v \rangle - \frac{1}{2}\langle T(x), x \rangle\right],$$

for all $x \in H$. Furthermore, the probability measure μ on H is nondegenerate if and only if $\ker(T) = \{0\}$.

For each $v \in H$ and each nonnegative trace-class operator T on H, there is a unique probability measure μ satisfying the equation in Proposition C.3, and so we can see that there are many Gaussian measures on H available.

We can use Gaussian measures on a Hilbert space to construct Gaussian measures on other Banach spaces.

THEOREM C.4. *Let X be a separable Banach space, and let $T : H \to X$ be a bounded linear operator from a separable Hilbert space H. If μ is a Gaussian measure on H, then the set function ν defined by*

$$\nu(A) = \mu(T^{-1}(A)),$$

for the Borel sets A in X, is a Gaussian measure on X.

In Theorem 7.25, we showed that a Lipschitz function $f : X \to Y$ between separable Banach spaces is Gâteaux differentiable outside of a Haar-null set when Y has the Radon–Nikodym property. We can make a similar statement about Gâteaux differentiability of Lipschitz functions with respect to Gaussian measure.

DEFINITION C.5. Let X be a separable Banach space. A Borel set A is said to be a *Gauss-null set* if $\mu(A) = 0$ for every nondegenerate Gaussian measure on X.

THEOREM C.6. *Suppose that X and Y are separable Banach spaces, and suppose Y has the Radon–Nikodym property. If $f : X \to Y$ is a Lipschitz function, then f is Gâteaux differentiable outside of a Gauss-null set.*

It turns out that Theorem C.6 is equivalent to Theorem 7.25 because Gauss-null sets coincide with Haar-null sets! The proof of this nontrivial (and non-obvious) fact can be found in Chapter 6 of [**13**]. (See Propositions 6.25, 6.27, and 6.32.)

Theorem C.6 allows us to give a beautiful proof of the density of Gâteaux differentiable functions in the set of Lipschitz functions between separable Banach spaces where the domain is separable and the codomain has RNP.

THEOREM C.7. *Let X be a separable Banach space and let Y be a Banach space with the Radon–Nikodym property. Any Lipschitz function $f : X \to Y$ can be uniformly approximated by Lipschitz functions that are Gâteaux differentiable everywhere.*

PROOF. Let μ be a nondegenerate Gaussian measure on X such that

$$\int_X \|v\|\, \mu(dv) < \infty.$$

For each $n \in \mathbb{N}$, define a function $g_n : X \to Y$ by

$$g_n(x) = \int_X f(x + 2^{-n} v)\, \mu(dv),$$

for each $x \in X$. For each $n \in \mathbb{N}$, the function g_n is a Lipschitz continuous function with $\mathrm{Lip}(g_n) \leq \mathrm{Lip}(f)$. Furthermore, since f is Gâteaux differentiable outside of a Gauss-null set (by Theorem C.6), it follows that g_n is Gâteaux differentiable everywhere. To complete the proof, we simply observe that

$$\|f(x) - g_n(x)\| = \left\| \int_X \left(f(x) - f(x + 2^{-n} v) \right) \mu(dv) \right\| \leq 2^{-n} \mathrm{Lip}(f) \int_X \|v\|\, \mu(dv).$$

Since this inequality is true for all $x \in X$ and $n \in \mathbb{N}$, we have completed the proof. $\square$

In particular, Theorem C.7 tells us that Lipschitz continuous functions that are Gâteaux differentiable everywhere are dense in the closed unit ball of $\mathrm{Lip}(X)$ whenever X is a separable Banach space (Theorem 6.9), since $\mathbb{R}$ certainly has the Radon–Nikodym property. This was a key step in our proof that separable Banach spaces have the isometric lifting property (Theorem 6.12).

In fact, Gaussian measures can be used to provide an alternate proof that separable Banach spaces have the isometric lifting property. Suppose X is a separable Banach space and μ is a nondegenerate Gaussian measure on X. Let $\delta_X : X \to \cancel{E}(X)$ be the Dirac lifting. Then the convolution $\mu * \delta_X$ is Gâteaux differentiable everywhere and $D(\mu * \delta_X)(0)$ is the desired linear lifting. We won't prove this here, but the details can be found in [**39**].

APPENDIX D

Notes on closest points

In the proof of Riesz's lemma (Theorem 8.9), we asserted that if E and F are finite dimensional subspaces of a strictly convex Banach space X, then for each $x \in \partial B_F$, there exists a unique closest element $\phi(x) \in E$. We wish here to provide some extra detail about why this is true. We start with a definition.

DEFINITION D.1. Let X be a metric space and let E be a subset of X. If for a given $x \in X$, there is some $e \in E$ such that $d(x, e) = d(x, E)$, then we call e a *closest element* in E to x.

We note that there is no uniqueness requirement in Definition D.1. It is possible to find a Banach space X with a subspace E in which an element x may have several closest points in E.

THEOREM D.2. *Let X be a normed linear space. If E is a nonempty compact subset of X, then each $x \in X$ has a closest point in E.*

PROOF. Let $x \in X$. We will show that there exists at least one $e \in E$ for which $\|x - e\| = d(x, E)$. (The closest point may not be unique.) Since
$$d(x, E) = \inf\{\|x - e\| : e \in E\}$$
is an infimum, we can find for each $n \in \mathbb{N}$ an element $e_n \in E$ such that
$$\|x - e_n\| < d(x, E) + \frac{1}{n}.$$
The set E is compact and $(e_n)_{n=1}^\infty$ is a sequence with terms in E, and so it has a subsequence $(e_{n_k})_{k=1}^\infty$ that converges to a point $e \in E$. Observe that for each $k \in \mathbb{N}$, we have (by the triangle inequality)
$$\|x - e\| \leq \|x - e_{n_k}\| + \|e_{n_k} - e\| < d(x, E) + \frac{1}{n_k} + \|e_{n_k} - e\|.$$
Now let $\epsilon > 0$ be given. There exists some $K \in \mathbb{N}$ so that $\|e_{n_k} - e\| < \frac{\epsilon}{2}$ whenever $k \geq K$. If we also choose K so large that $\frac{1}{n_k} < \frac{\epsilon}{2}$, then
$$\|x - e\| < d(x, E) + \frac{\epsilon}{2} + \frac{\epsilon}{2} = d(x, E) + \epsilon$$
whenever $k \geq K$. Since the choice of $\epsilon > 0$ in the above argument was arbitrary, it follows that $\|x - e\| \leq d(x, E)$. However, since $e \in E$, it must be true that $\|x - e\| \geq d(x, E)$, and so we conclude that $\|x - e\| = d(x, E)$, as required. □

If we impose more conditions on X, then we may relax the requirements on E.

THEOREM D.3. *Let X be a reflexive Banach space. If E is a nonempty closed subspace of X, then each $x \in X$ has a closest point in E.*

PROOF. Let $x \in X$. By definition, for each $n \in \mathbb{N}$, we can find an element $e_n \in E$ such that
$$\|x - e_n\| < d(x, E) + \frac{1}{n}.$$
Using the triangle inequality, we observe that for each $n \in \mathbb{N}$,
$$\|e_n\| \leq \|x - e_n\| + \|x\| < d(x, E) + \frac{1}{n} + \|x\| < d(x, E) + 1 + \|x\|.$$
Consequently, the sequence $(e_n)_{n=1}^\infty$ is bounded.

The subspace E is closed in X, and so closed in the weak topology on X. However, the space X is reflexive, and so E is closed in the weak* topology. Furthermore, when X is equipped with the weak* topology, it satisfies the Heine–Borel property. Therefore, the bounded sequence $(e_n)_{n=1}^\infty$ has a subsequence $(e_{n_k})_{k=1}^\infty$ that converges in the weak* topology. Let e be the weak* limit of $(e_{n_k})_{k=1}^\infty$.

Let $\epsilon > 0$ be given. For any x^* in X^*, we have that
$$x^*(x - e) = x^*(x - e_{n_k}) + x^*(e_{n_k} - e),$$
and so for each $x^* \in X^*$, there exists some $K \in \mathbb{N}$ (that depends in general on x^*) such that
$$|x^*(x - e)| \leq \|x^*\|\|x - e_{n_k}\| + \frac{\epsilon}{2} < \|x^*\|\left(d(x, E) + \frac{1}{n_k}\right) + \frac{\epsilon}{2},$$
whenever $k \geq K$. If we pick x^* to be nonzero, and if we choose K so large that $\frac{1}{n_K} < \frac{\epsilon}{2\|x^*\|}$, then
$$|x^*(x - e)| < \|x^*\|\left(d(x, E) + \frac{\epsilon}{2\|x^*\|}\right) + \frac{\epsilon}{2} = \|x^*\|d(x, E) + \epsilon.$$
Since the choice of $\epsilon > 0$ was arbitrary, we conclude that
$$|x^*(x - e)| \leq \|x^*\|d(x, E)$$
for each x^* in X^*. (We showed it for x^* nonzero, but it is trivially true if $x^* = 0$.) Consequently, we see that $\|x - e\| \leq d(x, E)$. However, since it is necessarily true that $\|x - e\| \geq d(x, E)$, we conclude that $\|x - e\| = d(x, E)$, as required. $\square$

COROLLARY D.4. *If X is a Banach space and E is a finite dimensional subspace, then each element x in X has a closest point in E.*

PROOF. Let $x \in X$ be chosen so that $x \notin E$. Let F be the closed linear span of x with E. Then F is a finite dimensional Banach space and $\dim(F) = \dim(E) + 1$. In particular, since F is finite dimensional, it is a reflexive Banach space. Since E is a finite dimensional subspace of F, it is a closed subspace of F. Thus, by Theorem D.3, there exists some $e \in E$ such that $d(x, e) = d(x, E)$, as claimed. $\square$

DEFINITION D.5. *If E is a subset of a metric space X for which each $x \in X$ has an element e of E that is closest to x, then E is called a proximinal set.*

With this new terminology, we see that Corollary D.4 states that in a Banach space X, any finite dimensional subspace E is a proximinal set. Shortly, we will show that if we strengthen our assumption on X to say that it is a *strictly convex* Banach space, then not only does every $x \in X$ have a closest point in the finite dimensional subspace E, but it has a *unique* closest point. First, we will introduce the related terminology.

DEFINITION D.6. Let X be a metric space. A subset E of X is called *Chebyshev* if for each $x \in X$, there is a unique $e \in E$ such that $d(x,e) = d(x,E)$.

In other words, the set E is a Chebyshev set in X if each $x \in X$ has exactly one closest element in E. The term "Chebyshev" in this context seems to be due to Efimov and Stechkin in 1958 [**31**].

We are now ready for our main result.

THEOREM D.7. *If X is a strictly convex Banach space, then every proximinal subspace of X is a Chebyshev set.*

PROOF. Assume that X is a strictly convex Banach space and let E be a proximinal subspace of X. That means that for each x in X, there is at least one e in E that is closest to x. We wish to show that there is at most one such closest point in E. Suppose that there are two closest points e_1 and e_2. Then

$$\|x - e_1\| = d(x, E) \quad \text{and} \quad \|x - e_2\| = d(x, E).$$

In particular, we are assuming that $e_1 \neq e_2$. Since X is strictly convex,

$$\left\| x - \frac{e_1 + e_2}{2} \right\| = \left\| \frac{x - e_1}{2} + \frac{x - e_2}{2} \right\| < \left\| \frac{x - e_1}{2} \right\| + \left\| \frac{x - e_2}{2} \right\| = d(x, E).$$

But $\frac{e_1 + e_2}{2}$ is an element of E, because E is a subspace of X. That means we have found an element of E such

$$d(x, E) \leq \left\| x - \frac{e_1 + e_2}{2} \right\| < d(x, E),$$

a contradiction. Consequently, there cannot be more than one closest point to x, as required. □

COROLLARY D.8. *If X is a strictly convex Banach space and E is a finite dimensional subspace of X, then each element x in X has a unique closest point in E.*

PROOF. From Corollary D.4, we know that E is a proximinal set in X. Since X is strictly convex and E is a proximinal subspace, Theorem D.7 tells us that E is a Chebyshev set in X, as required. □

Bibliography

[1] I. Aharoni, B. Maurey, and B. S. Mityagin, *Uniform embeddings of metric spaces and of Banach spaces into Hilbert spaces*, Israel J. Math. **52** (1985), no. 3, 251–265, DOI 10.1007/BF02786521. MR815815

[2] Israel Aharoni, *Every separable metric space is Lipschitz equivalent to a subset of c_0^+*, Israel J. Math. **19** (1974), 284–291, DOI 10.1007/BF02757727. MR511661

[3] Israel Aharoni and Joram Lindenstrauss, *Uniform equivalence between Banach spaces*, Bull. Amer. Math. Soc. **84** (1978), no. 2, 281–283, DOI 10.1090/S0002-9904-1978-14475-9. MR482074

[4] Israel Aharoni and Joram Lindenstrauss, *An extension of a result of Ribe*, Israel J. Math. **52** (1985), no. 1-2, 59–64, DOI 10.1007/BF02776080. MR815602

[5] Fernando Albiac and Nigel J. Kalton, *Topics in Banach space theory*, 2nd ed., Graduate Texts in Mathematics, vol. 233, Springer, [Cham], 2016. With a foreword by Gilles Godefory, DOI 10.1007/978-3-319-31557-7. MR3526021

[6] R. D. Anderson, *Hilbert space is homeomorphic to the countable infinite product of lines*, Bull. Amer. Math. Soc. **72** (1966), 515–519, DOI 10.1090/S0002-9904-1966-11524-0. MR190888

[7] Richard F. Arens and James Eells Jr., *On embedding uniform and topological spaces*, Pacific J. Math. **6** (1956), 397–403, http://projecteuclid.org/euclid.pjm/1103043959. MR0081458

[8] N. Aronszajn, *Differentiability of Lipschitzian mappings between Banach spaces*, Studia Math. **57** (1976), no. 2, 147–190, DOI 10.4064/sm-57-2-147-190. MR425608

[9] Patrice Assouad, *Remarques sur un article de Israel Aharoni sur les prolongements lipschitziens dans c_0 (Israel J. Math. 19 (1974), 284–291)*, Israel J. Math. **31** (1978), no. 1, 97–100, DOI 10.1007/BF02761384. MR511662

[10] Antonio Avilés, Félix Cabello Sánchez, Jesús M. F. Castillo, Manuel González, and Yolanda Moreno, *Separably injective Banach spaces*, Lecture Notes in Mathematics, vol. 2132, Springer, [Cham], 2016, DOI 10.1007/978-3-319-14741-3. MR3469461

[11] Steven F. Bellenot, *Transfinite duals of quasireflexive Banach spaces*, Trans. Amer. Math. Soc. **273** (1982), no. 2, 551–577, DOI 10.2307/1999928. MR667160

[12] Y. Benyamini and J. Lindenstrauss, *A predual of l_1 which is not isomorphic to a $C(K)$ space*, Israel J. Math. **13** (1972), 246–254 (1973), https://doi.org/10.1007/BF02762798. MR331013

[13] Yoav Benyamini and Joram Lindenstrauss, *Geometric nonlinear functional analysis. Vol. 1*, American Mathematical Society Colloquium Publications, vol. 48, American Mathematical Society, Providence, RI, 2000, DOI 10.1090/coll/048. MR1727673

[14] Yoav Benyamini, *The uniform classification of Banach spaces*, Texas functional analysis seminar 1984–1985 (Austin, Tex.), Longhorn Notes, Univ. Texas Press, Austin, TX, 1985, pp. 15–38. MR832247

[15] C. Bessaga and A. Pełczyński, *Spaces of continuous functions. IV. On isomorphical classification of spaces of continuous functions*, Studia Math. **19** (1960), 53–62, DOI 10.4064/sm-19-1-53-62. MR113132

[16] Vladimir I. Bogachev, *Gaussian measures*, Mathematical Surveys and Monographs, vol. 62, American Mathematical Society, Providence, RI, 1998, DOI 10.1090/surv/062. MR1642391

[17] K. Borsuk, *Drei Sätze über die n-dimensionale euklidische Sphäre*, Fund. Math. **20** (1933), 177–190.

[18] J. Bourgain, *Real isomorphic complex Banach spaces need not be complex isomorphic*, Proc. Amer. Math. Soc. **96** (1986), no. 2, 221–226, DOI 10.2307/2046157. MR818448

[19] J. Bourgain, *Remarks on the extension of Lipschitz maps defined on discrete sets and uniform homeomorphisms*, Geometrical aspects of functional analysis (1985/86), 1987, pp. 157–167, https://doi.org/10.1007/BFb0078143. MR907692

[20] J. Bourgain, H. P. Rosenthal, and G. Schechtman, *An ordinal L^p-index for Banach spaces, with application to complemented subspaces of L^p*, Ann. of Math. (2) **114** (1981), no. 2, 193–228, DOI 10.2307/1971293. MR632839

[21] Adam Bowers and Nigel J. Kalton, *An introductory course in functional analysis*, Universitext, Springer, New York, 2014. With a foreword by Gilles Godefroy, DOI 10.1007/978-1-4939-1945-1. MR3289046

[22] Robert Cauty, *Un espace métrique linéaire qui n'est pas un rétracte absolu* (French, with English summary), Fund. Math. **146** (1994), no. 1, 85–99, DOI 10.4064/fm-146-1-85-99. MR1305261

[23] Jens Peter Reus Christensen, *Measure theoretic zero sets in infinite dimensional spaces and applications to differentiability of Lipschitz mappings*, Publ. Dép. Math. (Lyon) **10** (1973), no. 2, 29–39. MR361770

[24] James A. Clarkson, *Uniformly convex spaces*, Trans. Amer. Math. Soc. **40** (1936), no. 3, 396–414, https://doi.org/10.2307/1989630. MR1501880

[25] Harry Corson and Victor Klee, *Topological classification of convex sets*, Proc. Sympos. Pure Math., Vol. VII, Amer. Math. Soc., Providence, RI, 1963, pp. 37–51. MR161119

[26] Mahlon M. Day, *Reflexive Banach spaces not isomorphic to uniformly convex spaces*, Bull. Amer. Math. Soc. **47** (1941), 313–317, DOI 10.1090/S0002-9904-1941-07451-3. MR3446

[27] J. Diestel and J. J. Uhl Jr., *The Radon-Nikodym theorem for Banach space valued measures*, Rocky Mountain J. Math. **6** (1976), no. 1, 1–46, DOI 10.1216/RMJ-1976-6-1-1. MR399852

[28] J. Diestel and J. J. Uhl Jr., *Vector measures*, Mathematical Surveys, No. 15, American Mathematical Society, Providence, RI, 1977. With a foreword by B. J. Pettis. MR453964

[29] Nelson Dunford and B. J. Pettis, *Linear operations on summable functions*, Trans. Amer. Math. Soc. **47** (1940), 323–392, https://doi.org/10.2307/1989960. MR0002020

[30] Rick Durrett, *Probability—theory and examples*, Cambridge Series in Statistical and Probabilistic Mathematics, vol. 49, Cambridge University Press, Cambridge, 2019. Fifth edition of [MR1068527], DOI 10.1017/9781108591034. MR3930614

[31] N. V. Efimov and S. B. Stečkin, *Some properties of Čebyšev sets* (Russian), Dokl. Akad. Nauk SSSR (N.S.) **118** (1958), 17–19. MR95445

[32] Per Enflo, *On the nonexistence of uniform homeomorphisms between L_p-spaces*, Ark. Mat. **8** (1969), 103–105, DOI 10.1007/BF02589549. MR271719

[33] P. Enflo, *Uniform structures and square roots in topological groups. I, II*, Israel J. Math. **8** (1970), 230–252; ibid. 8 (1970), 253–272, DOI 10.1007/bf02771561. MR263969

[34] Per Enflo, *Banach spaces which can be given an equivalent uniformly convex norm*, Israel J. Math. **13** (1972), 281–288 (1973), DOI 10.1007/BF02762802. MR336297

[35] P. Enflo, *On infinite-dimensional topological groups*, Séminaire sur la Géométrie des Espaces de Banach (1977–1978), École Polytech., Palaiseau, 1978, pp. Exp. No. 10–11, 11. MR520212

[36] T. Figiel, *On nonlinear isometric embeddings of normed linear spaces* (English, with Russian summary), Bull. Acad. Polon. Sci. Sér. Sci. Math. Astronom. Phys. **16** (1968), 185–188. MR231179

[37] Maurice Fréchet, *Les espaces abstraits* (French), Les Grands Classiques Gauthier-Villars. [Gauthier-Villars Great Classics], Éditions Jacques Gabay, Sceaux, 1989. Reprint of the 1928 original. MR1189135

[38] G. Godefroy, N. Kalton, and G. Lancien, *Subspaces of $c_0(\mathbf{N})$ and Lipschitz isomorphisms*, Geom. Funct. Anal. **10** (2000), no. 4, 798–820, DOI 10.1007/PL00001638. MR1791140

[39] G. Godefroy and N. J. Kalton, *Lipschitz-free Banach spaces*, Studia Math. **159** (2003), no. 1, 121–141, DOI 10.4064/sm159-1-6. Dedicated to Professor Aleksander Pełczyński on the occasion of his 70th birthday. MR2030906

[40] G. Godefroy, G. Lancien, and V. Zizler, *The non-linear geometry of Banach spaces after Nigel Kalton*, Rocky Mountain J. Math. **44** (2014), no. 5, 1529–1583, DOI 10.1216/RMJ-2014-44-5-1529. MR3295641

[41] E. Gorelik, *The uniform nonequivalence of L_p and l_p*, Israel J. Math. **87** (1994), no. 1-3, 1–8, https://doi.org/10.1007/BF02772978. MR1286810

[42] W. T. Gowers, *A solution to the Schroeder-Bernstein problem for Banach spaces*, Bull. London Math. Soc. **28** (1996), no. 3, 297–304, https://doi.org/10.1112/blms/28.3.297. MR1374409

[43] F. Grünbaum and E. H. Zarantonello, *On the extension of uniformly continuous mappings*, Michigan Math. J. **15** (1968), 65–74, http://projecteuclid.org/euclid.mmj/1028999906. MR0223506

[44] A.J. Guirao, V. Montesinos, and V. Zizler, *Renormings in banach spaces: A toolbox*, Monografie Matematyczne, Springer International Publishing, 2022, https://books.google.com/books?id=brSEEAAAQBAJ.

[45] S. Heinrich and P. Mankiewicz, *Applications of ultrapowers to the uniform and Lipschitz classification of Banach spaces*, Studia Math. **73** (1982), no. 3, 225–251, https://doi.org/10.4064/sm-73-3-225-251. MR675426

[46] Stefan Heinrich, *Ultraproducts in Banach space theory*, J. Reine Angew. Math. **313** (1980), 72–104, https://doi.org/10.1515/crll.1980.313.72. MR552464

[47] C. Ward Henson, *Nonstandard hulls of Banach spaces*, Israel J. Math. **25** (1976), no. 1-2, 108–144, https://doi.org/10.1007/BF02756565. MR461104

[48] J. R. Isbell, *Uniform neighborhood retracts*, Pacific J. Math. **11** (1961), 609–648, http://projecteuclid.org/euclid.pjm/1103037336. MR0141074

[49] Paata Ivanisvili, Ramon van Handel, and Alexander Volberg, *Rademacher type and Enflo type coincide*, Ann. of Math. (2) **192** (2020), no. 2, 665–678, DOI 10.4007/annals.2020.192.2.8. MR4151086

[50] W. B. Johnson, J. Lindenstrauss, and G. Schechtman, *Banach spaces determined by their uniform structures*, Geom. Funct. Anal. **6** (1996), no. 3, 430–470, DOI 10.1007/BF02249259. MR1392325

[51] W. B. Johnson, B. Maurey, and G. Schechtman, *Non-linear factorization of linear operators*, Bull. Lond. Math. Soc. **41** (2009), no. 4, 663–668, https://doi.org/10.1112/blms/bdp040. MR2521361

[52] W. B. Johnson and E. Odell, *Subspaces of L_p which embed into l_p*, Compositio Math. **28** (1974), 37–49. MR352938

[53] W. B. Johnson, H. P. Rosenthal, and M. Zippin, *On bases, finite dimensional decompositions and weaker structures in Banach spaces*, Israel J. Math. **9** (1971), 488–506, https://doi.org/10.1007/BF02771464. MR280983

[54] W. B. Johnson and M. Zippin, *On subspaces of quotients of $(\sum G_n)_{l_p}$ and $(\sum G_n)_{c_0}$*, Israel J. Math. **13** (1972), 311–316 (1973), https://doi.org/10.1007/BF02762805. MR331023

[55] M. I. Kadec, *A proof of the topological equivalence of all separable infinite-dimensional Banach spaces* (Russian), Funkcional. Anal. i Priložen. **1** (1967), 61–70. MR209804

[56] N. J. Kalton, *Linear operators on L_p for $0 < p < 1$*, Trans. Amer. Math. Soc. **259** (1980), no. 2, 319–355, https://doi.org/10.2307/1998234. MR567084

[57] N. J. Kalton, *Locally complemented subspaces and $\mathcal{L}_p$-spaces for $0 < p < 1$*, Math. Nachr. **115** (1984), 71–97, DOI 10.1002/mana.19841150107. MR755269

[58] N. J. Kalton, *Banach spaces embedding into L_0*, Israel J. Math. **52** (1985), no. 4, 305–319, https://doi.org/10.1007/BF02774083. MR829361

[59] N. J. Kalton, *Spaces of Lipschitz and Hölder functions and their applications*, Collect. Math. **55** (2004), no. 2, 171–217. MR2068975

[60] N. J. Kalton, *Coarse and uniform embeddings into reflexive spaces*, Q. J. Math. **58** (2007), no. 3, 393–414, DOI 10.1093/qmath/ham018. MR2354924

[61] N. J. Kalton, *Extending Lipschitz maps into $C(K)$-spaces*, Israel J. Math. **162** (2007), 275–315, https://doi.org/10.1007/s11856-007-0099-2. MR2365864

[62] Nigel J. Kalton, *The nonlinear geometry of Banach spaces*, Rev. Mat. Complut. **21** (2008), no. 1, 7–60, DOI 10.5209/rev_REMA.2008.v21.n1.16426. MR2408035

[63] N. J. Kalton, *Lipschitz and uniform embeddings into ℓ_∞*, Fund. Math. **212** (2011), no. 1, 53–69, DOI 10.4064/fm212-1-4. MR2771588

[64] N. J. Kalton, *The uniform structure of Banach spaces*, Math. Ann. **354** (2012), no. 4, 1247–1288, https://doi.org/10.1007/s00208-011-0743-3. MR2992997

[65] N. J. Kalton, *Examples of uniformly homeomorphic Banach spaces*, Israel J. Math. **194** (2013), no. 1, 151–182, https://doi.org/10.1007/s11856-012-0080-6. MR3047066

[66] N. J. Kalton and G. Lancien, *Best constants for Lipschitz embeddings of metric spaces into c_0*, Fund. Math. **199** (2008), no. 3, 249–272, https://doi.org/10.4064/fm199-3-4. MR2395263

[67] Nigel J. Kalton and N. Lovasoa Randrianarivony, *The coarse Lipschitz geometry of $l_p \oplus l_q$*, Math. Ann. **341** (2008), no. 1, 223–237, DOI 10.1007/s00208-007-0190-3. MR2377476

[68] L. Kantorovitch, *On the translocation of masses*, C. R. (Doklady) Acad. Sci. URSS (N.S.) **37** (1942), 199–201. MR9619

[69] Yitzhak Katznelson, *An introduction to harmonic analysis*, 3rd ed., Cambridge Mathematical Library, Cambridge University Press, Cambridge, 2004, DOI 10.1017/CBO9781139165372. MR2039503

[70] M. Kirszbraun, *Über die zusammenziehende und Lipschitzsche Transformationen* (German), Fundamenta Mathematicae **22** (1934), no. 1, 77–108, http://eudml.org/doc/212681.

[71] S. Kwapień, *Unsolved Problems: Problem 3*, Studia Math. **38** (1970), 469.

[72] H. Elton Lacey, *The isometric theory of classical Banach spaces*, Die Grundlehren der mathematischen Wissenschaften, Band 208, Springer-Verlag, New York-Heidelberg, 1974. MR493279

[73] J. Lindenstrauss and A. Pełczyński, *Absolutely summing operators in $\mathcal{L}_p$-spaces and their applications*, Studia Math. **29** (1968), 275–326, https://doi.org/10.4064/sm-29-3-275-326. MR231188

[74] J. Lindenstrauss and H. P. Rosenthal, *The $\mathcal{L}_p$ spaces*, Israel J. Math. **7** (1969), 325–349, https://doi.org/10.1007/BF02788865. MR270119

[75] Joram Lindenstrauss, *On nonlinear projections in Banach spaces*, Michigan Math. J. **11** (1964), 263–287. MR167821

[76] Joram Lindenstrauss and Lior Tzafriri, *Classical Banach spaces. I*, Ergebnisse der Mathematik und ihrer Grenzgebiete [Results in Mathematics and Related Areas], Band 92, Springer-Verlag, Berlin-New York, 1977. Sequence spaces. MR500056

[77] L. Lyusternik and L. Šnirel′man, *Topological methods in variational problems* (1930).

[78] Piotr Mankiewicz, *On the differentiability of Lipschitz mappings in Fréchet spaces*, Studia Math. **45** (1973), 15–29, DOI 10.4064/sm-45-1-15-29. MR331055

[79] Jiří Matoušek, *Using the Borsuk-Ulam theorem*, Universitext, Springer-Verlag, Berlin, 2003. Lectures on topological methods in combinatorics and geometry; Written in cooperation with Anders Björner and Günter M. Ziegler. MR1988723

[80] S. Mazur, *Une remarque sur l'homéomorphie des champs fonctionnels* (French), Studia Math. **1** (1929), no. 1, 83–85, http://eudml.org/doc/216983.

[81] S. Mazur, *Über konvexe Mengen in linearen normierten Räumen* (German), Studia Math. **4** (1933), no. 1, 70–84, http://eudml.org/doc/217865.

[82] S. Mazur and S. Ulam, *Sur les transformations isométriques d'espaces vectoriels normés* (French), C.R. Acad. Sci. Paris **194** (1932), 946–948.

[83] Robert E. Megginson, *An introduction to Banach space theory*, Graduate Texts in Mathematics, vol. 183, Springer-Verlag, New York, 1998, DOI 10.1007/978-1-4612-0603-3. MR1650235

[84] D. Milman, *On some criteria for the regularity of spaces of type (B)*, C. R. (Doklady) Acad. Sci. U.R.S.S **20** (1938), 243–246.

[85] George J. Minty, *On the extension of Lipschitz, Lipschitz-Hölder continuous, and monotone functions*, Bull. Amer. Math. Soc. **76** (1970), 334–339, DOI 10.1090/S0002-9904-1970-12466-1. MR254575

[86] James R. Munkres, *Topology*, Prentice Hall, Inc., Upper Saddle River, NJ, 2000. Second edition of [MR0464128]. MR3728284

[87] Leopoldo Nachbin, *A theorem of the Hahn-Banach type for linear transformations*, Trans. Amer. Math. Soc. **68** (1950), 28–46, DOI 10.2307/1990536. MR32932

[88] E. M. Nikišin, *Resonance theorems and superlinear operators* (Russian), Uspehi Mat. Nauk **25** (1970), no. 6(156), 129–191. MR296584

[89] E. M. Nikišin, *A resonance theorem and series in eigenfunctions of the Laplace operator* (Russian), Izv. Akad. Nauk SSSR Ser. Mat. **36** (1972), 795–813. MR343091

[90] Edward Odell and Thomas Schlumprecht, *The distortion problem*, Acta Math. **173** (1994), no. 2, 259–281, https://doi.org/10.1007/BF02398436. MR1301394

[91] J. Pelant, *Embeddings into c_0*, Topology Appl. **57** (1994), no. 2-3, 259–269, DOI 10.1016/0166-8641(94)90053-1. MR1278027

[92] A. Pełczyński, *Projections in certain Banach spaces*, Studia Math. **19** (1960), 209–228, https://doi.org/10.4064/sm-19-2-209-228. MR126145

[93] A. Pełczyński, *Linear extensions, linear averagings, and their applications to linear topological classification of spaces of continuous functions*, Dissertationes Math. (Rozprawy Mat.) **58** (1968), 92. MR227751

[94] V. G. Pestov, *Free Banach spaces and representations of topological groups* (Russian), Funktsional. Anal. i Prilozhen. **20** (1986), no. 1, 81–82. MR831059

[95] B. J. Pettis, *A proof that every uniformly convex space is reflexive*, Duke Math. J. **5** (1939), no. 2, 249–253, DOI 10.1215/S0012-7094-39-00522-3. MR1546121

[96] R. S. Phillips, *On linear transformations*, Trans. Amer. Math. Soc. **48** (1940), 516–541, https://doi.org/10.2307/1990096. MR0004094

[97] Gilles Pisier, *Martingales with values in uniformly convex spaces*, Israel J. Math. **20** (1975), no. 3-4, 326–350, DOI 10.1007/BF02760337. MR394135

[98] Krzysztof Przesławski and David Yost, *Lipschitz retracts, selectors, and extensions*, Michigan Math. J. **42** (1995), no. 3, 555–571, DOI 10.1307/mmj/1029005313. MR1357625

[99] N. Lovasoa Randrianarivony, *Characterization of quasi-Banach spaces which coarsely embed into a Hilbert space*, Proc. Amer. Math. Soc. **134** (2006), no. 5, 1315–1317, DOI 10.1090/S0002-9939-05-08416-9. MR2199174

[100] M. Ribe, *On uniformly homeomorphic normed spaces. II*, Ark. Mat. **16** (1978), no. 1, 1–9, DOI 10.1007/BF02385979. MR487402

[101] M. Ribe, *Existence of separable uniformly homeomorphic nonisomorphic Banach spaces*, Israel J. Math. **48** (1984), no. 2-3, 139 147, DOI 10.1007/BF02761159. MR770696

[102] Haskell P. Rosenthal, *On relatively disjoint families of measures, with some applications to Banach space theory*, Studia Math. **37** (1970), 13–36, DOI 10.4064/sm-37-1-13-36. MR270122

[103] Mary Ellen Rudin, *A new proof that metric spaces are paracompact*, Proc. Amer. Math. Soc. **20** (1969), 603, DOI 10.2307/2035708. MR236876

[104] Andrew Sobczyk, *Projection of the space (m) on its subspace (c_0)*, Bull. Amer. Math. Soc. **47** (1941), 938–947, DOI 10.1090/S0002-9904-1941-07593-2. MR5777

[105] A. H. Stone, *Paracompactness and product spaces*, Bull. Amer. Math. Soc. **54** (1948), 977–982, DOI 10.1090/S0002-9904-1948-09118-2. MR26802

[106] H. Toruńczyk, *Characterizing Hilbert space topology*, Fund. Math. **111** (1981), no. 3, 247–262, DOI 10.4064/fm-111-3-247-262. MR611763

[107] F. A. Valentine, *On the extension of a vector function so as to preserve a Lipschitz condition*, Bull. Amer. Math. Soc. **49** (1943), 100–108, DOI 10.1090/S0002-9904-1943-07859-7. MR8251

[108] F. A. Valentine, *A Lipschitz condition preserving extension for a vector function*, Amer. J. Math. **67** (1945), 83–93, https://doi.org/10.2307/2371917. MR0011702

[109] William A. Veech, *Short proof of Sobczyk's theorem*, Proc. Amer. Math. Soc. **28** (1971), 627–628, DOI 10.2307/2038025. MR275122

[110] Nik Weaver, *Lipschitz algebras*, World Scientific Publishing Co., Inc., River Edge, NJ, 1999, DOI 10.1142/4100. MR1832645

[111] Robert Whitley, *Mathematical Notes: Projecting m onto c_0*, Amer. Math. Monthly **73** (1966), no. 3, 285–286, DOI 10.2307/2315346. MR1533692

[112] M. Zippin, *The separable extension problem*, Israel J. Math. **26** (1977), no. 3-4, 372–387, https://doi.org/10.1007/BF03007653. MR0442649

[113] M. Zippin, *Extension of bounded linear operators*, Handbook of the geometry of Banach spaces, Vol. 2, 2003, pp. 1703–1741, https://doi.org/10.1016/S1874-5849(03)80047-5. MR1999607

Index

absolute Lipschitz retract, 2
absolutely continuous measure, 236
adjacent vectors, 215
Aharoni's theorem, 72, 85
almost Lipschitz isomorphic, 141
AL_p-space, 124
annihilator, 90
approximate metric midpoint, 162
Arens–Eells space, 37
asymptotic smoothness technique, 171

Banach lattice, 123
Banach–Mazur distance, 93, 105
Banach–Mazur theorem, 45, 71
barycenter, 40
basis, 110
basis constant, 45
binary intersection property (BIP), 5
biorthogonal functionals, 110
Bochner integral, 232
Bochner's theorem, 193
Bolzano–Weierstrass theorem, 74
Borel function, 58
Borsuk–Ulam theorem, 78
bounded convergence theorem, 111
bounded variation, 236
Brouwer fixed-point theorem, 103

c_0
   is a 2-ALR, 8
   not complemented in ℓ_∞, 10
center, 29
characteristic function, 231
Chebyshev center, 30
Chebyshev diameter, 30
Chebyshev radius, 30
Chebyshev set, 245
Chebyshev's inequality, 201, 211
CL-homeomorphism, 141
CL-type, 132, 141
Clarkson's inequalities, 125
closest element, 243
closest-point map, 13
coarse embedding, 199

coarse Lipschitz embedding, 162
coarse Lipschitz function, 127, 132
coarse Lipschitz homeomorphism, 141
coarsely continuous function, 127
coarsely homeomorphic, 129
complemented, 10, 42, 89
condition number, 93
convergence through a filter, 119
convolution, 55
coordinate functionals, 110
Corson–Klee lemma, 126
cotype, 223
countably additive vector measure, 235
crudely finitely representable, 86
cube measure, 56, 62

$\Delta_M(\epsilon)$, 220
dense cube measure, 56, 62
diameter, 29
Dirac lifting, 38
directional derivative, 61
directional difference quotient, 61
distortion constant, 93
Dunford integral, 234

Enflo type, 226
Enflo–Pisier theorem, 125
ϵ-net, 87
essentially separable range, 233
exact sequence, 89
extremally disconnected, 8

Figiel's Theorem, 31
filter, 119
finitely additive vector measure, 235
finitely representable, 86
Fréchet derivative, 49

Gâteaux derivative, 61
Gâteaux derivative, 33, 49
Gauss-null set, 240
Gaussian measure, 239
Gelfand–Fréchet space, 237
Gorelik principle, 104

Haar-null, 60
Hahn–Banach theorem, 21
Hahn-Banach theorem, nonlinear version, 2
Hamming distance, 171
Hausdorff metric, 19
Heinrich–Mankiewicz theorem, 71, 91
hermitian kernel, 186
hermitian matrix, 186
$\mathcal{H}(K)$, 19
   is a 8-ALR, 23
Hölder continuous function, 16

indicator function, 231
injective Banach space, 7
interlaced vectors, 215
invariant mean, 94, 198
isometric lifting property, 52

James space, 221
Johnson–Odell theorem, 161

Kalton–Randrianarivony theorem, 162
Khintchine's inequalities, 204
Kirszbraun's theorem, 14
Kronecker delta, 203

$L_0(\Omega, \mathbb{P})$, 185
lifting property, 52
limit through a filter, 120
ℓ_∞
   has BIP, 5
   is 1-ALR, 3
$\mathcal{L}_\infty$-space, 96
$\mathrm{Lip}(M)$, 37
Lipschitz constant, 126
Lipschitz continuous, 1, 126
Lipschitz dual, 37
Lipschitz embedding, 76
Lipschitz isomorphism, 1
Lipschitz isomorphism problem, 27
Lipschitz on large distances, 127
Lipschitz retraction, 4
Lipschitz splitting, 95
Lipschitz weak-star Kadec–Klee, 105
Lipschitz-free space, 37
locally complemented, 89, 90
λ-locally complemented, 90
locally finite cover, 25
$\mathcal{L}_p$-space, 97, 161
Lyusternik–Šnirel′man lemma, 79

Mazur map, 150, 222
Mazur maps, 150
Mazur's Theorem, 33
Mazur–Ulam Theorem, 27
mean, 94
metric midpoint, 27, 162
metrically convex, 1
Michael selection theorem, 25, 104

midpoint technique, 162
Milman–Pettis theorem, 125
Minty's theorem, 16
modulus of concavity, 200
modulus of continuity, 1, 126, 132
monotone basis, 45

natural isometric embedding into the Arens–Eells space, 38
negative-definite function, 188
negative-definite hermitian kernel, 187
negative-definite metric, 185
Nikišin's theorem, 200
non-principal ultrafilter, 119
nondegenerate Gaussian measure, 239
nonnegative mean, 94
norm, 200
norm-preserving function, 132

Odell–Schlumprecht theorem, 223
ω-dual, 222

p-stable, 194
paracompact, 25
partial sum projection, 45, 110
Pettis integral, 234
Pettis measurability theorem, 233
Phillips' lemma, 9
pointed metric space, 37
positive-definite function, 188
positive-definite hermitian kernel, 186
positive-definite matrix, 186
positively homogeneous function, 132
principal ultrafilter, 119
Principle of Local Reflexivity, 87
projection, 89
property $\Pi(\lambda)$, 75
property $\Pi_+(\lambda)$, 86
property $Q(\epsilon, \delta)$, 220
proximinal set, 244

quasi-Banach space, 200
quasi-norm, 200
quasi-reflexive, 221

$\mathbb{R}$ is a 1-ALR, 2
Rademacher cotype, 223
Rademacher random variable, 203, 211
Rademacher type, 210
Rademacher's theorem, 50, 236
radius, 29
Radon–Nikodym derivative, 236
Radon–Nikodym property, 236
Radon–Nikodym property (RNP), 57, 235
Radon–Nikodym theorem, 236
refinement, 25
Riesz space, 123
Riesz's lemma, 78

Schauder basis, 110

Schroeder–Bernstein Problem, 100
Schur space/property, 131
SCP, 42
second variation, 221
selection, 223
separable complementation property, 42
separably injective, 11
short exact sequence, 89
shortest path metric, 215
shrinking basis, 110
simple function, 231
splitting, 89, 95
Stonean space, 8
strict convexity, 27
strongly measurable function, 231
subordinate partition of unity, 25
super-reflexive, 125
support functional, 20

total variation measure, 235
trace, 239
trace-class, 239

ultrafilter, 119
ultrametric space, 227
ultraproduct, 121
unconditional basis, 223
uniform embedding, 185
uniform homeomorphism, 126, 185
uniformly almost Lipschitz isomorphic, 141
uniformly close, 141
uniformly continuous, 1, 126, 132
uniformly convex Banach space, 28, 124
uniformly homeomorphic spaces, 126, 185

vector lattice, 123
vector measure, 235

weak L_1, 200
weak L_p, 211

Zorn's lemma, 119

SELECTED PUBLISHED TITLES IN THIS SERIES

79 **Adam Bowers,** Nigel Kalton's Lectures in Nonlinear Functional Analysis, 2024
78 **Paul B. Larson,** Extensions of the Axiom of Determinacy, 2023
77 **Imre Bárány,** Combinatorial Convexity, 2021
76 **Mario Garcia-Fernandez and Jeffrey Streets,** Generalized Ricci Flow, 2021
75 **Raymond Cheng, Javad Mashreghi, and William T. Ross,** Function Theory and ℓ^p Spaces, 2020
74 **Leonid Polterovich, Daniel Rosen, Karina Samvelyan, and Jun Zhang,** Topological Persistence in Geometry and Analysis, 2020
73 **Armand Borel,** Introduction to Arithmetic Groups, 2019
72 **Pavel Mnev,** Quantum Field Theory: Batalin–Vilkovisky Formalism and Its Applications, 2019
71 **Alexander Grigor'yan,** Introduction to Analysis on Graphs, 2018
70 **Ian F. Putnam,** Cantor Minimal Systems, 2018
69 **Corrado De Concini and Claudio Procesi,** The Invariant Theory of Matrices, 2017
68 **Antonio Auffinger, Michael Damron, and Jack Hanson,** 50 Years of First-Passage Percolation, 2017
67 **Sylvie Ruette,** Chaos on the Interval, 2017
66 **Robert Steinberg,** Lectures on Chevalley Groups, 2016
65 **Alexander M. Olevskii and Alexander Ulanovskii,** Functions with Disconnected Spectrum, 2016
64 **Larry Guth,** Polynomial Methods in Combinatorics, 2016
63 **Gonçalo Tabuada,** Noncommutative Motives, 2015
62 **H. Iwaniec,** Lectures on the Riemann Zeta Function, 2014
61 **Jacob P. Murre, Jan Nagel, and Chris A. M. Peters,** Lectures on the Theory of Pure Motives, 2013
60 **William H. Meeks III and Joaquín Pérez,** A Survey on Classical Minimal Surface Theory, 2012
59 **Sylvie Paycha,** Regularised Integrals, Sums and Traces, 2012
58 **Peter D. Lax and Lawrence Zalcman,** Complex Proofs of Real Theorems, 2012
57 **Frank Sottile,** Real Solutions to Equations from Geometry, 2011
56 **A. Ya. Helemskii,** Quantum Functional Analysis, 2010
55 **Oded Goldreich,** A Primer on Pseudorandom Generators, 2010
54 **John M. Mackay and Jeremy T. Tyson,** Conformal Dimension, 2010
53 **John W. Morgan and Frederick Tsz-Ho Fong,** Ricci Flow and Geometrization of 3-Manifolds, 2010
52 **Marian Aprodu and Jan Nagel,** Koszul Cohomology and Algebraic Geometry, 2010
51 **J. Ben Hough, Manjunath Krishnapur, Yuval Peres, and Bálint Virág,** Zeros of Gaussian Analytic Functions and Determinantal Point Processes, 2009
50 **John T. Baldwin,** Categoricity, 2009
49 **József Beck,** Inevitable Randomness in Discrete Mathematics, 2009
48 **Achill Schürmann,** Computational Geometry of Positive Definite Quadratic Forms, 2008
47 **Ernst Kunz, David A. Cox, and Alicia Dickenstein,** Residues and Duality for Projective Algebraic Varieties, 2008
46 **Lorenzo Sadun,** Topology of Tiling Spaces, 2008
45 **Matthew Baker, Brian Conrad, Samit Dasgupta, Kiran S. Kedlaya, and Jeremy Teitelbaum,** p-adic Geometry, 2008

For a complete list of titles in this series, visit the
AMS Bookstore at **www.ams.org/bookstore/ulectseries/**.